SUPPLY CHAIN OPTIMIZATION

Applied Optimization

VOLUME 98

Series Editors:

Panos M. Pardalos
University of Florida, U.S.A.

Donald W. Hearn
University of Florida, U.S.A.

SUPPLY CHAIN OPTIMIZATION

Edited by

JOSEPH GEUNES
University of Florida, Gainesville, U.S.A.

PANOS M. PARDALOS
University of Florida, Gainesville, U.S.A.

Springer

Library of Congress Cataloging-in-Publication Data

Supply chain optimization/ edited by Joseph Geunes, Panos M. Pardalos.
p. cm. — (Applied optimization ; v. 98)
Includes bibliographical references.
ISBN 0-387-26280-6 (alk. paper) — ISBN 0-387-26281-4 (e-book)
1. Business logistics. 2. Delivery of goods. I. Geunes, Joseph. II. Pardalos, P.M. (Panos M.), 1954– III. Series.

HD38.5.S89615 2005
658.7′2—dc22

2005049768

AMS Subject Classifications: 90B50, 90B30, 90B06, 90B05

ISBN-10: 0-387-26280-6 ISBN-13: 978-0387-26280-2
e-ISBN-10: 0-387-26281-4 e-ISBN-13: 978-0387-26281-9

Printed in the United States of America.

9 8 7 6 5 4 3 2 1 SPIN 11498841

springeronline.com

Contents

Preface

The title of this edited book, *Supply Chain Optimization*, aims to capture a segment of recent research activity in supply chain management. This research area focuses on applying optimization techniques to supply chain management problems. While the general area of supply chain management research is broader than this scope, our intent is to compile a set of research papers that capture the use of state-of-the-art optimization methods within the field. Several researchers who initially expressed interest in contributing to this effort also expressed concerns that their work might not contain a sufficient degree of optimization. Others were uncertain as to whether the problems they proposed covered a broad enough scope in order to be considered as supply chain research. Our position has been that research that rigorously models elements of supply chain operations with a goal of improving supply chain performance (or the performance of some segment thereof) would fit under the umbrella of *supply chain optimization*. We therefore sought high-quality works from leading researchers in the field that fit within this general scope. We are quite pleased with the result, which has brought together a diverse blend of research topics and novel modeling and solution approaches for difficult classes of supply chain operations, planning, and design problems.

The book begins by taking an in-depth look at the role of information in supply chains. "Information Centric Optimization of Inventories in Capacitated Supply Chains: Three Illustrative Examples," by S. Gavirneni, considers how firms can best take advantage of the vast amounts of data available to them as a result of advanced information technologies. The author considers how capacity, inventory, information, and pricing influence supply chain performance, and provides strategies for leveraging information to enhance performance.

The second chapter, "An Analysis of Advance Booking Discount Programs between Competing Retailers," by K.F. McCardle, K. Rajaram, and C.S. Tang, considers a new mechanism for eliciting information from customers. The authors employ a strategy of providing discounts to customers who reserve a product in advance of a primary selling season. This information can be used by a supplier to reduce the uncertainty faced in the selling season, and the authors explore conditions under which equilibrium behavior among two retailers results in applying such a strategy.

In Chapter 3, A.M. Newman, C.A. Yano, and P.M. Kaminsky study a class of combined transportation and inventory planning problems faced by third-party logistics providers, who are becoming increasingly prevalent players in supply chains. This chapter, "Third Party Logistics Planning with Routing and Inventory Costs," considers route selection for full-truckload carriers contracted by manufacturers for repeated deliveries. The logistics provider faces a tradeoff between providing better service to customers through more frequent deliveries versus achieving the most cost-effective delivery pattern from a transportation cost perspective.

E. Bish addresses capacity investment and pricing decisions under demand uncertainty in Chapter 4, "Optimal Investment Strategies for Flexible Resources, Considering Pricing." While a number of past works have considered the problem of investing in flexible resources under uncertainty, this work explores how a firm's ability to set prices influences the value of resource flexibility. This work provides interesting insights on how pricing power can alter flexible resource capacity investment under different product demand correlation scenarios.

In "Multi-Channel Supply Chain Design in B2C Electronic Commerce" (Chapter 5), W.K. Chiang and D. Chhajed provide an interesting look at the challenges manufacturers face in simultaneously selling via traditional retail and direct on-line sales channels. Under a variety of scenarios and using a game-theoretic modeling approach, they provide insights on channel design strategy for both centralized and decentralized supply chains, when consumers have different preferences for direct and retail channels.

While a vast amount of literature applies game-theoretic modeling approaches to supply chain problems, J.J. Bartholdi III and E. Kemahlıoğlu-Ziya provide an innovative new model for sharing gains from cooperation in Chapter 6 ("Using Shapley Value to Allocate Savings in a Supply Chain"). They consider original equipment manufacturers (OEMs) with varying degrees of power who can influence whether a contract supplier may pool upstream inventories of common goods for multiple OEMs. By using the concept of Shapley value to create a mechanism for sharing the gains by allowing inventory pooling, the authors show that this method induces supply chain coordination and leads to a stable solution, although the resulting solution may still be perceived as "unfair" by some participants.

M.S. Pangburn and E. Stavrulaki consider an economic model of combined pricing, location, and capacity setting decisions in Chapter 7, "Service Facility Location and Design with Pricing and Waiting-Time Considerations." This model accounts for contexts where customers are sensitive to both transportation time and service waiting time that results from congestion effects. Customers will choose a facility if the associated utility (which accounts for distance and waiting-time costs) exceeds some reservation value. The authors address the implications of non-homogeneous customers, as well as equilibrium competitive behavior with two facilities.

Chapter 8 considers a recently emerging focus in supply chain design, where the robustness of the design under uncertainty is critical. In "A Conceptual Framework for Robust Supply Chain Design under Demand Uncertainty," Y. Mo and T.P. Harrison propose a modeling approach for addressing demand uncertainty in the design phase. The authors propose different robustness measures that incorporate various elements of risk and discuss different solution strategies, including the use of stochastic programming and sampling-based methods.

Staying with the supply chain design focus, Chapter 9, "The Design of Production-Distribution Networks: A Mathematical Programming Approach," by A. Martel, considers a wide range of decision factors in design. This chapter highlights important strategic factors, such as performance measures, planning horizon length and the associated uncertainty, process and product structure modeling, network flow modeling, modeling price, demand, and customer service, and facility layout options. The cost model accounts for various financial factors, such as tariffs, taxes, exchange rates, and transfer payments, in addition to transportation, inventory, and location costs. The result is a comprehensive large-scale nonlinear integer math programming model. The author discusses solution methods employed to develop a decision support system for supply chain design decisions.

Chapter 10, "Modeling & Solving Stochastic Programming Problems in Supply Chain Management Using *Xpress-SP*," by A. Dormer, A. Vazacopoulos, N. Verma, and H. Tipi, provides a further look at how to deal with uncertainty in supply chains. The authors identify various sources of risk in supply chains and how these affect performance. This chapter provides a nice discussion of stochastic programming problems in general, and in how to use the *Xpress-SP* package to model and solve these problems. Two illustrative examples of supply chain planning problems under uncertainty serve to illustrate the effective use of this tool for solving such problems.

Chapter 11 considers an operations-level planning problem facing logistics managers in container terminal operations. In "Dispatching Automated Guided Vehicles in a Container Terminal," Y.-L. Cheng, H.-C. Sen, K. Natarajan, C.-P. Teo, and K.-C. Tan study the problem of dispatching automated vehicles in a port terminal. Their model accounts for congestion effects in transportation using a deadlock prediction and avoidance scheme. They provide greedy and network flow–based heuristic solution approaches, and use a simulation model to validate the performance improvements as a result of the modeling and solution approaches they propose.

In the final chapter ("Hybrid MIP-CP techniques to solve a Multi-Machine Assignment and Scheduling Problem in Xpress-CP"), A. Vazacopoulos and N. Verma discuss hybrid constraint programming and mixed integer programming approaches for difficult multi-machine scheduling problems. While this model is motivated by the problem of scheduling jobs on different machines on a shop floor, it might also apply to the assignment of work to different facilities in a supply chain. The authors discuss the pros and cons of both constraint programming and mixed integer programming approaches, and consider hybrid approaches that combine the strengths of both of these methods. The authors illustrate the use of the *Xpress-CP* software package as a tool for implementing this hybrid approach, and compare the results obtained to prior results from the literature based on a common set of test problems.

This collection represents a set of stand-alone works that captures recent research trends in the application of optimization methods to supply chain operations, planning, and design problems. We are extremely grateful to the authors for their outstanding contributions and for their patience, which have led to a final product that far exceeded our expectations. All chapters were rigorously reviewed, and we would like to thank the anonymous reviewers for their quality reviews and responsiveness. We would also like to thank several graduate students in the ISE Department at the University of Florida for their help; in particular, we thank Ismail Serdar Bakal, Altannar Chinchuluun, and Yasemin Merzifonluoğlu for their contributions to this effort.

JOSEPH GEUNES AND PANOS PARDALOS

Chapter 1

INFORMATION CENTRIC OPTIMIZATION OF INVENTORIES IN CAPACITATED SUPPLY CHAINS: THREE ILLUSTRATIVE EXAMPLES

Srinagesh Gavirneni
Johnson Graduate School of Management
Cornell University
Ithaca, NY 14853

Abstract Recent enhancements in information technology have played a major role in the timely availability and accuracy of information across the supply chain. It is now cheaper to gather, store, and analyze vast amounts of data and this has presented managers with new opportunities for improving the efficiency of their supply chains. In addition, the latest developments in supply chain management have led everyone to believe that cooperation between members of a supply chain can lead to larger profits. While some gains have been realized from these developments, most organizations have failed to take the most advantage of them. To overcome this, there is a need to redesign a firm's supply chain with regards to its structure and modus operandi. This chapter illustrates this need for information-centric design and management of capacitated supply chains using three examples based on three different supply chain configurations.

1. Introduction

A supply chain is a group of organizations (including product design, procurement, manufacturing, and distribution) that are working together to profitably provide the right product or service to the right customer at the right time. Supply Chain Management (SCM) is the study of strategies and methodologies that enable these organizations to meet their objectives effectively. In the past few decades, people have realized that cooperation with other organizations in the supply chain can lead to significantly higher profits. As a result, industrial supplier-customer relations have undergone radical changes resulting in a certain level of co-operation, mainly in the area of information sharing, that was lacking before. The degree of co-operation varies significantly from one supply chain to another. The information sharing could range from generic (e.g. type of inventory control policy being used, type of production scheduling rules being used) to specific (e.g. day-to-day inventory levels, exact production schedules). There is a need for new models addressing these recent developments in information sharing because traditional models were developed under demand and informational assumptions that no longer universally hold in the manufacturing sector. In addition there have been reports, from industrial sources, of differing reactions to Electronic Data Interchange (EDI) benefits - while some were very happy with improved information, others were disappointed at the benefits (see Armistead and Mapes (1993) and Takac (1992)). The popular press is full of stories about companies disillusioned with their Enterprise Resource Planning (ERP) systems. It is estimated that 70% of all ERP implementations do not recoup their investments and are branded as failures (see *InfoWorld*, October 2001). While there could be many reasons for this high failure rate, the fact that companies are not adept at using the information provided by these ERP systems is a major factor. Since the availability and accuracy of information are the key contributions of such enterprise-wide systems, the organizations must position themselves to benefit from it.

While information will always be beneficial, it is important to know when it is most beneficial and when it is only marginally useful. In the latter case, some other characteristics of the system, such as end-item demand variance or supplier capacity may have to be improved before expecting significant benefits from information sharing. With regard to the benefits of information sharing and its dependence on the various supply chain characteristics (such as capacity, variance, service level, etc.), it is necessary to answer the following questions: (1) In the presence of Information Sharing, what is the optimal control policy?; (2) What is the benefit (in dollars) of Information flow?; and (3) How can the supply chain be changed in order to maximize this benefit? In an attempt to answer these questions, we (in Gavirneni, Kapuscinksi, and Tayur (1999)) studied a simple, yet representative, supply chain consisting of one supplier and one retailer using an (s, S) policy. In spite of its simple setup, this two stage supply chain provided valuable insights into managing more complex systems efficiently. The (s, S) policy dictates that the retailer will only order when her inventory level falls below s, and at that time she will order up-to S. Under this setting, we considered three situations: (1) a traditional model where there is no information, except from past data, to the supplier prior to a demand from the retailer; (2) the supplier has the information of the (s, S) policy used by the retailer as well as the end-item demand distribution; and (3) the supplier has full information about the state of the retailer. The availability of new retailer information about inventory policy (in situation 2) and inventory levels (in situation 3) presents new opportunities for the supplier. After formulating the appropriate decision problems at the supplier, we showed that order up-to policies continue to be optimal for models with information flow for the finite horizon, the infinite horizon discounted and the infinite horizon average cost cases. We developed efficient solution procedures for these three models and performed a detailed computational study to understand the relationships between capacity, inventory, and information at the supplier level and explain how they are affected by customer $(S - s)$ values and end-item demand distribution. In addition, we tabulated the benefits (averaging around 14% and ranging from 1% to 35%) of information sharing for this supply chain and made the following observations about their behavior: (1) Since information presents the supplier with more options, it is always beneficial; (2) More information generally results in larger savings; (3) The benefit of information flow is higher at higher capacities; (4) If the variance of the demand seen by the customer is small (high), we can expect the benefit of information flow to increase (decrease) with increase in penalty cost; (5) Information is most beneficial at moderate values of variance; and (6) Information is less beneficial at extreme values of $(S - s)$. These insights can lead to better management of projects that involve information sharing between members of a supply chain.

This study (Gavirneni, Kapuscinksi, and Tayur (1999)) was one of the first papers to be published on this topic and a number of articles have been published on this topic since then. Chen (1998) studied the benefits of information flow in a multi-echelon serial inventory system by computing the difference between the costs of using echelon reorder points and installation reorder points. He observed that information sharing reduced costs by as much as 9%, but averaged only 1.75%. Cachon and Fisher (2000) and Aviv and Federgruen (1998) studied the benefits of information flow in one warehouse multi-retailer systems. Both these studies observed that the benefits of information sharing under these settings were quite small, averaging around 2% in the case of Aviv and Federgruen and about 2.2% in the case of Cachon and Fisher. Gavirneni and Tayur (1999) studied the benefits of information in a setting where the retailer is using a target-reverting policy for placing orders. A target-reverting policy is one in which the retailer attempts to quickly get back to a previously published schedule in the event that the pre-determined schedule was not adhered to. In that situation, the benefits ranged from 6% to 28% and averaged around 11%. In Gavirneni (2001), I studied the benefits of information sharing in a one warehouse, multi-retailer setting and observed that savings could be as large as 27.5%, but averaged around 5%. While providing valuable insights into management of supply chains in the presence of information sharing, all these articles have failed to adequately answer an important question: *How should the supply chain structure and operating policies be changed in order to obtain the maximum benefit from these information flows?* The aforementioned studies incorporated information into the existing setup and none considered changing the structure and/or the operating procedures in order to make better use of the information. I believe that such a change must be considered if one wants to take full advantage of the information. There is a need for analysis of these supply chains centered on the inherent information flows. Such an information-centric design and management of capacitated supply chains will address the following issues:

1 How does one incorporate information flows into the decision making process?

2 How does one determine which information is useful and worth gathering? How much money can be invested in collecting the information?

3 How should the supply chain structure and operating policies be changed in order to make the best use of the information flows?

Supply chains come in many shapes and sizes. In addition, the operational characteristics (such as lead times, cost structures, yields, supplier capabilities) vary significantly from one to another. Supply chains in the retail industry tend to start at one place (distribution center or manufacturer) and diverge into many customer facing locations. Supply chains in the automotive or heavy equipment industry tend to involve a lot of assembly activities. As a result those supply chains have many suppliers shipping material into a central location. The pharmaceutical supply chains tend to have many stages, cross international boundaries, long leadtimes and also face many regulatory restrictions. Supply chains in the semi-conductor industry often involve complex, delicate manufacturing processes with significant yield losses and highly uncertain demands. Current knowledge in managing material, financial, and information flows in these supply chain leads us to believe that each of these supply chains should be treated individually. Rarely can observations on the benefits of information flows from one supply chain be extended to other supply chains. As a result, when studying the impacts of information, it is necessary to undertake research initiatives that encompass a wide variety of supply chain structures and operational characteristics.

I will demonstrate the benefits of information centric design and management of supply chains using three examples of different supply chain configurations. These examples were chosen to capture the presence of (i) significant setup or ordering costs; (ii) price fluctuations; and (iii) inventory allocation issues. These three characteristics of supply chains were identified by Lee, Padmanabhan, and Whang (1997) as the main reasons for information distortion. In section 2, I study a two stage supply chain with one supplier and one retailer facing end-customer demands. Due to the presence of a significant ordering cost, the retailer is using an (s, S) policy to manage inventories. Section 3 describes a two stage supply chain with a single supplier and a single retailer (facing i.i.d. end-customer demands) in which the supplier is charging the same price in every period. A single supplier, multi-retailer system is modeled and analyzed in section 4. For these three different supply chain configurations, I will propose, analyze, and compute the benefits of appropriate information centric policies that will significantly improve their performance. Section 5 contains ideas for future research and some closing remarks.

The models I study are discrete time periodic review non-stationary capacitated inventory control problems. The capacitated stationary inventory control problems were analyzed by Federgruen and Zipkin (1986a); Federgruen and Zipkin (1986b) and solution procedures for it were presented by Tayur (1993) and Glasserman and Tayur (1994); Glasserman and Tayur (1995). The capacitated non-stationary inventory control problem was the focus of articles by Kapuscinski and Tayur (1998), Gavirneni, Kapuscinksi, and Tayur (1999), and Scheller-Wolf and Tayur (1997). These three articles use Infinitesimal Perturbation Analysis (IPA) to solve these problems. I will use this approach as well and details on this method can be found in Glasserman (1991).

2. A two-stage supply chain with a retailer using (s, S) Policy

Consider a supply chain containing one capacitated supplier and a retailer facing i.i.d. demands for a single product. The supplier has finite production capacity, C. The end-customer demand distribution has cumulative distribution function (cdf) $\Psi(\cdot)$ and probability distribution function (pdf) $\psi(\cdot)$. The holding and penalty costs at the retailer are h_r and p_r respectively. They are h_s and p_s at the supplier. The costs and the demand distributions are known to both parties. There is a fixed ordering cost K between the retailer and the supplier. There are no lead times either at the retailer or at the supplier. The unsatisfied demands at the retailer are backlogged and the unsatisfied demands at the supplier are sent to the retailer using an expediting (e.g. overtime) strategy and p_s represents the cost of expediting. Thus, if needed, the retailer can order and receive an infinite quantity of the product in a period. All these assumptions are common in inventory control literature and in spite of its simple setup, this two stage supply chain can provide valuable insights into managing more complex systems efficiently. Cachon and Zipkin (1999), Gavirneni, Kapuscinksi, and Tayur (1999), and Gavirneni and Tayur (1999) have used settings similar to this one to understand the effect of cooperation on inventories in supply chains.

The sequence of events in this supply chain is as follows. (1) The supplier decides on her inventory level restricted by her production capacity. (2) The end-customer demands at the retailer are observed and the holding or penalty costs are incurred at the retailer. (3) The retailer places an order with the supplier, if necessary, to reach the desired inventory level. (4) The supplier satisfies (the product will be available at the retailer at the start of the next period) the retailer demands to the best of her abilities. (5) If there is inventory left at the supplier, she incurs holding costs and on the other hand if there is some unsatisfied demand, it is supplied by expediting and the costs of expediting are incurred.

For this supply chain I study two modes of operation at the retailer. In both models I assume that the retailer provides the supplier with information on the demands she is seeing in every period. In model 1, the retailer uses an (s, S) policy. That is, when her inventory falls below s, she orders up-to S; we know from Scarf (1962) that the (s, S) policy is optimal for the retailer in this case. Thus the retailer will not order every period, but provides information, to the supplier, on the end-customer demands she is experiencing. As these cumulative end-customer demands approach $S - s$, the supplier is able to predict more accurately whether she will receive an order from the retailer. She also will be able to better predict more accurately the size of demand if it would occur. Because of this predictability, her holding and penalty costs will decrease when compared to the situation in which the retailer did not provide this information. When the retailer is willing to provide this information I wish to ask the following questions: (1) is this the best way to manage this supply chain? and (2) are there ways to use the information to make the supply chain more efficient? For example, when the cumulative end-customer demand at the retailer is close to $S - s$, the supplier expects a demand and stocks inventory to meet it. If by chance, the next end-customer demand is very low and does not drop the retailer inventory below s, then the demand at the supplier is not realized and the supplier ends up incurring holding cost. There are ways to remove this uncertainty in timing of retailer demands and I formulate them in Model 2.

In model 2, the supplier and the retailer keep track of the cumulative end-customer demands since the previous retailer order. If at the end of a period, this cumulative demand is greater than a pre-specified value (denoted by δ), then the retailer must order after she has seen the next end-customer demand. In this case, the supplier knows for sure that there will be a demand in that period and can be better prepared to meet it. The supplier does not know the exact size of the order, but she knows the distribution from which it will be realized. For this model I will show that retailer uses an order up-to policy when she orders. I will also formulate the resulting non-stationary inventory control problem at the supplier and establish that her optimal policy is also order up-to, though the order up-to levels differ from one period to the next. In addition, I will choose (by exhaustive search) the δ value as the one with the lowest total cost. By using this policy I am removing some uncertainty at the supplier and this results in lower costs for her. But since an (s, S) policy is optimal for the retailer, moving to the operating policy in Model 2 is certain to increase her costs. In this paper, I want to study the relationship between these two opposing forces in the supply chain. I will show, via a detailed computational study, that if the δ value is chosen properly, the savings at the supplier are greater than the increase in costs at the retailer. Thus the total costs in the supply chain are reduced, making the supply chain more efficient.

2.1 The Models

In this section I analyze the two models described above. For each case I determine optimal policies for both the retailer and the supplier. I also present solution procedures for determining the optimal parameters.

2.1.1 Model 1 - The Traditional Model. Here the retailer is using the (s, S) policy that is optimal for her. The corresponding s and S values can be determined using an efficient solution procedure developed by Zheng and Federgruen (1991). In this setting the retailer does not order in every period, but informs the supplier about the end-customer demands. The non-stationary inventory control problem seen by the supplier was formulated and the relevant structural properties and solution procedures were described in detail in Gavirneni, Kapuscinksi, and Tayur (1999).

2.1.2 Model 2 - The Information Centric Model.

In this model I will consider a different operating policy at the retailer in the hope that the new operating policy will make better use of the information flow and thus improve the efficiency of the supply chain. Both the retailer and the supplier monitor the cumulative end-customer demand since the retailer last ordered. When this cumulative demand is greater than a predetermined value, δ, then the retailer must place an order after the next end-customer demand. Thus, the supplier knows a period ahead when demand is going to occur, but is not sure of the size of the order. She has a probability distribution from which this demand will be realized. Let us first look at the optimal retailer behavior under this strategy.

Retailer Behavior

To analyze the behavior of the retailer, it is necessary to pay close attention to the sequence of events. Let us assume that the problem is at the beginning of the first period in an n-period problem. Assume that the total end-customer demand since her last order is i and that she has y units of inventory on hand. Let $J_n(i, y)$ be the total cost of this n-period problem. If i is greater than δ, then she can place an order with the supplier at the end of this period after she has seen another end-customer demand. If i is less than δ, then she cannot place an order with the supplier and her next period will start with inventory $y - \xi$ and in state $i + \xi$ where ξ is the end-customer demand in this period. Thus:

$$J_n(i,y) = E_\xi[(y-\xi)^+ h_r + (y-\xi)^- p_r + V_{n-1}(i,\xi,(y-\xi))]$$

I use a^+ to represent $\max(0, a)$ and a^- to represent $\max(0, -a)$. In addition, E_a represents expectation with respect to the random variable a. $V_{n-1}(i, \xi, (y - \xi))$ is the optimal cost of an $n - 1$ period problem when the total end-customer demand until the previous period is i, the end-customer demand in this period is ξ and the inventory before the retailer orders, if she can, is $y - \xi$. This cost can be computed as follows:

$$\begin{aligned} V_{n-1}(i,\xi,(y-\xi)) &= J_{n-1}(i+\xi, y-\xi) \text{ If } i < \delta \\ &= \min_{x \geq (y-\xi)} J_{n-1}(0,x) \text{ Else} \end{aligned}$$

Starting with the initial condition $V_0(\cdot, \cdot, \cdot) = 0$, and using arguments (as detailed in Bertsekas (1988)) based on induction and convexity it can be shown that it is optimal for the retailer to order, when she is allowed to, up-to some fixed level y_*. This optimal order up-to level can be determined using IPA. When the retailer orders, she will be incurring a fixed ordering cost, but that cost does not figure in this optimization with a fixed δ. It will however, play a key role in determining the optimal δ value. Let y_*^δ be the optimal order up-to level corresponding to δ.

Property 1. If $\delta_1 < \delta_2$, then $y_*^{\delta_1} \leq y_*^{\delta_2}$.

Proof:. Let l_1 (l_2) be the number of periods it takes the cumulative end-customer demand to jump over δ_1 (δ_2). Clearly l_1 and l_2 are random and in addition $l_1 \leq_{st} l_2$. Let $G_1(\cdot)$ ($G_2(\cdot)$) be the distribution of end-customer demand over l_1+1 (l_2+1) periods. Since $G_1(\cdot) \leq_{st} G_2(\cdot)$, and $y^{\delta_1} = G_1^{-1}(\frac{p_r}{p_r+h_r})$ and $y^{\delta_2} = G_2^{-1}(\frac{p_r}{p_r+h_r})$, it follows that $y^{\delta_1} \leq y^{\delta_2}$. □

Once the optimal retailer behavior has been determined, I can analyze the inventory control problem at the supplier.

Supplier Behavior

The supplier faces a non-stationary inventory control problem which is defined below. In every period she is in one of many possible states. Her state is determined by the cumulative end-customer demand since the retailer last ordered. For all values of i less than δ, in state i she sees no demand in that period. On the other hand for i greater than δ, she sees a demand from the distribution with cdf $\Phi_i(\cdot)$ and pdf $\phi_i(\cdot)$. I know that these distributions are related to the end-customer demand distribution as follows:

$$\Phi_i(i+t) = \Psi(t)$$

Again by formulating the appropriate stochastic dynamic programming and using arguments of convexity and induction (see Bertsekas (1988)), I can show that the optimal policy at the supplier is a modified order up-to policy and these order up-to levels can be computed using IPA.

Thus for a given δ value, I know how to solve the problem in model 2. To find the optimal δ value that results in the lowest supply chain cost, I perform an exhaustive search over the possible set of values. I used this approach to perform a detailed study comparing the total supply chain costs of these two models. The results from this study are presented in the next section.

2.2 Computational Results

There are two principle objectives for this computational study: (1) Exhibit that using the strategy in the information centric model 2 does in fact result in a reduction in the total supply chain cost; and (2) Study how the reduction in cost is affected by various supply chain parameters such as capacity, fixed cost, holding and penalty costs, and demand variance. These sensitivity results should provide some insights into when the retailer should consider moving away from the locally optimal policy in order to realize a reduction in the total supply chain costs by enabling better use of information flows at the supplier.

The experimental setup for the study is as follows. The holding cost at the supplier is 1 while the penalty cost is allowed to take values 5, 8, and 11. The retailer was also setup similarly. The end-customer demand is assumed to have a mean of 20 and was sampled from distributions Exponential(20), Erlang(2,10), Erlang(4,5), Erlang(8,2.5), and Erlang(16,1.25). Thus the standard deviations of the end-customer demand were 20, 14.2, 10, 7.1, and 5 respectively. The production capacity at the supplier was allowed to take values 25, 45, and 65. Thus the capacity was always greater than the mean demand. For all these cases I computed the costs of models 1 and 2. Although in the previous section, I proposed an exhaustive search over all the possible values of δ, for ease of analysis I considered δ values from a smaller subset. When the setup cost was greater than or equal to 10, I used δ values ranging from 0 to 80 in multiples of 10. When the setup cost was lower than 10, I considered δ values from 0 to 10 in increments of 1. Using a more exhaustive search can only result in an improved performance for model 2. The difference between the costs of these two models can be attributed to better usage of the information flows. For each case, I computed the percentage reduction as follows:

$$\%\text{ reduction} = \frac{\text{traditional model cost} - \text{information centric model cost}}{\text{traditional model cost}} \times 100.$$

My observations from this computational study are detailed below. First I study the cost per period, then the optimal δ levels followed by the percentage reduction.

2.2.1 Cost per Period.

For both the models I observed that the cost per period increased with increase in demand variance, increased with increase in penalty cost, and decreased with increase in capacity. This behavior of the costs has been well documented in inventory control literature and thus I will not elaborate here. I also observed that in all but one of the 1215 cases, the cost of model 2 was lower than the cost of model 1. Thus I can conclude that, in general, model 2 makes better use of the information flows in this supply chain.

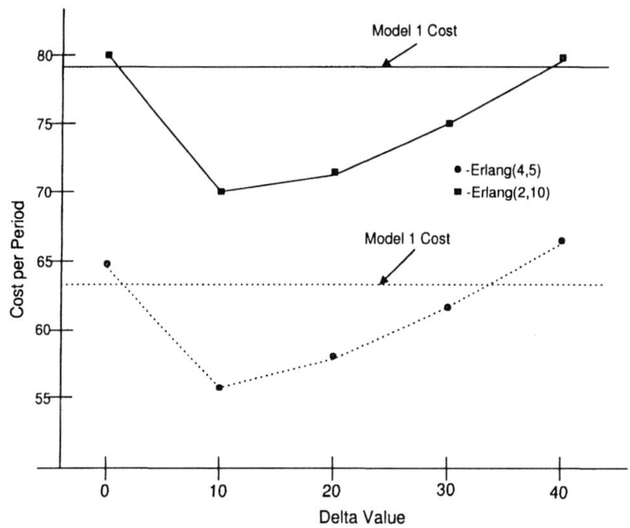

Figure 1.1. Cost per period as a function of δ value

2.2.2 Optimal δ Values. To determine the cost of model 2, I evaluated it under various values of δ and chose the δ value that resulted in the lowest cost. Figure 1.1 contains the plot of the cost per period as a function of δ for the Erlang(2,10) and Erlang(4,5) demand distributions. The retailer and supplier penalty costs were 5, the fixed ordering cost was 30, and the supplier capacity was 45. Notice that in both the cases, the optimal δ value was 10 and it was very easy to identify them. The situation was similar for all the other problem instances. The figure also contains the costs associated with model 1 for both the cases. It is worth noting that model 2 is more effective only when the δ values are chosen carefully. If they are selected arbitrarily, the performance of the supply chain could worsen. In addition I observed that the optimal δ values were (1) higher at higher capacities, (2) higher at higher fixed ordering costs, (3) lower at higher demand variances, and (4) lower at higher retailer or supplier penalty costs.

2.2.3 Percentage Reduction. In this section I will take a detailed look at the percentage reduction in cost realized by using model 2 in place of model 1. The percentage was positive in all but one (with setup cost 110, standard deviation of demand 20, capacity 25, supplier penalty cost 5, and retailer penalty cost 11) of the 1215 cases, and ranged from -0.44% to 33.7%, and averaged around 10.4%. This reduction is of significant size when compared to the savings, due to information shar-

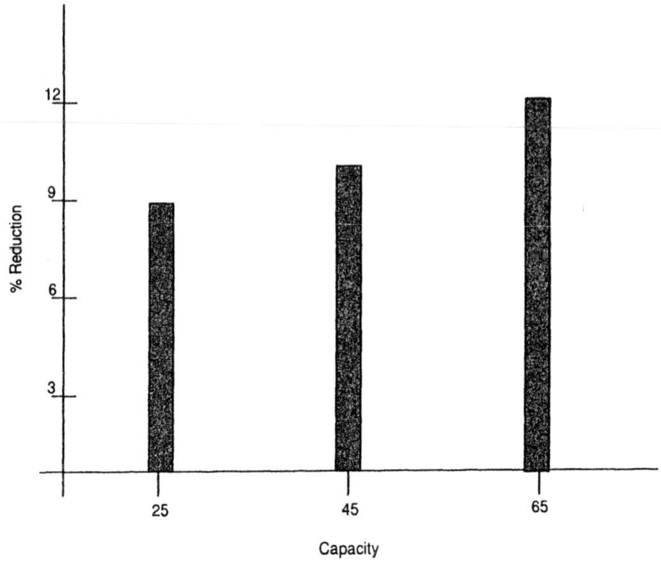

Figure 1.2. Percentage reduction as a function of capacity

ing, reported by Chen (1998), Cachon and Fisher (2000), and Aviv and Federgruen (1998). Thus in many cases, it would be better for both the supplier and the retailer to use the strategy in model 2. Clearly the retailer costs in model 2 will be higher than in model 1. But if the supplier was willing to share some of her savings, both the parties would be better off and the supply chain could be more efficient. However, if the setup cost or demand variance are extremely large, this strategy may not be effective. Let us take a closer look at how the supplier capacity, the penalty costs, and the demand variance affected the relative performance of model 2.

The Effect of Capacity

Figure 1.2 contains the average percentage reduction as a function of the supplier capacity. Model 2 was more effective at higher values of supplier capacity. The main reason for this behavior is the flexibility that additional capacity provides the supplier. If the supplier is not able to (due to tight capacity) react to the more effective information flows in model 2, there would be no reduction in cost. Thus when the supplier has higher capacity, she is able to use the information flows efficiently and reduce her costs more significantly. Thus, the strategy in model 2 makes the supply chain more efficient at larger supplier capacities.

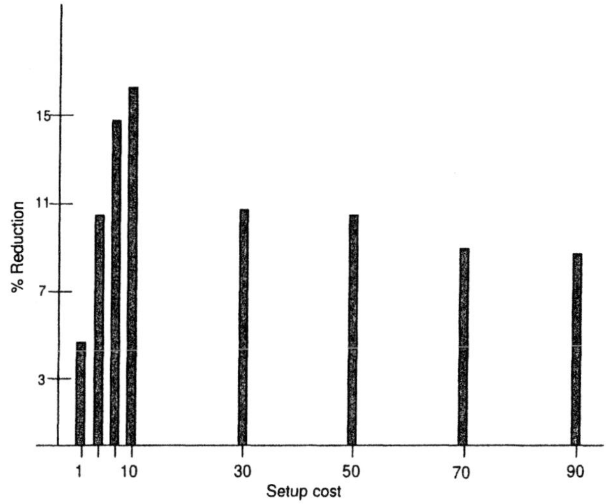

Figure 1.3. Percentage reduction as a function of setup cost

The Effect of Fixed Ordering Cost

The average relative performance of model 2 as a function of the fixed ordering cost is given in figure 1.3. The fixed cost K figures prominently in determining the optimal parameters for the two models. In model 1, the s and S are chosen in an optimal (at the retailer) fashion and for model 2, the fixed cost plays a role in determining the optimal δ value. I observed that, not surprisingly, at higher fixed costs, the optimal δ values were higher. The fact that savings in cost are lower at higher fixed costs can be explained as follows. At higher fixed ordering cost, the retailer orders (less frequently) larger amounts, and the presence of finite capacity requires the supplier to start producing well ahead of time. This reduces her ability to react to unexpected changes at the retailer and the effectiveness of model 2 is reduced. On the other hand, when the fixed costs are low, both models require that the retailer orders very frequently, thus reducing the difference in their performance. Thus, this strategy is most effective at moderate values of the fixed cost.

The Effect of Supplier and Retailer Penalty Costs

Figures 1.4 and 1.5 illustrate how the savings of model 2 are affected by the penalty costs at the supplier and the retailer respectively. Notice that model 2 performs better at higher supplier penalty costs and at lower retailer penalty costs. I observed this behavior consistently among all the distributions. The main reason for this behavior is the way the costs at the retailer and the supplier change under model 2. Recall that

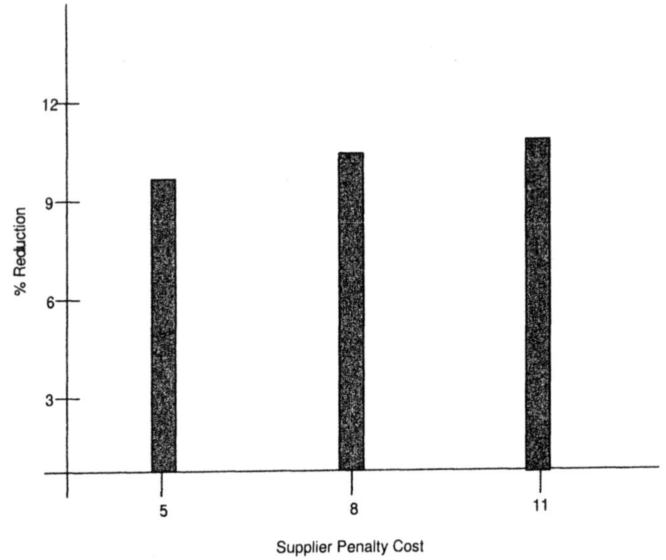

Figure 1.4. Percentage reduction as a function of supplier penalty cost

under model 2, the retailer is using a sub-optimal policy and her costs are increased while the costs at the supplier are decreased due to reduction in demand uncertainty. When her penalty costs are higher, the supplier realizes larger savings and the savings in model 2 are higher. On the other hand, when the penalty costs at the retailer are higher, her costs under model 2 increase more dramatically resulting in less effectiveness. Thus when the supplier penalty costs are high and the retailer penalty costs are low, the strategy in model 2 is more effective.

The Effect of Demand Variance

Figure 1.6 plots the average performance of model 2 as a function of the standard deviation of the end-customer demand. Notice that as the demand variance decreases the average performance of model 2 increases. Recall that while model 2 has no uncertainty about the timing of retailer demands, the quantity demanded is still uncertain. Thus when the end-customer demand has a high variance, the resulting uncertainty at the supplier is large even for model 2. Thus its performance is better at lower demand variances.

2.3 Conclusions

From the study of these two models, I conclude that using the information centric strategy defined in model 2, the information flows in

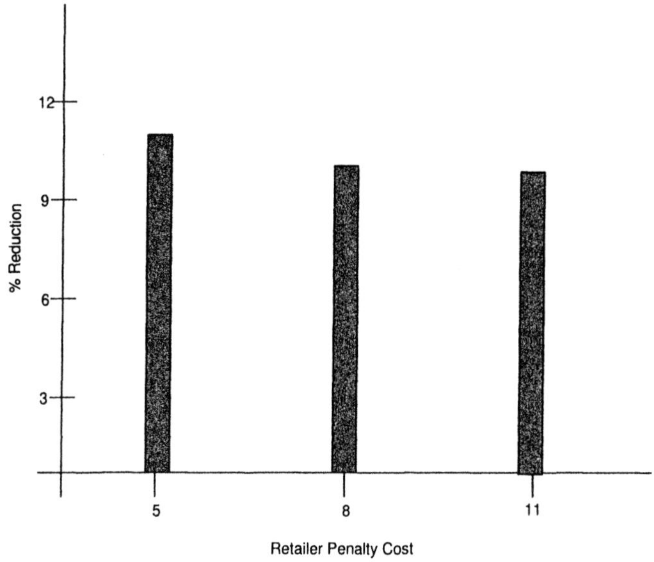

Figure 1.5. Percentage reduction as a function of retailer penalty cost

this two-stage supply chain can be better utilized resulting in an improvement (by as much as 34%) in the supply chain performance. This improvement is more dramatic when one or more of the following conditions hold: (1) the supplier capacity is high, (2) the fixed ordering cost is low, (3) the supplier penalty cost is high, (4) the retailer penalty cost is low, and (5) the demand variance is low.

3. Price Fluctuations and Supply Chain Performance

In this section, I consider the supply chain consisting of one supplier, one retailer, and one product. Existing research advocates that, in a decentralized setting, it is efficient that the retailer and the supplier use stationary order up-to policies. I show that in the presence of information sharing, the supply chain performance can be improved by the supplier offering fluctuating prices which in turn make the retailer and the supplier move away from stationary policies.

In the supply chain studied in this section, there is a single supplier with finite production capacity, C, supplying a single product to a newsvendor type retailer who is in turn facing independent and identically distributed demands (with cdf $\Psi(\cdot)$ and pdf $\psi(\cdot)$) from end-customers. The holding and penalty costs are respectively h_r and p_r at the retailer and h_s and p_s at the supplier. The costs and the demand

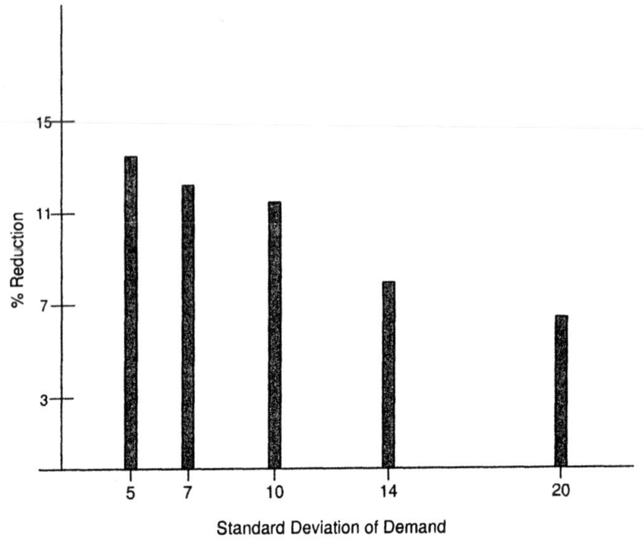

Figure 1.6. Percentage reduction as a function of standard deviation

distributions are known to both parties. There are no fixed ordering costs or lead times either at the retailer or the supplier. The unsatisfied demands at the retailer are backlogged and the unsatisfied demands at the supplier are sent to the retailer using an expediting strategy and p_s represents the cost of expediting. Thus, if needed, the retailer can order and receive an infinite quantity of the product in a period. All these assumptions are common in inventory control literature and most of them, except the one on ordering costs, can be relaxed without significantly changing the general behavior of the system. Cachon and Zipkin (1999) studied a setting similar to this one. They used game theoretic models to study the impact on inventory levels of competition and cooperation between the retailer and the supplier.

I study this supply chain under a periodic setting and the sequence of events in every period is as follows: (1) The supplier decides (restricted by her capacity) how much to produce. The product is available immediately; (2) The retailer faces the end-customer demand and satisfies it to the best of her abilities. Unsatisfied demands are backlogged; (3) The retailer decides how much to order from the supplier; (4) The supplier satisfies the retailer's demand to the best of her abilities. Unsatisfied demands are supplied through the expedited source. The product is available to the retailer at the beginning of the next period; (5) The holding and penalty costs at both the retailer and the supplier are computed and the problem goes to the next period. I measure the performance of this supply chain using the total holding and penalty costs at both the retailer and the supplier. Since the purchase costs between the retailer and the supplier are internal to the supply chain, they are not explicitly included in the total supply chain cost. The objective here is to study the effect of price fluctuations (at the supplier) and information sharing (between the retailer and the supplier) on the performance of this supply chain.

I study the interaction between these two strategies in this supply chain by formulating and analyzing the retailer and supplier behavior in two different models. In Model 1 (the everyday low price (EDLP) Model), the supplier charges the retailer the same price (c dollars per unit) in every period. In this setting, it is optimal for the retailer to use a stationary order up-to policy with the order up-to level z in every period. Thus the end-customer demands at the retailer are transmitted to the supplier without any change and the supplier sees i.i.d. demands in every period. In every period, the supplier is completely aware of the inventory level at the retailer and there is no need for the retailer to provide additional information. In Model 2 (the HI-LO pricing Model), the supplier alternates the selling price between c' and $c' - \epsilon$ from one period to the next. This leads to the retailer using an ordering pattern that repeats every two periods. In every cycle of two periods, the first period has an order up-to level z' while the second period has the order up-to level $z' + \Delta_\epsilon$. Under this retailer inventory policy, the demands seen by the supplier are no longer i.i.d. I characterize the information (retailer inventory policy parameters in setting 1 and retailer inventory levels in setting 2) available to the supplier and formulate the resulting non-stationary inventory control problem she faces. Though the variance of demands seen by the supplier is increased, the benefits realized from the associated information flow will result in lower costs at her location. In addition, I will show that this reduction in costs at the supplier far outweighs the increase in the retailer's costs. Thus, if the supplier is willing to share some of the benefit she realizes with the retailer, the retailer may be willing to provide the inventory information and the whole supply chain will be more efficient.

While the ways in which the prices at the supplier can be made to fluctuate are numerous, I restrict my attention to fluctuations that repeat every two periods. This is very similar to the *HI-LO* pricing popular among many suppliers. As will be seen later in the section on the computational results, I further assume that these fluctuations are symmetric around the price offered in the constant pricing scheme. Under these assumptions, to determine the optimal fluctuating pricing scheme, one needs only to search over the possible values of the ϵ value. I develop an efficient procedure to determine the optimal supplier and retailer behavior for a given value of ϵ and use that to search over the set of the possible ϵ values to determine the optimal fluctuating pricing scheme.

Designing efficient supply contracts has recently been a favorite topic of many in the supply chain management research community. Anupindi and Bassok (1999), Cachon (1999), and Lariviere (1999) are excellent sources of information on this topic. It is not surprising that pricing plays an important role in designing good supply contracts. Pasternack (1985) and Lee, et al. (2000) showed that price protection (a method for compensating the retailer for excess inventory at her location) is a strategy that the supplier can use to achieve channel coordination. Chen, Federgruen, and Zheng (2001) have shown that in order to achieve channel coordination in a supply chain with non-identical retailers, a discounting scheme based on three quantities (namely annual sales volume, order quantity, and order frequency) is necessary. Munson and Rosenblatt (2001) explored the benefits of using quantity discounts in a three level supply chain and showed that savings can be significant. However few researchers have specifically looked at price fluctuations and the role they play in improving supply chain performance. Iyer and Ye (2000) studied price fluctuations at the retailer and their effect on grocery supply chains. They observed that if the supplier obtains information about the price fluctuations at the retailer, in some cases she can use that information to improve her performance. In this paper, I focus on the effect of price fluctuations at the supplier and their impact on the performance of the whole supply chain.

3.1 Two Models

In this section I study two inventory control problems which differ in the way the supplier prices the product for the retailer. I establish the corresponding optimal policies for the retailer and the supplier and develop efficient solution procedures for computing the optimal parameters.

3.1.1 Model 1 - EDLP Model. In this model the supplier charges the retailer c dollars per unit in every period. Under this setting, it is optimal for the retailer to use a stationary order up-to policy with the order up-to level z. Based on the assumption of expediting at the supplier, z is the newsvendor solution for the retailer. Thus

$$z = \Psi^{-1}\left(\frac{p_r}{h_r + p_r}\right)$$

Under this retailer ordering policy the demands seen by the supplier are i.i.d. with cumulative distribution function $\Phi(\cdot)$ and density function $\phi(\cdot)$. In addition, the distribution $\Phi(\cdot)$ is exactly equal to the distribution $\Psi(\cdot)$. Based on Federgruen and Zipkin (1986a); Federgruen and Zipkin (1986b) a modified order up-to policy is optimal for the supplier. The optimal order up-to level, y, while not available in closed form, can be computed using IPA. Details on IPA validation and implementation can be found in Glasserman and Tayur (1994); Glasserman and Tayur (1995).

When the retailer uses a stationary order up-to policy, it presents a stable environment for the supplier. Since the retailer starts at her optimal order up-to level in every period and the end-customer demand is transmitted unaltered to the supplier, the supplier is fully aware of the inventory position at the retailer. There is no additional information that can be exchanged between the two.

3.1.2 Model 2 - HILO Pricing Model.

In this section, I model the case in which the supplier charges the retailer fluctuating prices from one period to the next. Specifically, I will assume that the supplier alternates the selling price between c' and $c' - \epsilon$ from one period to the next. Under this setting I will study the optimal retailer and supplier behavior.

3.1.3 Retailer Behavior: Model 2.

Property 2. When the unit selling price at the supplier alternates between c' and $c' - \epsilon$, the retailer optimal order up-to level alternates between z' and $z' + \Delta_\epsilon$.

Proof. This policy with cyclic order up-to levels follows from Karlin (1960) and Zipkin (1989) as a special case of cycle length equal to 2. □

When the retailer uses this ordering policy, the demands seen by the supplier are no longer i.i.d. In the next section I formulate the corresponding non-stationary inventory control model at the supplier and determine her optimal policy.

3.1.4 Supplier Behavior: Model 2.

For this model, I will analyze the supplier behavior for two specific settings. In setting 1, the supplier is only aware of the retailer inventory policy parameters (namely z' and $z' + \Delta_\epsilon$) and in setting 2, in addition to knowing the retailer policy parameters, the supplier obtains information about the day-to-day inventory levels at the retailer.

3.1.5 Setting 1: Information on Retailer Policy Parameters.

Since the retailer ordering policy follows a two period pattern, the demands at the supplier also will exhibit a cyclic pattern with a cycle time of two periods. In the first period, the demand, d, at the supplier is either zero (if ξ_1 is less than Δ_ϵ) or $\xi_1 - \Delta_\epsilon$ (if ξ_1 is greater than Δ_ϵ) where ξ_1 is the end-customer demand seen at the retailer. Let us call the state the supplier is in as state 1. In the next period, she is in one of two possible states. I will say that she is in state 2 if the demand from the retailer was zero in the previous period. On the other hand, if the retailer order in the previous period was non-zero, then I will say that the supplier is in state 3. In state 2, the demand seen by supplier is $\xi_1 + \xi_2$ where both ξ_1 and ξ_2 are end-customer demands from the distribution $\Psi(\cdot)$ and ξ_1 is less than Δ_ϵ. In state 3, the supplier sees demand $\xi_2 + \Delta_\epsilon$. For ease of presentation, I will say that $\Phi_i(\cdot)$ ($\phi_i(\cdot)$) is the cdf (pdf) of the distribution of demand seen by the supplier in state i. Clearly $\Phi_1(\cdot) \leq_{st} \Phi_2(\cdot) \leq_{st} \Phi_3(\cdot)$. From states 2 and 3, the supplier transitions into state 1 in the next period. The transition probability from state 1 to 2 is $\Psi(\Delta_\epsilon)$ and the transition probability from state 1 to 3 is $1 - \Psi(\Delta_\epsilon)$.

I first present some structural properties for this problem and also discuss ways for computing the optimal solutions.

Structural Properties: Setting 1

Property 3. For finite horizon and infinite horizon (discounted and average cost) a modified order up-to policy is optimal.

Proof. Let $L_i(y)$ be the one-period cost in state i with inventory level y. It can be computed as follows:

$$L_i(y) = h \int_0^y (y-t)\phi_i(t)dt + p_s \int_y^\infty (t-y)\phi_i(t)dt$$

Let $V_n^i(x)$ be the optimal cost of an n-period problem starting in state i with inventory level (before production) x. $V_n^i(x) = \min_{y \in [x, x+C]} J_n^i(y)$ where

$$J_n^1(y) = L_1(y) + \Psi(\Delta)V_{n-1}^2(y) + \int_\Delta^\infty V_{n-1}^3(y-t+\Delta)^+\psi(t)dt$$

$$J_n^2(y) = L_2(y) + \int_0^\infty V_{n-1}^1(y-t)^+\phi_2(t)dt$$

$$J_n^3(y) = L_3(y) + \int_0^\infty V_{n-1}^1(y-t)^+ \phi_3(t)dt$$

Since there are no salvage costs at the end of horizon, I have $V_0^i() \equiv 0$ for all i. Since $L_i(y)$ is convex in y, it is easy to establish (see Kapuscinski and Tayur (1998), and Scheller-Wolf and Tayur (1997)) via induction that for all values of n and i:

1 $J_n^i(y)$ is convex in y;

2 $V_n^i(x)$ is convex in x; and

3 a modified order up-to policy is optimal.

Let z_n^i denote the optimal order up-to level for an n-period problem starting in state i. Based on these convexity properties and using results from pages 210-212 (for the discounted case) and pages 310-313 (for the average case) of Bertsekas (1988), I can establish that for each state i, the infinite horizon optimal policy is modified order up-to y_i^*. □

Property 4. The order up-to levels are ordered as follows: (1) $y_1^* \leq y_3^*$, and (2) $y_2^* \leq y_3^*$.

Proof. I use $'$,$''$ to denote first and second derivatives respectively.

$$\begin{aligned}
J_n^{'1}(y) &= L_1'(y) + \Psi(\Delta)V_{n-1}^{'2}(y) + \int_\Delta^{y+\Delta} V_{n-1}^{'3}(y-t+\Delta)^+ \psi(t)dt \\
J_n^{'2}(y) &= L_2'(y) + \int_0^y V_{n-1}^{'1}(y-t)\phi_2(t)dt \\
J_n^{'3}(y) &= L_3'(y) + \int_0^y V_{n-1}^{'1}(y-t)\phi_3(t)dt \\
V_n^{'i}(x) &= \max\{J_n^{'i}(x), \min\{0, J_n^{'i}(x+C)\}\};
\end{aligned}$$

Since $\Phi_1(\cdot) \leq_{st} \Phi_2(\cdot) \leq_{st} \Phi_3(\cdot)$, I know that $L_1'(y) \geq L_2'(y) \geq L_3'(y)$ for all values y. Using this relation and starting from the initial condition $V_0^{'i} = 0\ \forall i$, using induction I can show that $V_n^{'1}(x) \geq V_n^{'3}(x)$ and $V_n^{'2}(x) \geq V_n^{'3}(x)$ for all values of n and x. This leads us to conclude that $z_n^1 \leq z_n^3$ and $z_n^2 \leq z_n^3$ for all n. This ordering must also hold for infinite horizon order up-to levels. □

The order up-to level in state 1 is lower than the order up-to level in state 2 and the order up-to level in state 2 will be lower than that in state 3. This follows from the stochastic ordering of the demand distributions. Let y_i^C be the optimal order up-to level in state i when the supplier capacity is C. At lower capacities, the order up-to levels will be higher.

Property 5. For two different capacities C_1 and C_2 such that $C_1 \leq C_2$, $y_i^{C_1} \geq y_i^{C_2}$ for all i in $\{1,2,3\}$.

Proof. Let ${}^{C_1}J_n^i(.)$, ${}^{C_1}V_n^i(.)$, ${}^{C_1}z_n^i$ and ${}^{C_2}J_n^i(.)$, ${}^{C_2}V_n^i(.)$, ${}^{C_2}z_n^i$ be the quantities defined above when the capacities are C_1 and C_2 respectively. First I prove that if $C_1 \leq C_2$ then

1 ${}^{C_1}J_n'^i(x) \leq {}^{C_2}J_n'^i(x)$;

2 ${}^{C_1}V_n'^i(x) \leq {}^{C_2}V_n'^i(x)$; and

3 ${}^{C_1}z_n^i \geq {}^{C_2}z_n^i$.

I first prove (1) and (2) by induction. They are obviously true for n=0. Assume they are true for $n-1$. After comparing ${}^{C_1}J_n'^i(x)$ and ${}^{C_2}J_n'^i(x)$, and using (2) for $n-1$, it is easily established that ${}^{C_1}J_n'^i(x) \leq {}^{C_2}J_n'^i(x)$. Furthermore from the convexity of J_n and inductional assumption ${}^{C_1}J_n'^i(x+C_1) \leq {}^{C_1}J_n'^i(x+C_2) \leq {}^{C_2}J_n'^i(x+C_2)$. Using the expression for $V_n'^i(x)$ given above and the observation that if $A \geq B$ then $\min(0,A) \geq \min(0,B)$ and $\max(0,A) \geq \max(0,B)$, it is easily established that ${}^{C_1}V_n'^i(x) \leq {}^{C_2}V_n'^i(x)$. So, by induction parts (1) and (2) of the property are true for all n. Since both ${}^{C_1}V_n^i(x)$ and ${}^{C_2}V_n^i(x)$ are convex and ${}^{C_1}V_n'^i(x) \leq {}^{C_2}V_n'^i(x)$, I have ${}^{C_1}z_n^i \geq {}^{C_2}z_n^i$. This proves part (3). This relation will be valid for the infinite horizon order up-to levels as well. □

Let $y_i(\Delta)$ be the optimal order up-to level in state i when the retailer ordering policy is $\{z', z'+\Delta\}$.

Property 6. For values Δ_1 and Δ_2 such that $\Delta_1 \leq \Delta_2$, $y_i^*(\Delta_1) \leq y_i^*(\Delta_2)$ for $i \in \{2,3\}$, and $y_1^*(\Delta_1) \geq y_1^*(\Delta_2)$

Proof. Let ${}^{\Delta_1}\Phi^i(.)$, ${}^{\Delta_1}L_i(.)$, ${}^{\Delta_1}J_n^i(.)$, ${}^{\Delta_1}V_n^i(.)$, ${}^{\Delta_1}z_n^i$ and ${}^{\Delta_2}\Phi^i(.)$, ${}^{\Delta_2}L_i(.)$, ${}^{\Delta_2}J_n^i(.)$, ${}^{\Delta_2}V_n^i(.)$, ${}^{\Delta_2}z_n^i$ be the quantities defined above corresponding to two different values Δ_1 and Δ_2 such that $\Delta_1 \leq \Delta_2$. It is easy to establish that ${}^{\Delta_1}\Phi^1(.) \geq_{st} {}^{\Delta_2}\Phi^1(.)$ and ${}^{\Delta_1}\Phi^i(.) \leq_{st} {}^{\Delta_2}\Phi^i(.)$ for $i \in \{2,3\}$. It follows that ${}^{\Delta_1}L_1'(.) \leq {}^{\Delta_2}L_1'(.)$ and ${}^{\Delta_1}L_i'(.) \geq {}^{\Delta_2}L_i'(.)$ for $i \in \{2,3\}$. Starting with the initial condition ${}^{\Delta_1}V_0^i(.) = {}^{\Delta_2}V_n^i(.) = 0$, and using induction I can establish that ${}^{\Delta_1}V_n'^1(.) \leq {}^{\Delta_2}V_n'^1(.)$ and ${}^{\Delta_1}V_n'^i(.) \geq {}^{\Delta_2}V_n'^i(.)$ for $i \in \{2,3\}$. This leads us to conclude that

${}^{\Delta_1}z_n^1(.) \geq {}^{\Delta_2}z_n^1(.)$ and ${}^{\Delta_1}z_n^i(.) \leq {}^{\Delta_2}z_n^{\prime i}(.)$ for $i \in \{2,3\}$. Since these relations hold for all n, they must hold for the infinite horizon order up-to levels as well. □

In states 2 and 3, myopic order up-to levels that minimize the cost of a single period are upper bounds on the optimal order up-to levels. Since $\Delta_\epsilon + \Psi^{-1}(\frac{p_s}{h_s+p_s})$ is the myopic solution for state 3, based on property 4, I can say that:

Property 7. For all states $i \in \{1,2,3\}$, $y_i^* \leq \Delta_\epsilon + \Psi^{-1}(\frac{p_s}{h_s+p_s})$.

Proof. Observe that $L_i'(\Delta_\epsilon + \Psi^{-1}(\frac{p_s}{h_s+p_s})) \geq 0$ for all i. This leads, via induction, to the fact that $V_n^{\prime i}(\Delta_\epsilon + \Psi^{-1}(\frac{p_s}{h_s+p_s})) \geq 0$ for all i implying that z_n^i must be smaller than $\Delta_\epsilon + \Psi^{-1}(\frac{p_s}{h_s+p_s})$ for all values of n and i. Thus the infinite order up-to levels y_i must be smaller than $\Delta_\epsilon + \Psi^{-1}(\frac{p_s}{h_s+p_s})$. □

Since the unsatisfied demands at the supplier are lost (due to expediting), when the supplier capacity is greater than $\Delta_\epsilon + \Psi^{-1}(\frac{p_s}{h_s+p_s})$, I can assume that the supplier is uncapacitated.

Property 8. When $C \geq \Delta_\epsilon + \Psi^{-1}(\frac{p_s}{h_s+p_s})$, there exists $\Delta^u = \Psi^{-1}(\frac{p_s}{h_s+p_s})$ such that for all values of $\Delta_\epsilon \geq \Delta^u$, $y_1(\Delta_\epsilon) = 0$.

Proof. Define Δ^u such that $\Psi(\Delta^u) = \frac{p_s}{h_s+p_s}$ which will result in the fact that for any $\Delta \geq \Delta^u$, $L_1'(0) \geq 0$. Since $C \geq \Delta_\epsilon + \Psi^{-1}(\frac{p_s}{h_s+p_s})$, based on Property 7 I know that the optimal order up-to level can always be reached leading to the relation $V_n^i(0) = 0$ for all $i \in \{2,3\}$. Combining this with the fact that $L_1'(0) \geq 0$ will lead us to conclude that $V_n^{\prime 1}(0) \geq 0$. Thus the optimal order up-to level in state 1, z_n^1, must be at most zero. Given that it cannot be smaller than zero, it must be equal to zero. □

These properties will be useful in developing solution procedures for computing the optimal order up-to levels.

3.1.6 Solution Procedures: Setting 1. In this section I propose efficient solution procedures for this non-stationary inventory control problem. I do this in two stages. First, I develop a procedure for the uncapacitated situation. This procedure, while not applicable for the capacitated case, will be helpful in providing starting solutions and developing heuristic procedures.

Uncapacitated Situation. When the production capacity is infinite and the demand distribution is stationary, the optimal order up-to level is given by the newsvendor formula. When the demands are non-stationary, the optimal order up-to levels are generally not available in closed form. However efficient procedures for computing them are available. Song and Zipkin (1993) present solution procedures for the continuous time non-stationary problem under the additional assumption of Poisson demands. Karlin (1960) and Zipkin (1989) present solution procedures for the discrete time problem when the demands are cyclic. However, these solution procedures are not useful here since the state in a period sometimes is dependent on the demand observed in the previous period. Gavirneni and Tayur (2001) have a solution procedure that is applicable here. Their approach is based on recursively estimating the derivatives of the cost function and has been shown to be very efficient. I will use their procedure for the uncapacitated situation.

Capacitated Situation. Capacitated inventory control problems are hard to solve even when demands are stationary. In the presence of non-stationarities, they are particularly hard, and closed form solutions do not exist for these order up-to levels. So I resort to IPA. IPA is a simulation-based solution procedure. During the simulation run, while computing the costs of the system, IPA also computes the derivatives of the costs with respect to the order up-to levels. Using these derivatives in a gradient based search procedure, I can iteratively compute the optimal order up-to levels. Glasserman (1991) provides a good introduction to this technique. For its application in multi-echelon inventory control models, the reader is referred to Glasserman and Tayur (1994); Glasserman and Tayur (1995). To use the IPA procedure I need to establish that the derivatives estimated from the simulation are in fact valid. This validation can be achieved here by using arguments similar to those found in Kapuscinski and Tayur (1998) and Scheller-Wolf and Tayur (1997).

3.1.7 Setting 2: Information on Retailer Inventory Levels.

In this section, I consider the situation when the retailer does not order in a particular period (i.e., the end-customer demand is less than Δ_ϵ), she informs the supplier about her inventory level. This information can be used by the supplier to accurately predict her demand. She now knows that in the next period the demand she sees will be $\xi_1 + \xi_2$ where ξ_1 is known and ξ_2 is from the distribution $\Psi(\cdot)$. Using this information, she can further reduce her costs (we know for sure that her costs cannot increase) while the costs at the retailer, when compared to those in setting 1, are not affected. Thus the supply chain will probably be more efficient in setting 2 than in setting 1.

The non-stationarity of the demands at the supplier can be formulated as follows. In the first period, the demand, d, at the supplier is either zero (if ξ_1 is less than Δ_ϵ) or $\xi_1 - \Delta_\epsilon$ (if ξ_1 is greater than Δ_ϵ) where ξ_1 is the end-customer demand seen at the retailer. Let us call this state 0 for the supplier. In the next period, she is in one of Δ_ϵ possible states. I will say that she is in state ξ_1 (which is known to the supplier) if the demand from the retailer was zero in the previous period. In this case I know that ξ_1 is less than Δ_ϵ. On the other hand, if the retailer order in the previous period was non-zero, then I will say that the supplier is in state Δ_ϵ. In state i, the demand seen by supplier is $i + \xi_2$ where ξ_2 is an end-customer demand from the distribution $\Psi(\cdot)$. For convenience in notation I have assumed that the end-customer demands have been discretized and are strictly positive. I will say that $\Phi_i(\cdot)$ ($\phi_i(\cdot)$) is the cdf (pdf) of the distribution of demand seen by the supplier in state i. Clearly $\Phi_i(\cdot) \leq_{st} \Phi_{i+1}(\cdot)$ for all values of i in $\{0, 1, 2, \ldots, \Delta_\epsilon\}$. From a state $i > 0$, the supplier transitions into state 0 in the next period. The transition probabilities from state 0 to other states can also be appropriately determined. Using arguments of convexity and induction, I can show that:

Property 9. For finite horizon and infinite horizon (discounted and average cost) a modified order up-to policy is optimal. □

The structural properties established for setting 1 can be appropriately extended to cover setting 2. In addition, the optimal policy parameters can be computed efficiently using IPA. Once the optimal solutions have been characterized, and efficient solution procedures have been determined, it is time to determine the extent of the benefits that are realizable by these strategies.

3.2 Computational Results

The experimental setup for this study is as follows. The holding cost at the supplier is 1 while the penalty cost is allowed to take values 5, 8, and 11. The end-customer demand is assumed to have a mean of 20 and was sampled from distributions Exponential(20), Erlang(10,2), Erlang(5,4), and Erlang(2.5,8). Thus the standard deviations of the end-customer demand were 20, 14.1, 10, and 7.1 respectively. The production capacity at the supplier was allowed to take values 25, 45, and 65. So, the capacity was always greater than the mean demand. For model 1, the cost at the supplier was kept constant at five dollars per unit. In model 2, I let the cost at the supplier alternate between $5.0 + K * 0.25$ and $5.0 - K * 0.25$. I computed the total supply chain costs for values of K ranging from 0 to 5 and chose the value of K that resulted in the lowest total supply chain costs. Since the case $K = 0$ represents the case of stationary policies, I know that this optimization can never result in increased supply chain costs. Computation results also demonstrate that, in many cases, the total costs of the supply chain were reduced. It is also possible to use a finer grid for the purchasing costs by changing the factor 0.25 to 0.1. Since the key factor here is the difference in costs, the fact that I only consider symmetric fluctuations in selling prices does not significantly affect the results.

The difference between the costs of these two models can be attributed to price fluctuations and information sharing. For each case, I computed the percentage benefit of these strategies as follows:

$$\% \text{ benefit} = \frac{\text{EDLP model cost} - \text{HILO model cost}}{\text{EDLP model cost}} \times 100.$$

Our observations from this computational study are detailed below.

3.2.1 Relationship Between Δ and ϵ. In an infinite horizon inventory control problem, the relationship between ϵ and Δ is independent of c' as long as it is greater than ϵ. Clearly, when $\epsilon = 0$, the value of Δ corresponding to it is also zero. As ϵ increases, the Δ value increases (see Figure 1.7). For end-customer demand distributions that are reasonably well behaved (continuous and differentiable), this relationship between ϵ and Δ is also well behaved (see Figure 1.7) and for a given value of ϵ, I can easily determine the corresponding Δ_ϵ. For the experiments that resulted in Figure 1.7, I assumed that the holding and penalty costs (h_r and p_r) at the retailer were 1 and 5 respectively.

3.2.2 Cost per Period. I observed that, in both settings, the cost per period increased with an increase in demand variance, increased with an increase in penalty cost, and decreased with an increase in capacity. This behavior of the costs has been well documented in inventory control literature and thus I will not elaborate on this here.

3.2.3 Supplier Order up-to Levels: Setting 1. Figure 1.8 contains the plot of the optimal supplier order up-to levels (for the case Erlang(10,2), $C = 65, p_s = 11$) for the three states as a function of the Δ value. Notice that these order up-to levels satisfy the properties 5, 6,

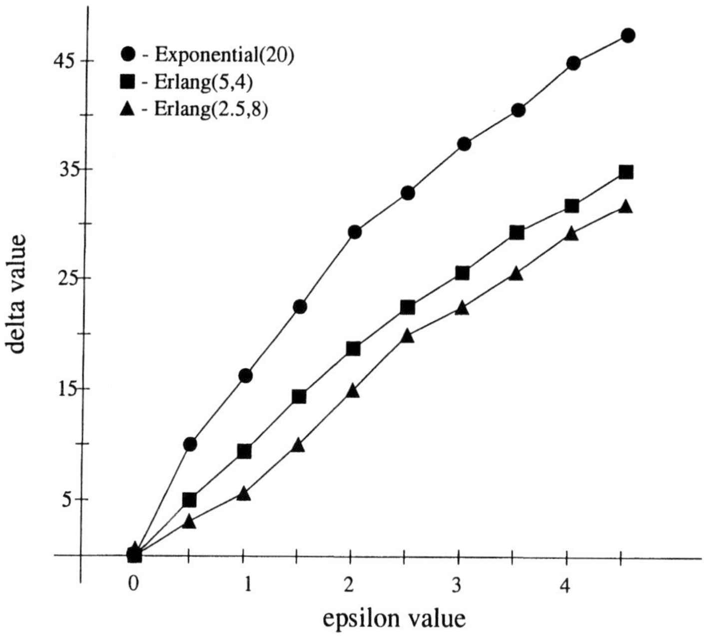

Figure 1.7. The plot of Δ as a function of ϵ

and 7 established in the previous section. The order up-to level for state 3 was always greater than the order up-to levels for states 1 and 2. The order up-to level for state 1 decreased with increase in the Δ value while the order up-to levels for states 2 and 3 increased. At a large Δ value, the order up-to level in state 1 drops to zero and remains there for all higher values.

3.2.4 Reduction in Total Supply Chain Costs. In this section, I present the results on percentage benefit gained when these strategies are implemented. First, I detail the results for setting 1 in which the supplier is only aware of the retail inventory policy parameters.

Setting 1: Information on Retailer Policy Parameters

I observed that while the total supply chain costs were not reduced for the demand distributions of Erlang(10,2), Erlang(5,4), and Erlang(2.5,8), I was able to reduce the total supply chain costs when the end-customer demand had the Exponential(20) distribution. The reasons for this were twofold: (1) The information flows associated with the fluctuating prices were more beneficial when the demand variance was high; and (2) The expected per-period holding and penalty costs at the retailer are less

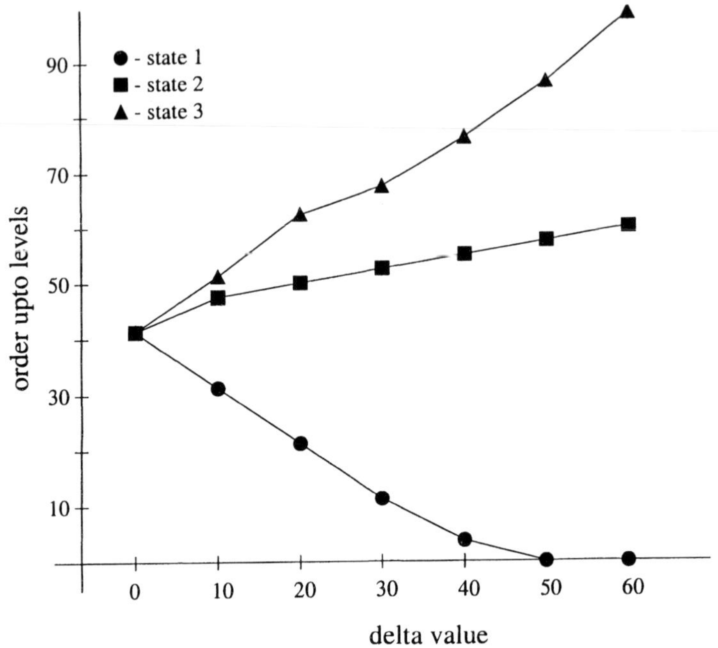

Figure 1.8. Order up-to levels as a function of Δ value

sensitive to the inventory levels when the demands have higher variance. Thus when the end-customer demand variance is high, I am able to reduce the total supply chain costs by using price fluctuations and the information flows associated with them. However, the reductions observed were quite small and ranged from 0.00% to 0.98%.

Figure 1.9 presents the plot of the percentage reduction in the total costs in the supply chain as a function of capacity for the Exponential(20) demand distribution. Notice that reductions were higher at higher capacities. The main reason for this behavior is that information flows are more beneficial at higher capacities and thus at higher capacities the supplier realizes higher savings while the retailer costs are not affected.

Figure 1.10 illustrates the relationship between the percentage reduction in the total costs and the penalty cost for the Exponential(20) demand distribution. Notice that reductions were higher at lower penalty costs. The main reason for this behavior is that at higher penalty costs, the increased variability in the retailer ordering process increases the supplier costs dramatically and the little information that is available to her is not effective in reducing her costs. Thus the benefit of these strategies is lower at higher penalty costs.

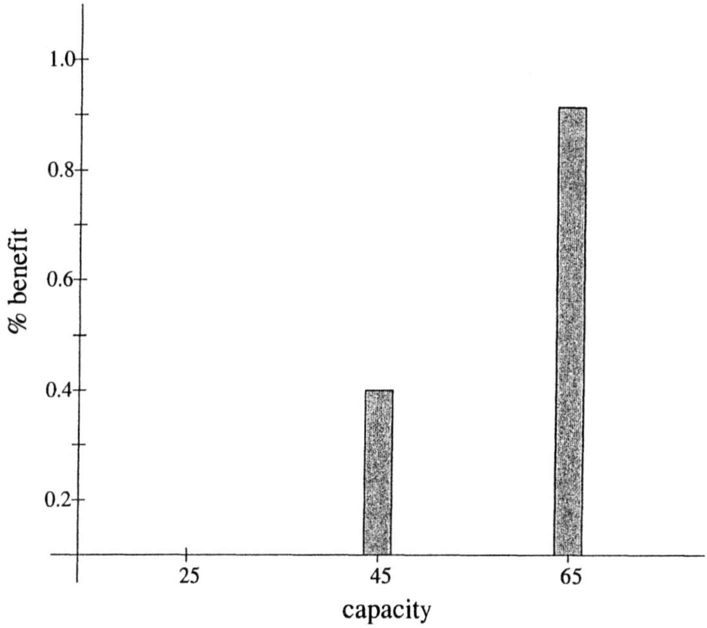

Figure 1.9. Percentage benefits as a function of supplier capacity: Setting 1

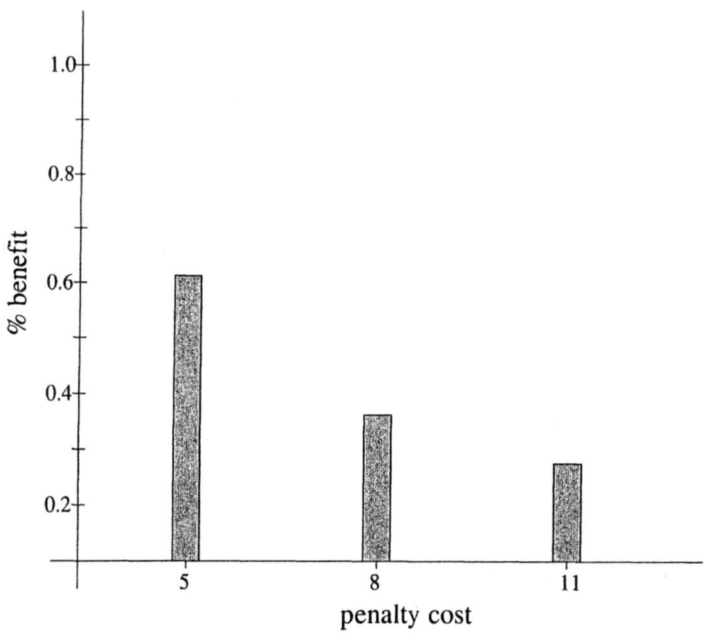

Figure 1.10. Percentage benefit as a function of supplier penalty cost: Setting 1

When the end-item demand distribution has very high variance, the supplier capacity is not restrictive, and the supplier penalty cost is low, I am able to reduce the total supply chain costs by considering non-stationary policies at the retailer even though the end-customer demand distribution is stationary and i.i.d. Admittedly, the reductions observed here are not large ($< 1\%$), but the most notable observation is that such a reduction is indeed possible.

Setting 2: Information on Retailer Inventory Levels

In this section, I report on the reduction in total supply chain costs when the supplier has information about the retailer inventory levels. In this setting, the total supply chain costs reduced by as much as 16.3% (average of 5.0%). Let us now study how the supplier capacity, supplier penalty cost, and end-customer demand variance affect these benefits.

Effect of Supplier Capacity. Figure 1.11 presents the plot of average percentage reduction as a function of supplier capacity. Notice that the benefits are consistently increasing as the capacity increases. This is because, when her capacity is not restrictive, the supplier is able to react to the information flows from the retailer. This enables her to realize higher benefits from these information flows, thus far eclipsing the inefficiencies (at the retailer) caused by the price fluctuations.

Effect of Supplier Penalty Cost. The average percentage reduction is plotted as a function of the supplier penalty cost in Figure 1.12. It is clear that the percentage benefit is higher at higher penalty costs. This is in contrast to the behavior observed for setting 1 (see Figure 1.10). The reason for this is as follows: when there is tighter cooperation between the supplier and the retailer, the expediting necessary at the supplier is drastically reduced. Such a reduction has a higher benefit when the supplier penalty cost is high.

Effect of End-Customer Demand Variance. The plot of average percentage reduction as a function of the standard deviation of end-customer demand is given in Figure 1.13. It is evident that when the end-customer demand is more variable, the percentage reduction is higher. This is due to the fact that when the demand has a variance, the information flows from the retailer are more beneficial. They are able to quickly alert the supplier of large swings in the end-customer demand.

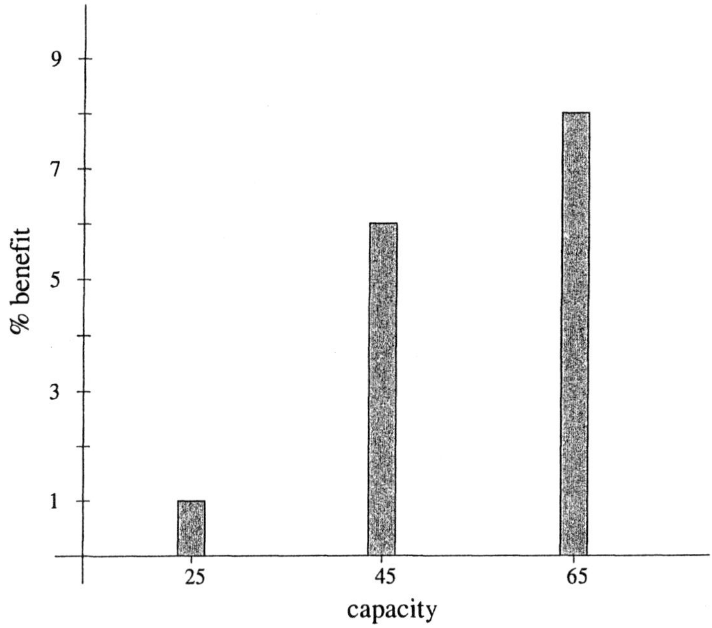

Figure 1.11. Percentage benefit as a function of supplier capacity: Setting 2

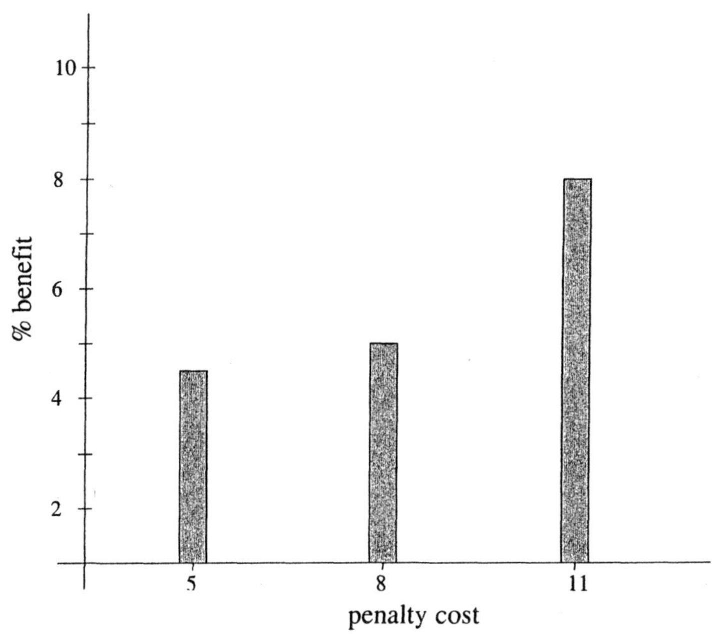

Figure 1.12. Percentage benefit as a function of supplier penalty cost: Setting 2

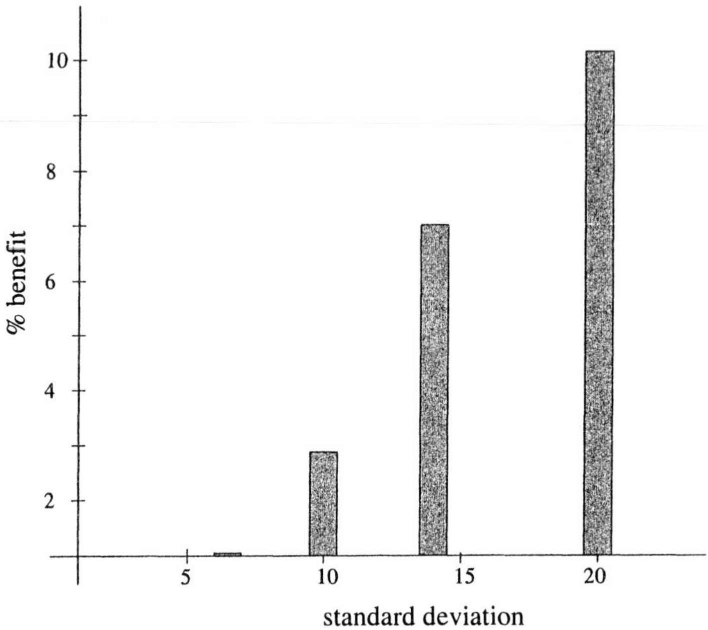

Figure 1.13. Percentage benefit as a function of standard deviation of end-customer demand: Setting 2

3.3 Conclusions

From the results of this study I conclude that a significant reduction (as much as 16%) in total supply chain costs can be realized when the supplier fluctuates her selling price and the retailer is willing to provide information about her inventory levels. I further observed that these reductions are larger at higher supplier capacities, supplier penalty costs, and end-customer demand variance.

4. Scheduled Ordering Policies with Information Sharing

In this section, I analyze a capacitated supplier, following a make-to-stock policy, providing a single product to n retailers who are facing i.i.d. (in time) demands from the end-customers. The supplier has a finite production capacity, C, but has access to an alternate (possibly using overtime) costlier source that has infinite capacity. The difference in costs between these two modes of production is captured by her penalty cost p_s, and her holding cost h_s. The retailers are all identical, face the same end-customer demand distribution, and have holding and penalty costs h_r and p_r respectively. There are no leadtimes, productions costs, or fixed ordering costs. The end-customer demand distribution, at each of the retailers, has cdf (pdf) $\Psi(\cdot)$ $(\psi(\cdot))$ with mean μ . Most of these assumptions are common in supply chain management and I believe they (except the one on fixed ordering costs) can be relaxed with little impact on the results.

For this supply chain, I consider two models. In model 1, every retailer is allowed to order every period. Since there are no fixed costs between the retailers and the supplier, in every period, the supplier faces random demands from each of the retailers. If possible, she satisfies these demands from stock, and the unsatisfied demands are supplied (by incurring the penalty cost) from the alternate source. In model 2, the supplier specifies that retailer j is allowed to order only in period $in + j$ for $i \in \{0, 1, 2, \ldots\}$. In the periods that she is not allowed to order, she will receive a fixed quantity η. For this model, I consider two different possibilities: (1) the retailers are not providing information about their inventory levels to the supplier; and (2) the retailers are sharing information with the supplier. When the retailers provide information to the supplier, she uses this information, especially from the retailers that are going to order in the immediate future, to determine the inventory level she wishes to maintain.

It is possible to obtain balanced ordering with more than one retailer ordering in every period. For example, if there are four retailers, one could balance the ordering having two retailers order in every period. In this paper, I do not consider those possibilities and restrict attention to cases in which only one retailer orders in every period. It is however not difficult to extend the models to capture those possibilities.

I identify optimal policies and compute optimal costs for these models and attribute the difference in costs to scheduled ordering policies with or without information sharing. The main objective of this study is to determine whether these strategies are effective in reducing the total supply chain costs. It is quite clear that a transition from model 1 to model 2 will increase the retailers' costs and may decrease the supplier's costs. However, I want to determine the conditions under which the reduction in cost at the supplier is greater than the increase in costs at the retailers. Only in those situations will the total supply chain cost decrease.

Similar supply chains, i.e., with one supplier and many retailers, were analyzed by Cachon (1999) and Aviv and Federgruen (1998). In both of these papers, the presence of either batch sizes or fixed ordering costs necessitated that a retailer does not order in every period. They studied the effect of *balanced* (or *staggered*) ordering policies on the supply chain costs. Cachon demonstrated a significant reduction in supply chain costs when moving from a *synchronized* ordering pattern to a *balanced* ordering pattern, but failed to demonstrate a significant reduction between *randomized* ordering pattern and *balanced* ordering pattern. He also did not consider the possibility of information sharing between the retailers and the supplier. Aviv and Federgruen also demonstrated a significant reduction in costs between *peak* ordering pattern and *smooth* staggered ordering pattern. In addition, they studied the effect of information sharing and noted that its effect was small (around 2%) to insignificant (around 1%).

I noticed that when the retailers were willing to share information, these policies reduced, even when the demands across the retailers were independent, the supply chain costs by as much as 10.7%, but in some cases increased the costs by as much as 4.3%. When the demands across the retailers were positively correlated, scheduled ordering policies with information sharing were uniformly effective (by 3.4% to 25.32%) in reducing the supply chain cost. In addition, I found that these strategies were more effective when there are a few customers (2 or 3), the supplier capacity is high, the end-customer demand variance is high, or the supplier penalty cost is, relative to that of the retailers, high.

4.1 The Models

In this section I describe in detail the two models that I introduced in the previous section. In both the models, the sequence of events in every period is as follows: (1) The supplier decides (bound by the capacity restriction) her production quantity; (2) End-customer demands at the retailers are realized and satisfied. Unsatisfied demands are backlogged; (3) The retailers incur holding or penalty costs; (3) The retailers (in model 2, if they are allowed) place their orders with the supplier; (4) The supplier ships product to the retailers (from stock or via expediting) and the product will be available to them in the next period; (5) The supplier incurs holding or penalty costs.

4.1.1 Model 1. In this model, every retailer is allowed to order in every period. Since each retailer, in effect, can order and receive any amount of the product, the optimization problem at the retailer is the standard newsvendor problem. It is well known that for this problem an order up-to policy is optimal and the optimal order up-to level z_r can be computed as follows:

$$z_r = \Psi^{-1}\left(\frac{p_r}{h_r + p_r}\right)$$

When the retailers use stationary order up-to policies, the end-customer demands are transmitted unchanged to the supplier. Thus in every period the supplier faces i.i.d. demands that are cumulative of n i.i.d. demands from the distribution $\Psi(\cdot)$. Since she is faced with finite production capacity, from Federgruen and Zipkin (1986a); Federgruen and Zipkin (1986b), I know that her optimal policy is modified order up-to. However, this optimal order up-to level is not available in closed form and a procedure using IPA (see Glasserman and Tayur (1994); Glasserman and Tayur (1995)) may be used to compute it efficiently.

4.1.2 Model 2. In this model, retailer j is allowed to order only in period $in+j$ for $i \in \{0,1,2,\ldots\}$. In the other periods, she will be shipped a fixed quantity η. I will perform a search over reasonable values of η to determine the best possible value. I must however determine how to solve this problem for a given value of η. First I consider the case in which the retailers are willing to share information with the supplier.

The Retailers' Problem

Since the retailers are all identical, the inventory control problem at each retailer is the same and can be formulated as follows. Let $J_t^j(y)$ be the cost at retailer j at the beginning of the tth period, when her inventory level is y. Let $V_t^j(x)$ be the optimal cost at retailer j when her inventory, before ordering, is x and I am at the end of the tth period.

$$J_t^j(y) = E_\xi[h_r(y-\xi)^+ + b_r(\xi - y)^+ + V_t^j(y-\xi)]$$

Recall that retailer j is allowed to order only in period $in+j$ and that in all other periods she receives a default shipment of η units. Thus I have:

$$\begin{aligned} V_t^j(x) &= \min_{y \geq x} J_{t-1}(y) \quad \text{if} \quad t = in+j \\ &= J_{t-1}(x+\eta) \quad \text{otherwise} \end{aligned}$$

For this problem, I establish some analytical results and structural properties. Since these results follow from standard arguments (see Kapuscinski and Tayur (1998), Scheller-Wolf and Tayur (1997)) in inventory control literature, I do not provide detailed proofs.

Property 10. $J_t^j(y)$ and $V_t^j(x)$ are convex in y and x respectively for all values of t and j.

Proof. Starting with the initial condition $J_0^j(x) \equiv 0 \ \forall x, j$, this property follows from standard arguments based on induction. □

Property 11. For average cost criterion as well as the discounted cost criterion, the infinite horizon optimal policy at the retailer is order up-to.

Proof. This property follows from the convexity of the cost functions and the properties of dynamic programs detailed in Bertsekas (1988). □

Thus, in a period in which retailer j is allowed to order, it is optimal for her to order up-to a level Z_r. If her inventory level is above Z_r, it is optimal for her not to order. Though this optimal level, Z_r, is not available in closed form, it can be computed efficiently using specialized procedures (see Gavirneni and Tayur (2001)). I will however use IPA.

The Supplier Problem

In this section I formulate and analyze the inventory control problem at the supplier. For ease of analysis, I assume that the end-customer demands are discrete. The supplier keeps track of the inventory levels at the retailers in every period. I define the state of the supplier using two indices. I say that she is in state (i, j) if the retailer that is supposed to order next has inventory level $Z_r + i$ and the retailer that is supposed to order after her has inventory $Z_r + j$. Notice that either i or j or both can be negative. When $n \geq 3$, this representation is an incomplete representation of the state of the supplier. However, I decided to use this approach to prevent the state space from becoming too large. The probability matrix that determines the transition among these states can be easily determined from the end-customer demand distribution. As I do not require an explicit representation of this matrix, I will not discuss how to compute it. In state (i, j) the supplier sees demand from a distribution $\Phi_i(\cdot)$. The demand distribution depends only on the state of the retailer that is to order next. Demand in state (i, j), $\Phi_i(\cdot)$, is related to the end-customer demand $\Psi(\cdot)$ as follows:

$$\begin{aligned}\Phi_i(0) &= \Psi(i) \\ \phi_i(x) &= \psi(x+i) \text{ if } x > 0\end{aligned}$$

From this relation it follows that for $i_1 \geq i_2$, $\Phi_{i_1}(\cdot) \leq_{st} \Phi_{i_2}(\cdot)$. That is, when the retailer has a higher inventory, she will probably order less. In addition, I can show that:

Property 12. For the discounted and average cost criterion, the optimal policy is modified order up-to.

Proof. This follows from arguments based on induction and convexity of cost functions. □

The optimal order up-to level depends on the state of the period. Let $Z_s^{(i,j)}$ be the optimal order up-to level in state (i, j). These order up-to levels satisfy:

Property 13. For states $(i_1, j_1), (i_2, j_2)$ such that $i_1 \leq i_2, j_1 \leq j_2$, $Z_s^{(i_1,j_1)} \geq Z_s^{(i_2,j_2)}$.

Proof. This follows from the monotonicity of demands. □

Because of the restriction on the production capacity, these order up-to levels are generally not available in closed form. They can however be computed efficiently using IPA. Since the validation and implementation of the IPA procedure follows closely along the lines of Kapuscinski and Tayur (1998) and Scheller-Wolf and Tayur (1997), I do not present them in detail here.

No Information Sharing

The scenario when the retailers do not provide information about their inventory levels is a special case of the analysis presented above. In this situation, the supplier state does not change (since retailers are identical and the ordering pattern is balanced) from one period to the next and it is optimal for her to use a stationary order up-to policy. This optimal order up-to level can also be computed using IPA.

Once I have determined ways to solve the problem in model 2 for a given value of η, I propose to determine the best value of η by an exhaustive search over all the reasonable values. I have implemented these procedures for both the models and performed an extensive computational study to understand the effects of scheduled ordering policies with information sharing.

4.2 Computational Results

In this section I report results from a detailed computational study. There are two main objectives for performing this study. First, I want to determine the effect of scheduled ordering policies with and without information sharing in supply chains. Do these strategies decrease or increase the total supply chain cost? Second, I want to study the effect of various supply chain parameters such as supplier capacity, supplier penalty cost, variance of end-customer demands, and the number of retailers on this behavior. Specifically, I want to understand when it is prudent to implement these strategies in supply chains.

The experimental setup is as follows. The holding costs at the supplier and the retailers are set to 1. The penalty costs at the retailers were equal to 9 while the penalty cost at the supplier was either 19, 29, or 39. The end-customer demands have a mean equal to 20, and were allowed to follow the distributions: Exponential(20), Erlang(2,10), Erlang(4,5), and Erlang(8,2.5). Thus, demand variances were 400, 200, 100, and 50 respectively. The number of retailers was either 2, 3, or 4. The supplier capacity, when there were n retailers, was set equal to $a \times n \times 20$ and a was allowed to take values 1.5, 2.0, and 2.5. I also study two different conditions on the end-customer demands. In one case, the end-customer demands are assumed to be independent across the retailers. In another case, these demands are assumed to be highly ($\rho = 1$) positively correlated. One can easily guess that these scheduled ordering policies will be more beneficial when the demands at the retailers are positively correlated. All the same I decided to study the case of correlated demands to (1) observe how effective, in the best case, these strategies can be; and (2) get a better understanding of the effect of various system parameters on the effectiveness of these strategies. In addition, I study these cases with and without information sharing between the retailers and the supplier. For each problem instance in model 2, I determined the best possible η value by performing an exhaustive search over values ranging from 0 to 40 in increments of 5. In all the cases the best η value was either 20 or 25, i.e., very close to the mean. I detail my observations from this computational study. First, I study the cost per period and then the effectiveness of these strategies.

4.2.1 Cost per Period. In all the cases the cost per period increased when the number of retailers increased, the supplier penalty cost increased, or the end-customer variance increased. In addition, when the supplier capacity decreased, the cost per period went up. This behavior is expected and the reasons for it have been well documented in inventory control literature. So I will not elaborate here. I will however focus on the change in cost per period when moving from model 1 to model 2.

4.2.2 Percentage Difference. We compute the percentage difference between these two models as follows:

$$\% \text{ Difference} = \frac{\text{cost of model 1} - \text{cost of model 2}}{\text{cost of model 1}} \times 100$$

We detail below my observations on the behavior of this percentage difference. First I study the case of no information sharing followed by the case of information sharing.

No Information Sharing

When there was no information sharing and the demands were independent, the scheduled order policies resulted in an increase in the total supply chain costs in all of the cases I studied. This increase in cost ranged from 7.4% to 37.7%. However, when the demands at the retailers were positively correlated, there was a significant (from 4.5% to 16.8%) reduction in the total supply chain cost in all the cases. This result is not surprising and the reasoning for it is as follows. When the demands are independent, using scheduled ordering policies results in a loss of the risk pooling that is realized when all the retailers order in every period. However when the demands are highly correlated, the benefits from such risk pooling are minimal and it is actually better for the supplier to separate the customer orders in order to avoid major fluctuations in demand. I noticed that in general, these strategies were more effective when (1) the supplier capacity was high, (2) the supplier penalty cost was high, (3) the number of retailers was small, and (4) the end-customer demand variance was neither too high nor too low. Since this behavior was similar to the case of information sharing, I have deferred the explanation of these results to the next section. I conclude that *scheduled ordering policies without information sharing are beneficial only when the demands are significantly positively correlated across the retailers.*

Information Sharing

In this section I detail the effect of scheduled ordering policies with information sharing. For the case of independent demands, this approach was effective in reducing the supply chain cost in 50% of the problem instances. While in some cases the supply chain cost reduced by as much as 10.7%, in other cases it increased by as much as 14.9%. The average difference was an increase of 1.6%. In the case of correlated demands, the supply chain cost recorded a decrease (ranging from 10.9% to 32.9% and averaging around 21.8%) in all the cases. Thus I conclude that *these strategies are beneficial in some cases of independent demands and their effectiveness grows when the demands are positively correlated.*

Effect of Supplier Capacity. Figures 1.14 and 1.15 give the plots of percentage difference as a function of the supplier capacity parameter, a, for independent and correlated demands respectively. Notice that in

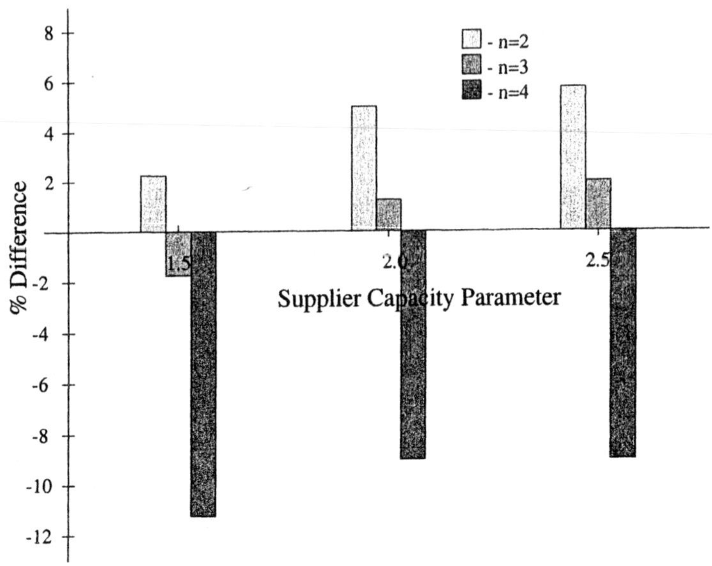

Figure 1.14. % Difference as a function of supplier capacity for independent demands (information sharing)

both the cases the percentage difference is increasing with increase in capacity. The principal reason for this behavior is that the supplier, when she has excess capacity, is more flexible to react to the information provided by the retailers. I conclude that *these strategies are more useful when the supplier has excess capacity.*

Effect of Supplier Penalty Cost. The plots of the percentage difference as a function of the supplier penalty cost for independent and correlated demands are given in figures 1.16 and 1.17 respectively. Observe that, in both the cases, as the supplier penalty cost increases percentage difference also increases. The reason for this behavior is that when the supplier penalty costs are high (relative to retailer penalty costs), the savings realized at the supplier using these strategies is greater than the resulting cost increases at the retailer. Thus I conclude that *these strategies are more beneficial when the supplier penalty costs are high relative to those at the retailers.*

Effect of End-customer Demand Variance. The percentage difference as a function of the end-customer demand variance for independent and correlated demands is given in figures 1.18 and 1.19 respectively. The effect of variance is not immediately obvious. For instance,

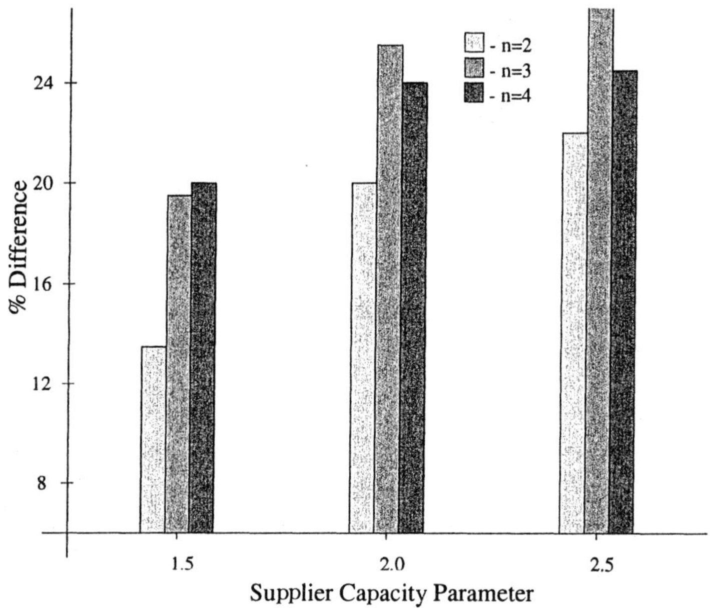

Figure 1.15. % Difference as a function of supplier capacity for correlated demands (information sharing)

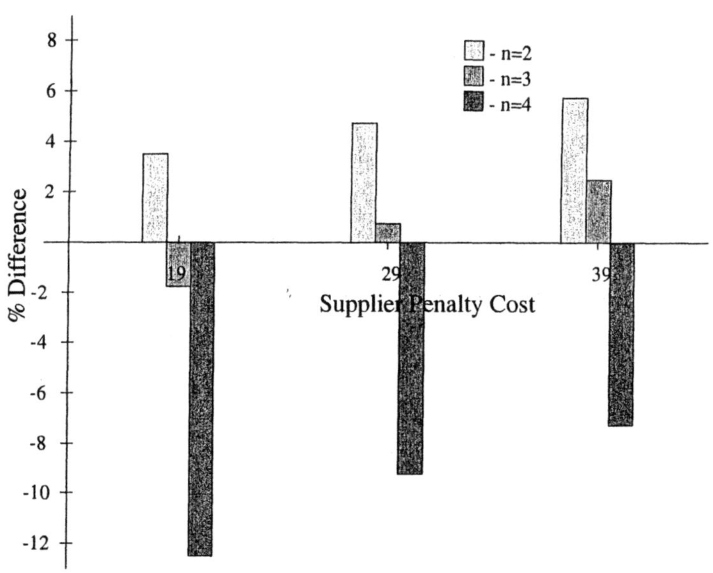

Figure 1.16. % Difference as a function of supplier penalty cost for independent demands (information sharing)

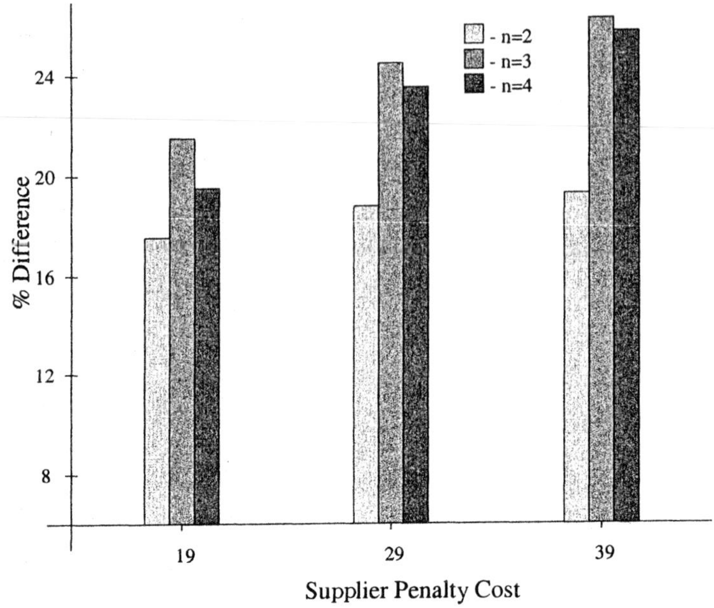

Figure 1.17. % Difference as a function of supplier penalty cost for correlated demands (information sharing)

when the demands were independent, the percentage difference consistently decreased with decrease in the demand variance. On the other hand, in the presence of correlation between demands, the percentage difference appears to be highest at medium values of variance. When the demands are independent, the loss of benefits from risk pooling far outweigh the benefits from information sharing. Thus, there is a reduction in percentage difference as the demand variance decreases. When the demands are highly correlated, I know that loss of benefits from risk pooling is minimal and I must look at reduction in costs due to information sharing (which is lower at lower variances) and increase in costs due to infrequent retailer ordering (which is lower at extreme values of variance).

Thus I conclude that *these strategies are beneficial at higher variances when the demands are independent and at medium variances when the demands are correlated.*

Effect of Number of Retailers. In figures 1.14 to 1.19, the percentage difference for the cases of 2, 3, or 4 retailers has been separately identified. Notice that in figures 1.14, 1.16, and 1.18 the percentage difference is highest when there are two retailers and it decreases as the

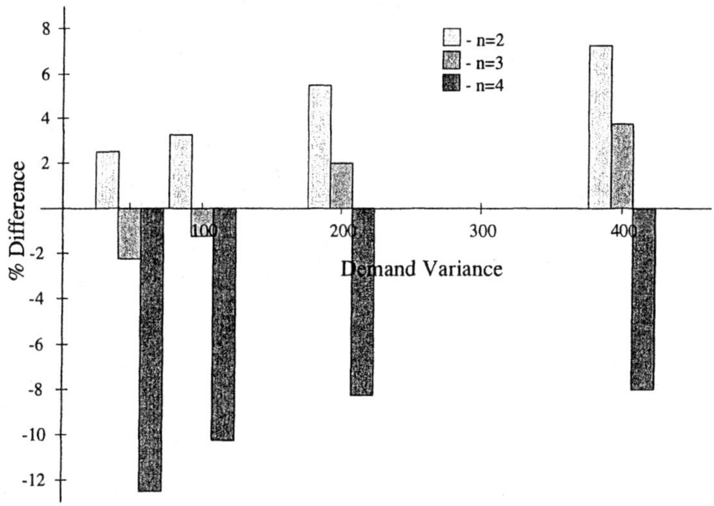

Figure 1.18. % Difference as a function of demand variance for independent demands (information sharing)

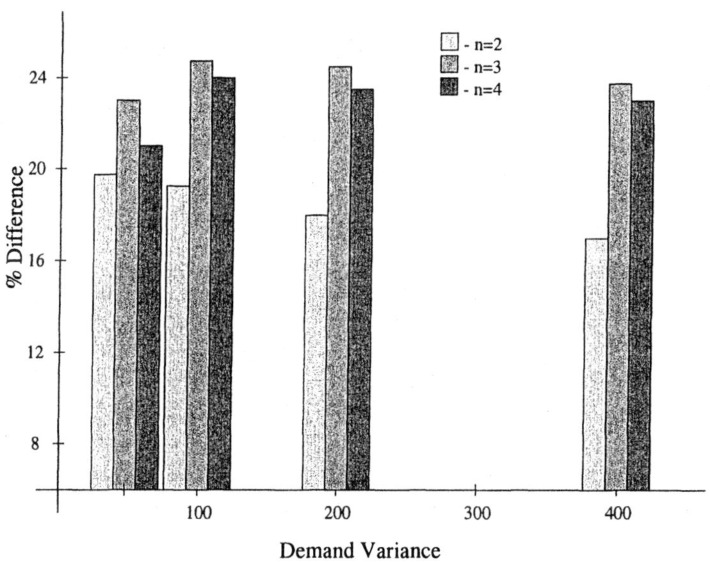

Figure 1.19. % Difference as a function of demand variance for correlated demands (information sharing)

number of retailers increases to 3 and then to 4. In figures 1.15, 1.17, and 1.19 observe that the percentage difference is highest when there are three customers. For the case when the demands are independent, the benefits lost from risk pooling are higher when the number of retailers is high. Thus in that case, these strategies are not as beneficial when there are more retailers. When the demands are correlated, there is not as much loss from the lack of risk pooling, but in the presence of a large number of retailers, the lowered frequency in the retailer ordering results in higher costs. Thus these strategies tend to lose their effectiveness when the number of retailers is high. So I conclude that *these strategies are effective when the number of retailers is small (2 or 3)*.

Benefits of Information Sharing
In this section, I take a closer look at the benefits of information sharing, i.e., reduction in total supply chain cost due to the retailers providing information about their inventory levels. From the computational study I identified the reduction in costs due to information sharing and analyzed it with respect to various system parameters. Figures 1.20 to 1.22 contain the relevant graphs. Observe that the benefits of information sharing were always positive. This is not surprising since the availability of information provides the supplier with more alternatives and thus can not reduce her cost. I observed that these benefits were higher at higher supplier capacities. When she has excess capacity, the supplier is able to react to information and thus realize a higher benefit. I also observed that the benefits were higher at higher supplier penalty costs and this is due to the greater reduction, enabled by the flow of information, in penalty costs. The benefits were higher when the number of retailers was higher and at higher demand variance. This was because when the number of retailers is high or the demand variance is high, the information is more effective in reducing the uncertainty in the system.

4.3 Conclusion

This study of scheduled ordering policies with information sharing has demonstrated that benefits of information centric design and management of supply chains can be extended to distribution intensive supply chains. While the resulting benefits were lower when the demands were independent among the retailers, they were significantly large when the demands were positively correlated.

5. Future Research

In this chapter, via the use of three examples, I illustrated the benefits of information centric design and management of supply chains. Computational results from simulations of these supply chains have shown that supply chain performance can be significantly improved by the ap-

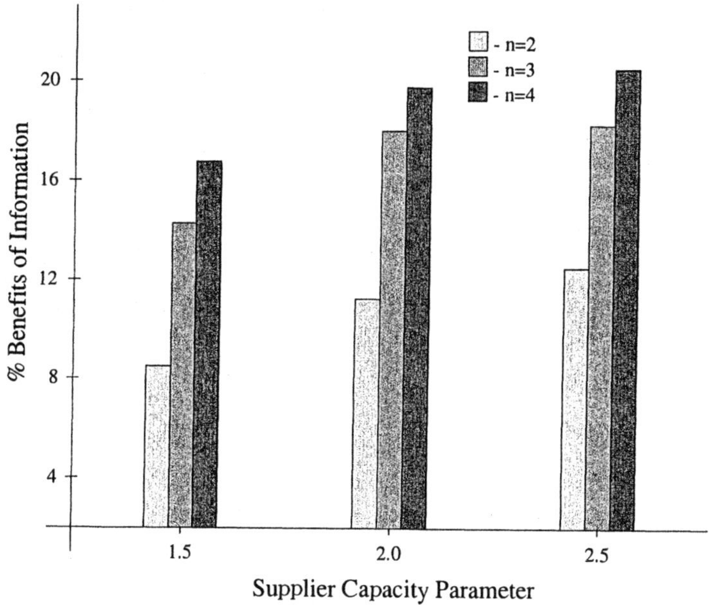

Figure 1.20. % Benefits of information sharing as a function of supplier capacity

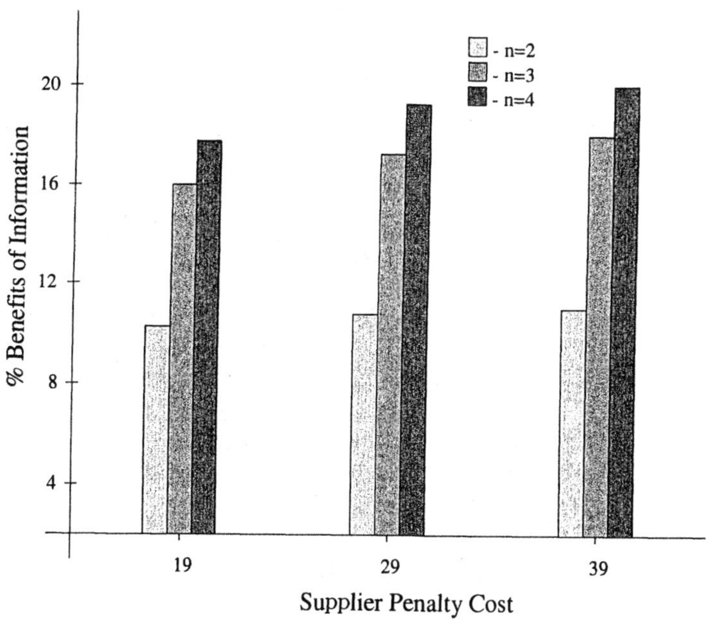

Figure 1.21. % Benefits of information sharing as a function of supplier penalty cost

propriate use of these strategies. While the supply chains studied here are representative of a wide array of supply chains, they by no means

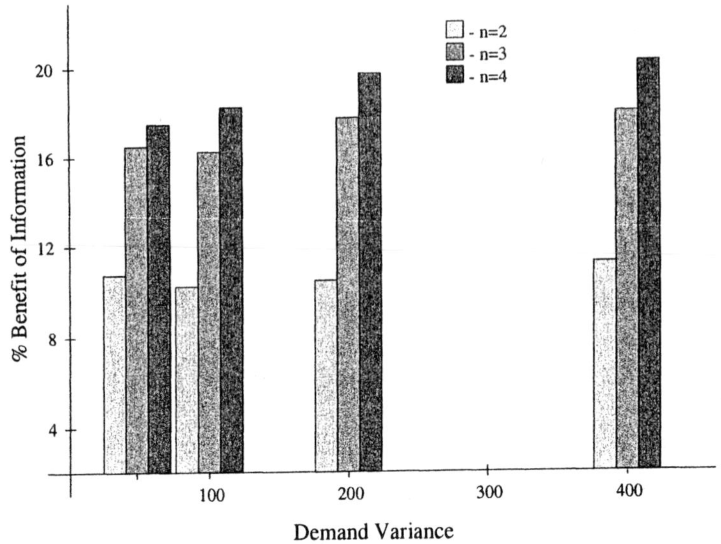

Figure 1.22. % Benefits of information sharing as a function of demand variance

capture the complete spectrum. As a result, significant research activity is still needed in order to show that information centric strategies can be universally beneficial. I am currently involved in the following related research projects intended to further expand the understanding behind managing information intensive supply chains:

1 In a multi-stage distribution supply chain, it is widely believed that well placed distribution centers can be effective in risk pooling, thus reducing uncertainties and improving supply chain performance. I wish to determine whether distribution centers continue to play an important role in supply chains in which information is readily available. I believe and wish to show that distribution centers do not carry such benefits in information-centric supply chains.

2 Based on my experience in the semiconductor industry, where the lead times are very long and the yields are often low and random, I wish to study the effect of information, on the current status of the production process, from supplier to retailer. In practice, the yields associated with a certain batch are only visible to the retailer when he/she receives the shipment associated with the batch. However, the supplier has known all along how this particular batch is in the manufacturing process. In this project, I will model this flow of information from the supplier to the retailer and determine the appropriate decision making strategies for the retailer.

3 In a typical assembly supply chain, many suppliers ship parts or modules to a single location at which these parts or modules are assembled into the final product. Such supply chains are commonly found in automotive and heavy equipment manufacturing. Traditionally, in such supply chains, a supplier has a localized perspective and is not aware of the status of the other suppliers. This often leads to a lack of coordination and results in major inefficiencies in the system. It is possible for the various suppliers to share information, so that the decisions are made in a coordinated manner. However, it is not easy to determine which information should be provided to whom and how this information should be used. This project will explore these possibilities and come up with strategies for managing these supply chain efficiently.

References

Anupindi, R. and Y. Bassok, 1999. "Supply Contracts with Quantity Commitments and Stochastic Demand," in *Quantitative Models for Supply Chain Management,* S. Tayur, R. Ganeshan, and M. Magazine (eds.), Kluwer Academic Publishers.

Armistead, C.G. and Mapes, J., 1993. "The impact of supply chain integration on operating performance," *Logistics Information Management* 6(4), 9-14.

Aviv, Y. and A. Federgruen, 1998. "The Operational Benefits of Information Sharing and Vendor Managed Inventory (VMI) Programs," Olin School of Business, Washington University.

Bertsekas, D.P., 1988. *Dynamic programming: Deterministic and stochastic models,* Prentice-Hall, Englewood Cliffs, NJ.

Cachon, G. and M. Fisher, 2000. "Supply Chain Inventory Management and the Value of Shared Information," *Management Science* 46(8), 1032-1048.

Cachon, G. and P. Zipkin, 1999. "Competitive and Cooperative Inventory Policies in a Two Stage Supply Chain," *Management Science* 45(7), 936-953.

Cachon, G., 1999. "Competitive Supply Chain Inventory Management," in *Quantitative Models for Supply Chain Management,* S. Tayur, R. Ganeshan, and M. Magazine (eds.), Kluwer Academic Publishers.

Chen, F., 1998. "Echelon Reorder Points, Installation Reorder Points, and the Value of Centralized Demand Information," *Management Science*, 44(12), 221-234.

Chen, F., Federgruen, A., and Y.S. Zheng, 2001. "Coordination Mechanisms for a Distribution System with One Supplier and Multiple Retailers," *Management Science* 47(5), 693-708.

Federgruen, A. and P. Zipkin, 1986a. "An inventory model with limited production capacity and uncertain demands I: The average-cost criterion," *Mathematics of Operations Research* 11(2), 193-207.

Federgruen, A. and P. Zipkin, 1986b. "An inventory model with limited production capacity and uncertain demands II: The discounted-cost criterion," *Mathematics of Operations Research* 11(2), 208-215.

Gavirneni, S., Kapuscinski, R. and S. Tayur, 1999. "Value of Information in Capacitated Supply Chains," *Management Science* 45(1), 16-24.

Gavirneni, S., 2001. "Benefits of Cooperation in a Production-Distribution Environment," *The European Journal of Operational Research* 130(3), 164-174.

Gavirneni, S and S. Tayur, 2001. "An Efficient Procedure for Nonstationary Inventory Control," *IIE Transactions* 33(2), 83-89.

Gavirneni, S., and S. Tayur, 1999. "Managing a Single Customer using a Target Reverting Policy," *Manufacturing & Service Operations Management* 1(2), 157-173.

Glasserman, P., 1991. "Gradient estimation via perturbation analysis," Kluwer Academic Publishers, Boston.

Glasserman, P. and S. Tayur, 1994. "The stability of a capacitated, multi-echelon production-inventory system under a base-stock policy," *Operations Research* 42(5), 913-925.

Glasserman, P. and S. Tayur, 1995. "Sensitivity analysis for base-stock levels in multi-echelon production-inventory systems," *Management Science* 42(5), 263-281.

Iyer, A. and J. Ye, 2000. "Vendor Managed Inventory in a Promotional Retail Environment," *M&SOM* 2(2), 128-143.

Kapuscinski, R. and S. Tayur, 1998. "A capacitated production-inventory model with periodic demand," *Operations Research* 46(6), 899-911.

Karlin, S., 1960. "Optimal policy for dynamic inventory process with stochastic demands subject to seasonal variations," *J. SIAM* 8, 611-629.

Lariviere, M., 1999. "Supply Chain Contracting and Coordination with Stochastic Demand," in *Quantitative Models for Supply Chain Management*, S. Tayur, R. Ganeshan, and M. Magazine (eds.), Kluwer Academic Publishers.

Lee. H., Padmanabhan. P., and S. Whang, 1997. "Information Distortion in a Supply Chain: The Bullwhip Effect," *Management Science* 43(4), 546-558.

Lee. H., Padmanabhan. P., Taylor, T.A. and S. Whang, 2000. "Price Protection in the Personal Computer Industry," *Management Science* 46(4), 467-482.

Munson, C.L. and M.J. Rosenblatt, 2001. "Coordinating a three-level supply chain with quantity discounts," *IIE Transactions* 33, 371-384.

Pasternack, B.A., 1985. "Optimal pricing and return policies for perishable commodities," *Marketing Science* 4(2), 166-176.

Scarf, H., 1962. "The Optimality of (s, S) policies in the dynamic inventory problem," in K.J. Arrow, S. Karlin, and P. Suppes (editors), Mathematical Methods in Social Sciences, Stanford University Press.

Scheller-Wolf, A. and S. Tayur, 1997. "Reducing International Risk through Quantity Contracts," *GSIA working paper*, Carnegie Mellon University, Pittsburgh, PA, April 1997.

Song, J., and P. Zipkin, 1993. "Inventory control in a fluctuating demand environment," *Operations Research* 41(2), 351-370.

Takac, P.F., 1992. " Electronic data interchange (EDI): an avenue to better performance and the improvement of trading relationships?," *International Journal of Computer Applications in Technology* 5(1), 22-36.

Tayur, S., 1993. "Computing the optimal policy for capacitated inventory models," *Stochastic Models* 9(4), 585-598.

Zheng, Y.S. and A. Federgruen, 1991. "Finding Optimal (s, S) Policies is About as Simple as Evaluating a Single Policy," *Operations Research* 39(4), 654-665.

Zipkin, P., 1989. "Critical number polices for inventory models with periodic data," *Management Science* 35(1), 71-80.

Chapter 2

AN ANALYSIS OF ADVANCE BOOKING DISCOUNT PROGRAMS BETWEEN COMPETING RETAILERS*

Kevin F. McCardle, Kumar Rajaram, Christopher S. Tang**
Anderson Graduate School of Management, UCLA
Los Angeles, CA 90095-1481

Abstract As product demand uncertainty increases and life cycles shorten, retailers respond by developing mechanisms for more accurate demand forecasting and supply planning to avoid over-stocking or under-stocking a product. We model the situation in which two retailers consider launching one such mechanism, known as the 'Advance Booking Discount' (ABD) program. In this program customers are enticed to pre-commit their orders at a discount price prior to the regular selling season. However, these pre-committed orders are filled during the selling season. While the ABD program enables the retailers to lock in a portion of the customer demand and use this demand information to develop more accurate forecasts and supply plans, the advance booking discount price reduces profit margin. We analyze the four possible scenarios wherein each of the two firms offer an ABD program or not, and establish conditions under which the unique equilibrium calls for launching the ABD program at both retailers. We also provide a detailed numerical example to illustrate how these conditions are affected by the level of demand uncertainty, demand correlation, market share, and fixed costs for instituting an ABD program.

1. Introduction

In several service industries, customers are encouraged to purchase various types of services at a time prior to the actual consumption. For

*This chapter is an expanded version of McCardle, Rajaram, and Tang (2004).

**Christopher Tang's research was partially supported by the UCLA James Peters Research Fellowship.

instance, AT&T's pre-paid calling cards, UCLA's Bruin smart cards, and Sheraton's pre-paid vouchers for hotels enable customers to pay for the service in advance at a discount price (c.f., Lollar, 1992; McVea, 1997). This advance selling strategy reduces the risk of under-utilized capacity because the service provider lengthens the selling season. In addition, this strategy improves the cash flow of the service provider because the customers pay for the service in advance and redeem the service at a later period. In a recent paper, Xie and Shugan (2001) develop a single-firm model to analyze the conditions under which the firm should sell the service in advance at a discount price. They also determine the optimal advance price as well as the optimal capacity available for the sales in advance.

In addition to these examples in the telephone and hotel service industry, we have observed that many retailers are now launching 'Advance Booking Discount' (ABD) programs when selling physical goods such as books, CDs, electronic toys, cakes, Christmas trees, etc. Under an ABD program, the retailer offers a product at a price discount prior to the selling season. If customers accept this offer, then they 'reserve' the product to be picked up (or delivered) during the selling season by pre-paying the entire discounted price prior to the selling season. While no order cancellation or refund is permitted, the retailer guarantees product availability for pre-committed orders. If customers decline the ABD offer, then they can purchase the product at the regular price during the selling season, though the retailer does not guarantee product availability. Retailers typically implement such advance booking discount programs when selling perishable products consumed during a well defined and concentrated selling season (such as pumpkin pies or fresh turkeys during Thanksgiving, moon cakes during the Chinese mid-Autumn Festival, or Christmas trees during Christmas), or certain kinds of new durable products with a short selling season and high demand uncertainty (such as new music CDs, video games, etc.). Examples of such retailers include Maxim's Bakery in Hong Kong, Amazon.com, Movies Unlimited, Toys-R-Us, and Electronics Boutique.

When selling a physical product, the ABD program offers three major benefits. First, this program extends the selling season without the need for immediate delivery. This enables the retailer to entice more customers to buy the product over a longer period of time without being constrained by the production capacity. Second, the ABD program offers an opportunity for the retailer to utilize the pre-committed orders received during the advance selling period to generate a better demand forecast prior to the start of the selling season. Such improved forecasts allow a more accurate order to be placed at the start of the selling season, which in turn reduces the overstock and understock costs.[1] Third, the ABD program allows the retailer to improve cash flows because the retailer receives the payments in advance during the advance selling period. Tang et al. (2004) present a single-firm model that quantifies the benefits of the ABD program in addition to characterizing the optimal discount price. Weng and Parlar (1999) examine a single-firm model for analyzing the ABD program that is based on different underlying assumptions. Because an ABD program can be considered to be a type of discount promotion, the reader is referred to Tang et al. (1999) for a review of the marketing and operations management literature that deals with discount promotion.

To the best of our knowledge, all of the existing research that examines ABD programs considers single-firm models or models in which only one firm in an industry can adopt an ABD program. These models solve for the optimal (monopolist) discount price and analyze the benefits of the ABD program. In this chapter, we extend these models to incorporate the competitive nature of retailing by developing a duopoly model to analyze ABD programs under competition. The two retailers in our model sell the same product within a short and concentrated sales season. They each may or may not launch an ABD program – resulting in 4 separate scenarios. This competitive extension significantly expands and complicates the analysis. Within this competitive context our specific contributions to the literature are as follows. First, we develop a consumer response function to competitive ABD programs which captures the impact of price competition across firms and also captures the willingness of customers to switch retailers in order to partake in the ABD program. Second, in each of the competitive scenarios when ABD is offered, we calculate the optimal discount coefficients under equilibrium within the scenario and analyze how these values change with changes in demand, product, and market characteristics. Third, we determine the optimal choice of scenarios of each retailer, determining, in effect, the equilibrium of the meta-game across scenarios. We analyze how the fixed cost of instituting an ABD program, along with product-demand uncertainty, market share, and degree of demand correlation affect the

[1] In this chapter, we shall consider the case in which the retailer can place exactly one order at the beginning of the selling season. This situation occurs when the replenishment lead-time is longer than the selling season. However, when the replenishment lead time is sufficiently short, the retailer can use the sales data of the early part of the selling season to improve the forecast even further and place an additional order at the middle of the selling season. This specific scenario has been examined by Fisher, Rajaram and Raman (1999) and tested at a catalog retailer.

meta-game equilibrium. In the next section, we present the basic modeling framework of the ABD program under retail competition.

2. The Model

Consider a situation in which two retailers A and B sell the same (or a similar) product during a short selling season. The unit cost, selling price, and salvage value of this product are c, p, and s, respectively. We consider the case in which the consumer market consists of two segments: one segment intends to buy from retailer A and the other segment intends to buy from retailer B. We assume that each customer buys no more than one unit of the product.[2] The joint distribution of the anticipated demands for retailers A and B, denoted by D_A and D_B, is assumed to be bivariate normal with means μ_A and μ_B, standard deviations σ_A and σ_B, and correlation coefficient $\rho \in (-1, 1)$.[3]

To simplify the exposition, we assume that the anticipated demands D_A and D_B have the same coefficient of variation θ, where $\theta = \sigma_A/\mu_A = \sigma_B/\mu_B$. This seems reasonable since both firms are selling the same product and will consequently have similar degrees of demand uncertainty. Let μ be the expected total market demand, where $\mu = \mu_A + \mu_B$. Let $\alpha \in (0, 1)$ be the market share of Brand A, where $\alpha = \mu_A/\mu$. Given the definition of α and θ, we have $\mu_A = \alpha\mu$, $\mu_B = (1 - \alpha)\mu$, $\sigma_A = \theta\alpha\mu$ and $\sigma_B = \theta(1 - \alpha)\mu$.

In this chapter, we assume that both retailers charge a fixed price p per unit during the regular season. This assumption seems reasonable since there is usually an advertised Manufacturers Suggested Retail Price (MSRP) for the durable and perishable types of products we consider in the ABD program. In addition, the prices for these products are set to the MSRP and held constant in the regular season. This is because of the short duration of the regular season and because customers are typically very sensitive to the timing of the purchase of the product and consequently are more willing to pay full price to avail the product during the season.

2.1 The Advance Booking Discount Program

We now present the ABD program and model its impact on the demand for each retailer. Consider the case in which retailer i, where

[2]This assumption seems reasonable across a wide variety of durable products such as music CD's, books, toys, and video games, or perishable items such as Christmas trees that are consumed during a well-defined and concentrated sales season. Customers receive the product only during the season.

[3]The bivariate normal distribution is degenerate for $\rho = -1$ and $\rho = 1$.

$i = A, B$, launches the ABD program by offering a discount price $x_i p$ per unit of the product prior to the beginning of the season.[4] The discount coefficient is equal to x_i with $0 \leq x_i \leq 1$. Notice that firm i can launch an ABD program with $x_i = 1$, corresponding to the case in which firm i allows customers to pre-order but offers no discount on the regular season price. If customers accept the ABD offer, then they can pre-commit their orders by pre-paying $x_i p$ per unit prior to the selling season and pick up their orders during the season. If customers decline the ABD offer, they can always attempt to purchase the product during the regular selling season by paying the regular price p per unit; however, the availability of the product will not be guaranteed.

Based on the description of the ABD program, it is conceivable that the ABD program affects the 'anticipated' demand and the retailer's supply planning in several ways. First, by offering a discount price $x_i p$ prior to the beginning of the season, retailer i can entice segment i's customers to pre-commit their order. Second, if retailer i's discount price $x_i p$ is lower than that of retailer j or if retailer j does not offer an ABD program, then some customers from segment j may switch to retailer i by pre-committing their orders. Third, the ABD program enables the retailers to lock-in these pre-committed orders and to use this information to develop more accurate forecasts and supply plans. These observations enable us to model the impact of the ABD program on the anticipated demand D_i for each retailer i. Suppose both retailers launch ABD programs under which retailer i's discount price is $x_i p$ and retailer j's discount price is $x_j p$. Each customer in segment i has to select one of the following options: (1) pre-commit the order with retailer i by paying discount price $x_i p$; (2) pre-commit the order with the competing retailer j by paying discount price $x_j p$; or (3) purchase the product from retailer i at regular price p during the selling season. Figure 2.1 depicts these 3 options for the customers in segment i.

2.2 Consumer Response Functions

Let $R_{ie}(x_A, x_B) \in [0, 1]$ represent the fraction of customers in segment i who opt for option (1) and pre-order from retailer i. Let $R_{is}(x_A, x_B) \in [0, 1]$ be the fraction of customers in segment i who opt for option (2) and pre-order from competing retailer j. The remainder, $[1 - R_{ie}(x_A, x_B) -$

[4]To simplify the exposition at this point, we do not include a fixed cost that may be incurred when a retailer launches the ABD program. This fixed cost is included, however, in a later section.

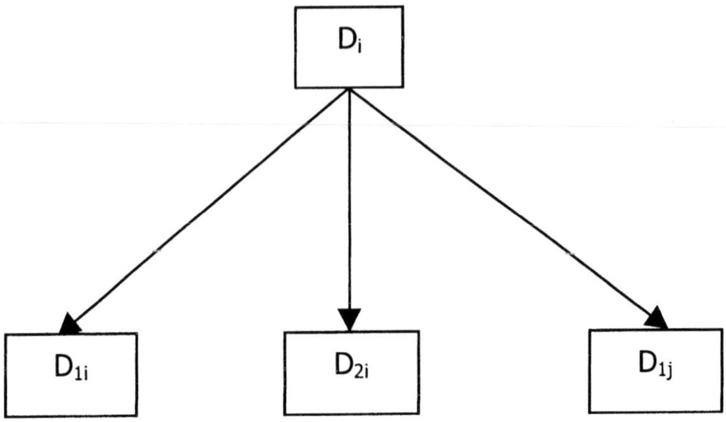

Figure 2.1. The Impact of the ABD Program on Segment i's Purchasing Behavior

$R_{is}(x_A, x_B)$], wait for the regular selling season.[5] Note that subscript e corresponds to early order, while subscript s corresponds to a switch from one retailer to the other. In order to obtain tractable analytical results, we develop a specific functional form for the consumer response functions $R_{ie}(x_A, x_B)$ and $R_{is}(x_A, x_B)$. To do so, let us consider the situation faced by the customers prior to the selling season. As depicted in Figure 2.1, prior to the selling season, each consumer in segment i has to compare the discount prices x_Ap and x_Bp and then decide whether to (1) pre-commit to retailer i, (2) pre-commit with retailer j, or (3) wait for the regular selling season. This situation is akin to the case in which the customer has to choose between 3 products with different prices. While marketing researchers have considered various functional forms for the consumer's choice function, we employ a modified version of the functional form proposed by Raju et al. (1995) as follows.[6] Specifically,

[5]It is conceivable that aggregate demand may increase as a result of the price discount offered through the ABD program. However, because the products we consider are either durable or perishable in nature and customers receive the product only during the selling season, it seems reasonable to assume that customers do not consume more during the selling season and that consumption does not increase with the level of the discount.

[6]We choose this specific functional form for the following reasons: (a) it has a precursor in the marketing literature (c.f., Narasimhan (1984), Achabal, et. al. (1990), Smith and Achabal (1998), and Bhardwaj and Sismeiro (2001)); (b) it is linear in x_A and x_B, as widely modeled in the marketing literature (c.f., Choi (1991), Eliashberg and Steinberg (1991) and Lal (1990)); (c) it is consistent with individual utility maximizing behavior (c.f., Shubik and Levitan, (1980)); and (d) tractability.

we assume that the response functions possess the following form:

$$R_{ie}(x_A, x_B) = \begin{cases} 0 \text{ if retailer } i \text{ does not launch ABD} \\ (r_{ie}/2)[(1 - ax_i) + b(x_j - x_i)], \quad \text{for } i = A, B \end{cases} \quad (2.1)$$

$$R_{is}(x_A, x_B) = \begin{cases} 0 \text{ if retailer } j \text{ does not launch ABD} \\ (r_{is}/2)[(1 - ax_j) + b(x_i - x_j)], \quad \text{for } i = A, B \end{cases} \quad (2.2)$$

Suppose neither firm offers an ABD program. Then (1) and (2) imply that $R_{ie} = R_{is} = R_{je} = R_{js} = 0$, i.e., there are no early orders.

Next, consider the case in which firm i does not launch an ABD program but firm j does. It can be seen from (1) and (2) that $R_{ie} = 0$ and $R_{js} = 0$, respectively. To complete the specification of the response functions in this case, we set $x_i = 1$, yielding

$$\begin{aligned} R_{je}(x_A, x_B) &= (r_{je}/2)[(1 - ax_j) + b(1 - x_j)], \quad \text{and} \\ R_{is}(x_A, x_B) &= (r_{is}/2)[(1 - ax_j) + b(1 - x_j)]. \end{aligned}$$

Similarly, when firm j does not launch an ABD program but firm i does, (1) and (2) imply that $R_{is} = R_{je} = 0$. To complete the specification of the response functions in this case, we set $x_j = 1$, yielding

$$\begin{aligned} R_{ie}(x_A, x_B) &= (r_{ie}/2)[(1 - ax_i) + b(1 - x_i)], \quad \text{and} \\ R_{js}(x_A, x_B) &= (r_{js}/2)[(1 - ax_i) + b(1 - x_i)]. \end{aligned}$$

As explained in Raju et al. (1995), the parameter $a \in (0, 1)$ in equations (1) and (2) represents the price sensitivity of customers, while $b \in (0, 1)$ represents the cross price sensitivity.[7] This interpretation fits our situation quite well. First, observe from the first equation that the proportion of customers in segment i who will pre-commit to retailer i (i.e., $R_{ie}(x_A, x_B)$) increases when retailer i offers a lower discount price $x_i p$ or when retailer i offers a discount price lower than that of retailer j; i.e., when $(x_j - x_i)p > 0$. Second, observe from the second equation that the proportion of customers in segment i who will 'switch' and pre-commit to retailer j (i.e., $R_{is}(x_A, x_B)$) increases when retailer j offers a lower discount price $x_j p$ or when retailer j offers a lower discount price than that of retailer i; i.e., when $(x_i - x_j)p > 0$.

Two features of our response functions that distinguish them from Raju, et al. (1995) are the parameters r_{ie} and r_{is}. These parameters

[7] In order to guarantee that $R_{ie}(x_A, x_B)$ and $R_{is}(x_A, x_B)$ are both non-negative, we need to impose $a + b < 1$.

allow us to capture different levels of sensitivity for a customer in segment i pre-committing to retailer i versus switching and pre-committing to retailer j. Their inclusion is essential in a competitive ABD model. The first parameter, r_{ie}, represents the consumer's risk aversion to a stockout. The larger r_{ie}, the greater is the chance that a customer will accept the ABD offer to avoid a stockout during the selling season. The second parameter, r_{is}, represents the degree of brand *dis*loyalty. The larger r_{is}, the greater the chance that a customer will switch from their regular firm and accept an ABD offer from the competing firm. Notice from above that the parameters r_{ie} and r_{is} are bounded between 0 and $1/(1+b)$ so as to guarantee that the response functions are bounded between 0 and 1/2. It is essential to bound these response functions between 0 and 1/2 to ensure $R_{is} + R_{ie} \leq 1$, ensuring consistency with our assumption that discounts do not increase consumption.

Note that the specification of the response function also allows for the case when both retailers offer an advanced booking program without providing a pre-season discount, i.e., AB without D. In this case, the ABD program serves as an early reservation system. To determine the response functions in this case we set $x_i = x_j = 1$ in (1) and (2), so that $R_{ie}(x_A, x_B) = (r_{ie}/2)(1-a)$ and $R_{is}(x_A, x_B) = (r_{is}/2)(1-a)$. Since $a < 1$, this implies that $R_{ie}(x_A, x_B) > 0$ when $r_{ie} > 0$. This shows that when customers are risk averse, they will use the ABD program as an early reservation system. Observe that as r_{ie} increases, customers are more averse to a stockout and switch to the early reservation system with their regular firm. Similarly, when $r_{is} > 0$, $R_{is}(x_A, x_B) > 0$. This is because a guarantee of product availability always induces some switchers. As r_{is} increases, customers are more disloyal and switch to the early reservation with the competing firm.

2.3 Effective Demands

Since each retailer may or may not launch an ABD program, we need to consider 4 scenarios. In scenario I, neither retailer offers the ABD program. In scenario II, retailer A offers the ABD program while retailer B does not. In scenario III, retailer A does not offer the program while retailer B does. In scenario IV, both retailers offer the ABD program. Based on the impact of an ABD program on the purchasing behavior of each segment as depicted in Figure 2.1, the effective demand (including the pre-committed orders prior to the selling season and the demand occurring during the selling season) for each retailer associated with scenario k, where $k = I, II, III, IV$, can be depicted in Figure 2.2.

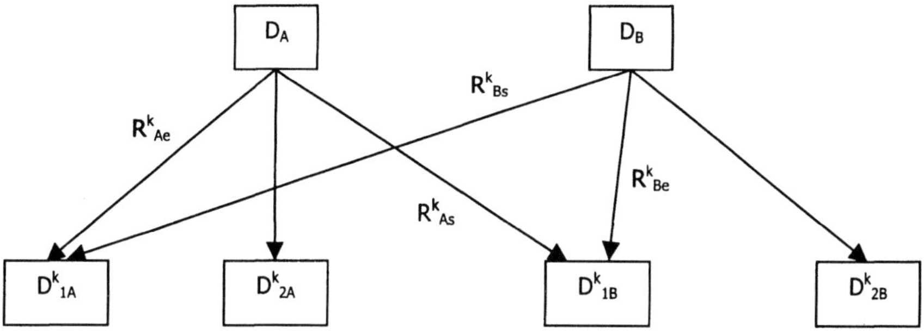

Figure 2.2. Effective Demands under Scenario $k, k = I, II, III, IV$

As shown in Figure 2.2, retailer i, $i = A, B$, faces two types of 'effective' demands in scenario k: the pre-committed orders placed prior to the season $D^k_{1i}(x_A, x_B)$, and the demand that occurs during the regular selling season $D^k_{2i}(x_A, x_B)$.[8] In this case, it is easy to check from Figure 2.2 that the 'effective' demands associated with retailer i can be expressed as follows:

$$\begin{aligned} D^k_{1i}(x_A, x_B) &= R^k_{ie}(x_A, x_B)D_i + R^k_{js}(x_A, x_B)D_j, \quad \text{for } j \neq i, \text{ and} \\ D^k_{2i}(x_A, x_B) &= [1 - R^k_{ie}(x_A, x_B) - R^k_{is}(x_A, x_B)]D_i. \end{aligned} \tag{2.3}$$

In scenario k, the total 'effective' demand obtained by retailer i is equal to $D^k_{1i}(x_A, x_B) + D^k_{2i}(x_A, x_B) = D_i + R^k_{js}(x_A, x_B)D_j - R^k_{is}(x_A, x_B)D_i$. Therefore, retailer i will gain additional demand in scenario k when more customers switch from retailer j to retailer i than those who switch from retailer i to retailer j; i.e., when $R^k_{js}(x_A, x_B)D_j - R^k_{is}(x_A, x_B)D_i > 0$.

Recall that (D_A, D_B) has a bivariate normal distribution. Because $D^k_{1i}(x_A, x_B)$ and $D^k_{2i}(x_A, x_B)$ are linear functions of D_A and D_B, it follows that $(D^k_{1i}(x_A, x_B), D^k_{2i}(x_A, x_B))$ also has a bivariate normal distribution. Firm i observes D^k_{1i} before placing the order prior to the selling season (period 2); thus, we are interested in the distribution of $(D^k_{2i}|D^k_{1i})$, which is also normal. It follows from (3) that the parameters of these

[8] For tractability, we assume that each retailer's demand depends only one their own customer pool, so that each retailer does not need to consider the unsatisfied demand from the other retailer.

normally distributed demand distributions for Firm A, for example, are:

$$\mu_{D^k_{1A}(x_A,x_B)} = R^k_{Ae}(x_A,x_B)\alpha\mu + R^k_{Bs}(x_A,x_B)(1-\alpha)\mu \tag{2.4}$$

$$\sigma^2_{D^k_{1A}(x_A,x_B)} = (R^k_{Ae}(x_A,x_B))^2[\theta\alpha\mu]^2 + (R^k_{Bs}(x_A,x_B))^2[\theta(1-\alpha)\mu]^2 + 2R^k_{Ae}(x_A,x_B)R^k_{Bs}(x_A,x_B) \times \rho(\theta\alpha\mu)(\theta(1-\alpha)\mu) \tag{2.5}$$

$$\mu_{D^k_{2A}(x_A,x_B)} = [1 - R^k_{Ae}(x_A,x_B) - R^k_{As}(x_A,x_B)]\alpha\mu \tag{2.6}$$

$$\sigma^2_{D^k_{2A}(x_A,x_B)} = [1 - R^k_{Ae}(x_A,x_B) - R^k_{As}(x_A,x_B)]^2[\theta\alpha\mu]^2 \tag{2.7}$$

$$Cov(D^k_{1A}(x_A,x_B), D^k_{2A}(x_A,x_B)) = \alpha R^k_{Ae}(x_A,x_B) + \rho(1-\alpha)R^k_{Bs}(x_A,x_B) \tag{2.8}$$

$$\mu_{(D^k_{2A}(x_A,x_B)|D^k_{1A}(x_A,x_B)=d)} = \mu_{D^k_{2A}(x_A,x_B)} + Corr\,(D^k_{1A}(x_A,x_B), D^k_{2A}(x_A,x_B)) \times \left(\frac{\sigma_{D^k_{2A}}}{\sigma_{D^k_{1A}}}\right)[d - \mu_{D^k_{1A}(x_A,x_B)}] \tag{2.9}$$

$$\sigma^2_{(D^k_{2A}(x_A,x_B)|D^k_{1A}(x_A,x_B)=d)} = \sigma^2_{D^k_{2A}(x_A,x_B)}[1 - Corr^2\,(D^k_{1A}(x_A,x_B), D^k_{2A}(x_A,x_B))]. \tag{2.10}$$

Because both firms have the same coefficient of variation θ for the primary demand, deriving the parameters for Firm B amounts to substituting A for B and $1-\alpha$ for α in the above set of equations; we omit the details.

Notice from (2.10) that the variance of period-two demand (i.e., the demand that occurs during the regular selling season) conditioned on observed period-one demand (i.e., the pre-committed orders placed prior to the selling season) is less than the variance of the unconditioned period-two demand (2.7). This is exactly as should be expected: the information content in period-one demand reduces the uncertainty for period two. Indeed, this reduction in uncertainty is one of the major benefits of the ABD program as it allows the firm to more accurately gauge total demand, reducing both shortage and spoilage costs. Furthermore, it can be shown that $\sigma^2_{(D^k_{2A}(x_A,x_B)|D^k_{1A}(x_A,x_B)=d)}$ is concave and decreasing in ρ for $\rho > 0$. Thus, when the primary demands, D_A and D_B, of the two firms are positively correlated, Firm A can use the information content not only from its own early purchasers, $R^k_{Ae}(x_A,x_B)$, but also the information content in the switchers from Firm B, $R^k_{Bs}(x_A,x_B)$, to reduce period-two demand uncertainty.

In the analysis that follows, substitutions from (1) and (2) will be made for the consumer response functions in the parameter equations (2.4) – (2.10) in each of four specific cases. The comparative statics in Section 3 will be developed from these.

2.4 The Analysis Framework

By utilizing the effective demands associated with retailer i in scenario k (i.e., $D_{1i}^k(x_A, x_B)$ and $D_{2i}^k(x_A, x_B)$), we can determine the optimal discount price that maximizes the retailer's expected profit as follows. First, notice that the order is placed at the start of the selling season. Therefore, retailer i can order the exact amount to fulfill the pre-committed orders $D_{1i}^k(x_A, x_B)$ observed prior to the selling season. The profit generated from those pre-committed orders is equal to $(x_i p - c) D_{1i}^k(x_A, x_B)$. Second, retailer i can utilize the information about $D_{1i}^k(x_A, x_B)$ to update the distribution of $D_{2i}^k(x_A, x_B)$. The retailer orders additional quantity Q (in addition to $D_{1i}^k(x_A, x_B)$) so as to cover the demand during the selling season. The profit generated from the demand $D_{2i}^k(x_A, x_B)$ is equal to $\{p \cdot \min\{Q, D_{2i}^k(x_A, x_B)\} + s[Q - D_{2i}^k(x_A, x_B)]^+ - cQ\}$. Given the discount prices $x_A p$ and $x_B p$, retailer i's expected profit in scenario k can be written as:

$$\begin{aligned} \pi_i^k(x_A, x_B) = E_{D_{1i}^k(x_A, x_B)} \Big\{ & (x_i p - c) D_{1i}^k(x_A, x_B) \\ & + \text{Max}_Q E_{[D_{2i}^k(x_A, x_B) | D_{1i}^k(x_A, x_B)]} \{p \ \min\{Q, D_{2i}^k(x_A, x_B)\} + \\ & s[Q - D_{2i}^k(x_A, x_B)]^+ - cQ\} \Big\}. \end{aligned} \tag{2.11}$$

Since retailer A selects x_A and retailer B selects x_B, the optimal expected profits attained by retailers A and B in equilibrium in scenario k must satisfy the following equations simultaneously:

$$\pi_A^k = Max_{x_A} \ \pi_A^k(x_A, x_B) \tag{2.12}$$

$$\pi_B^k = Max_{x_B} \ \pi_B^k(x_A, x_B). \tag{2.13}$$

Given the optimal equilibrium expected profits in each of the scenarios, we can construct the normal form of the ABD competitive game that lists each retailer's strategies and the payoff associated with each scenario k, where $k = I, II, III, IV$. Table 2.1 presents the normal form of the ABD competitive game, including the discount coefficients x_A^k and x_B^k associated with each scenario k. For instance, in scenario II, retailer A launches an ABD program while retailer B does not. Therefore, only Firm A needs to determine the optimal discount coefficient x_A^{II}. The scenario II payoffs for retailers A and B are π_A^{II} and π_B^{II}, respectively.

Firm A \ Firm B	Does Not Launch ABD Program	Launches ABD Program
Does Not Launch ABD Program	Scenario I: $(\pi^{I}_{A}, \pi^{I}_{B})$	Scenario III: $(\pi^{III}_{A}, \pi^{III}_{B})$ Discount coefficient: x^{III}_{B}
Launches ABD Program	Scenario II: $(\pi^{II}_{A}, \pi^{II}_{B})$ Discount coefficient: x^{II}_{A}	Scenario IV: $(\pi^{IV}_{A}, \pi^{IV}_{B})$ Discount coefficients: x^{IV}_{A}, x^{IV}_{B}

Table 2.1. Normal Form of the ABD Competitive Game.

In this section, we defined the consumer response functions $R_{ie}(x_A\, x_B)$ and $R_{is}(x_A, x_B)$, the effective demands $D^k_{1i}(x_A, x_B)$ and $D^k_{2i}(x_A, x_B)$, the profit function $\pi^k_i(x_A, x_B)$, and the optimal profit π^k_i for retailer i in scenario k. We now preview the structure of the remainder of this chapter. In Section 3, we analyze the optimal discount coefficient x^k_i and the optimal profit π^k_i for retailer i in scenario k, where $k = I, II, III, IV$. Section 4 utilizes the results obtained in Section 3 to construct the payoff matrix as presented in Table 2.1. Also, we characterize the conditions under which both retailers will offer the ABD program at equilibrium. Moreover, we shall discuss how the fixed cost, demand uncertainty, demand correlation, and market share will affect the optimal profit of the retailers and the equilibrium. Detailed numerical results and managerial implications are contained in Section 5. Finally, we present some concluding remarks in Section 6.

3. Analysis

With the basic model in hand, our analysis proceeds by consideration of the four separate scenarios:

Scenario I: Neither firm offers ABD.

Scenario II: Firm A offers ABD and Firm B does not.

Scenario III: Firm B offers ABD and Firm A does not.

Scenario IV: Both firms offer ABD.

The analysis of the individual scenarios does not include the decision of whether or not to offer ABD – the equilibrium analysis in Section 4

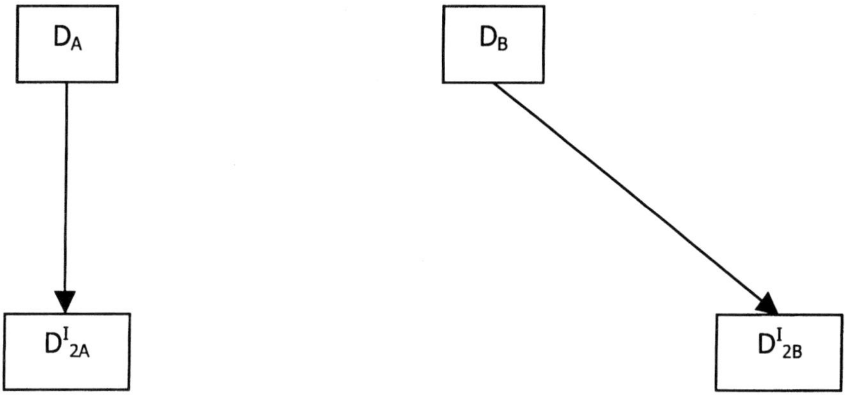

Figure 2.3. Effective Demands under Scenario I: Neither Firm Launches ABD Program

deals with this issue. Rather, the analysis in the current section solves for the optimal price-discount coefficient and the optimal order quantity assuming ABD is offered or not based on the scenario. To simplify the exposition we present our analysis as follows. In Scenarios I and IV our analysis will focus on the behavior of Firm A: unless otherwise noted, the behavior of Firm B is symmetric to that of Firm A. The behavior of both firms will be analyzed in Scenario II: Scenario III is symmetric to Scenario II.

3.1 Scenario I: Neither Firm Offers ABD

Because neither firm offers ABD, there is no switching of consumers from one retailer to the other, and there is no period-1 demand: $R^I_{Ae} = R^I_{As} = R^I_{Bs} = R^I_{Be} = 0$, and $D^I_{1A} = D^I_{1B} = 0$. All demand arises in period 2: $D^I_{2A} = D_A$, and $D^I_{2B} = D_B$. In this scenario there is no competitive interaction between the firms: each firm behaves as a monopolist. See Figure 2.3. The retailer charges p for each unit during the selling season and receives a salvage value s for each unit after the season, where $s < c < p$. The retailer needs to determine the optimal order quantity Q^I_i that maximizes the total expected profit. Let π_i be the optimal expected profit, where

$$\pi^I_i = \max_{Q_i \geq 0} E_{D_i}\{p \min\{Q_i, D_i\} + s[Q_i - D_i]^+ - cQ_i\}, i \in (A, B). \quad (2.14)$$

The above problem is the news-vendor problem with normally distributed demand. It is well known that the optimal order quantity Q^I_i and the optimal expected profit π_i^I for retailer i are given by:

$$Q_i^I = \mu_i + w\sigma_i, \quad i \in (A, B), \tag{2.15}$$
$$\pi_i^I = (p-c)\mu_i - (p-s)\phi(w)\sigma_i, \quad i \in (A, B), \tag{2.16}$$

where $w = \Phi^{-1}(\frac{p-c}{p-s})$, and $\Phi(\cdot)$ and $\phi(\cdot)$ are the cumulative distribution and the density functions of the standard normal distribution, respectively (see Silver, Pyke and Peterson 1998). Notice from (2.16) that the term $(p-s)\phi(w)\sigma_i$ can be rewritten as $[(p-c)+(c-s)]\phi(w)\sigma_i$, which corresponds to the sum of the expected overstock and understock costs associated with the optimal order quantity Q_i^I. From (2.16), we see that it is desirable for the retailer to reduce demand variance σ_i^2 using mechanisms such as the ABD program.

Rewriting the profit function for retailer A with $\mu_A = \alpha\mu$ and $\sigma_A = \theta\alpha\mu$, we get:

$$\pi_A^I = (p-c)\mu_A - (p-s)\phi(w)\sigma_A = \alpha\mu[(p-c) - (p-s)\phi(w)\theta]. \tag{2.17}$$

Similarly, we can determine the profit function for retailer B. As one would expect in a market-share model with fixed prices, where retailer A has share α and retailer B has share $1-\alpha$:

$$\pi_B^I = [(1-\alpha)/\alpha]\pi_A^I. \tag{2.18}$$

3.2 Scenario II: Only Firm A Offers ABD

When Firm A offers the item at discounted price $x_A p$ in period 1, Firm A gets period-1 demand both from Firm B's customers who switch, $R_{Bs}^{II}(x_A)D_B$, and from its own customers who order early $R_{Ae}^{II}(x_A)D_A$. Since Firm B does not offer ABD, Firm B has no pre-committed (i.e., period-1) orders. Thus, as depicted in Figure 2.4, we have:

$$D_{1A}^{II} = R_{Ae}^{II}(x_A)D_A + R_{Bs}^{II}(x_A)D_B, \tag{2.19}$$
$$D_{1B}^{II} = 0, \tag{2.20}$$
$$D_{2A}^{II} = [1 - R_{Ae}^{II}(x_A)]D_A, \tag{2.21}$$
$$D_{2B}^{II} = [1 - R_{Bs}^{II}(x_A)]D_B. \tag{2.22}$$

Here we will consider the behavior of both firms, starting with the non-ABD offering Firm B. Firm B, which faces demand only in period 2, has a classic single-period newsvendor problem to solve. From (2.22), Firm B's demand is normally distributed with mean and variance:

$$\mu_{D_{2B}^{II}} = [1 - R_{Bs}^{II}(x_A)]\mu_B = [1 - R_{Bs}^{II}(x_A)](1-\alpha)\mu \tag{2.23}$$
$$\sigma_{D_{2B}^{II}}^2 = [1 - R_{Bs}^{II}(x_A)]^2\sigma_B^2 = [1 - R_{Bs}^{II}(x_A)]^2[\theta(1-\alpha)\mu]^2. \tag{2.24}$$

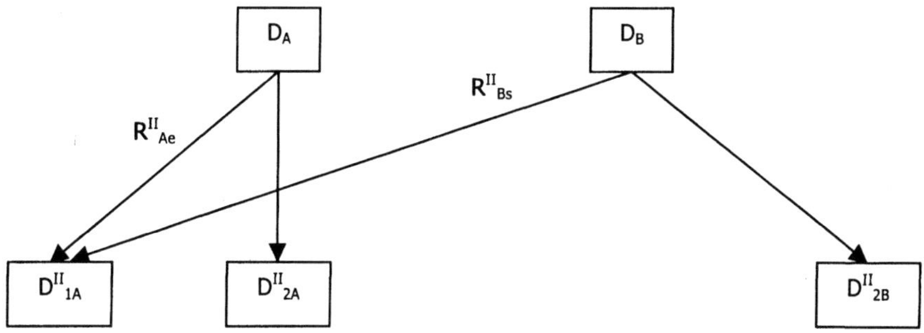

Figure 2.4. Effective Demands under Scenario II: Only Firm A Launches ABD Program

By using the same approach (the single-period newsvendor solution) as described in section 3.1 for Scenario I, we can determine the optimal expected profit π_B^{II} for Firm B:

$$\pi_B^{II} = (p-c)E(D_{2B}^{II}) - (p-s)\phi(w)\sigma_{D_{2B}^{II}},$$

which can be simplified as:

$$\pi_B^{II} = [1 - R_{Bs}^{II}(x_A)](1-\alpha)\mu[(p-c)-(p-s)\phi(w)\theta]. \tag{2.25}$$

From the definition of $R_{Bs}^{II}(x_A)$ [see (2)], it follows that Firm B's optimal profit in scenario II is increasing in x_A, Firm A's first-period discount coefficient. That is, the higher the price charged by Firm A in its ABD program, the higher the expected profits for Firm B when it does not offer ABD. Comparing the optimal expected profits for Firm B in scenario II, (2.25), with those in scenario I, (2.18), it is easy to see that

$$\pi_B^{II} \leq \pi_B^{I}. \tag{2.26}$$

That is, when only Firm A offers ABD, Firm B is worse off than the case where neither firm offers ABD.

We now turn our attention to Firm A, whose scenario-II actions are independent of those of Firm B. That is, Firm A takes it as given that Firm B does not offer ABD and optimizes accordingly. Firm A chooses a period-one price discount coefficient x_A and period-two order quantity

Q_A^{II} to maximize total expected profits π_A^{II}, where

$$\begin{aligned}\pi_A^{II} &= \max_{x_A}\Big\{E_{D_{1A}^{II}}\Big\{(x_A \cdot p - c)D_{1A}^{II} \\ &+ \max_{Q_A} E_{[D_{2A}^{II}|D_{1A}^{II}]}\{p \cdot \min\{Q_A, D_{2A}^{II}\} \\ &+ s[Q_A - D_{2A}^{II}]^+ - cQ_A\}\Big\}\Big\}. \end{aligned} \tag{2.27}$$

Substituting the definitions of the consumer response functions (1) and (2) into the equations for the parameters of the demand distributions, (2.4) – (2.10), and then optimizing (2.27) yields the first order condition for Firm A's period-one discount coefficient x_A^{II}. It can be easily shown that (2.27) is concave in x_A^{II}. Therefore, these first order conditions are sufficient for the optimality of x_A^{II}. By simplifying terms, we have:

$$x_A^{II} = \frac{p\overline{r}(1+b) + U_2 r_{Ae}(a+b) + c(a+b)\overline{r}}{2(a+b)p\overline{r}} \tag{2.28}$$

where

$$\overline{r} = r_{Ae}\alpha + r_{Bs}(1-\alpha) \tag{2.29}$$

$$U_2 = (p-c)\alpha - (p-s)\phi(w)\theta\alpha T_2 \tag{2.30}$$

$$\begin{aligned} T_2 &= \sqrt{1 - Corr(D_{1A}^{II}, D_{2A}^{II})} \\ &= (1-\alpha)r_{Bs}\sqrt{1-\rho^2} \times \\ &\quad \sqrt{\frac{1}{\alpha^2 r_{Ae}^2 + (1-\alpha)^2 r_{Bs}^2 + 2r_{Ae}r_{Bs}\rho\alpha(1-\alpha)}}. \end{aligned} \tag{2.31}$$

PROPOSITION 2.1 *In Scenario II, the optimal Firm-A ABD discount coefficient, x_A^{II}, is increasing in the cost c, salvage value s, and the correlation ρ when $\rho > 0$. Furthermore, x_A^{II} is decreasing in the coefficient of variation θ.*

The interpretation of this proposition is straightforward. The price charged by Firm A in period 1 is $x_A^{II}p$. As the amount of demand uncertainty θ increases, the firm values the information content of the early orders more; hence it lowers its first-period price to increase period-1 sales. As the cost c increases, the firm has less leeway to reduce price, hence the optimal first-period price increases. As the salvage value s increases, there is less downside risk to excess inventory, hence the value of the information in the early orders decreases and the firm increases period-1 price. Finally, as the demand correlation $\rho > 0$ increases, the information content of each early order increases, and Firm A can get the same level of information with fewer period-1 sales, hence can raise the period-1 price. In addition, the variance of period-2 demand decreases. The reduced period-two variance and the higher first-period price increase overall profits.

Substituting (2.28) into (2.27) yields the optimal expected Firm-A profit in scenario II:

$$\pi_A^{II} = (x_A^{II}p-c)\frac{1}{2}\mu\bar{r}(1+b-(a+b)x_A^{II})+U_2\mu\Big(1-\frac{r_{Ae}}{2}(1+b)+x_A^{II}\frac{r_{Ae}}{2}(a+b)\Big). \tag{2.32}$$

As previously noted, if Firm B does not offer ABD, Firm B is better off if Firm A also does not offer ABD. As might be expected, but somewhat less easy to see, the opposite is true for Firm A: Firm A is better off by offering an ABD program when Firm B does not offer an ABD program. We prove this claim in two steps.

LEMMA 2.2 *If Firm B does not offer an ABD program, then offering an ABD program with discount coefficient 1 is at least as good for Firm A than not having an ABD program, i.e.,* $\pi_A^{II}(1) \geq \pi_A^I$.

PROOF: Substitute in equations (2.4)–(2.10) to get:

$$\begin{aligned}
E(D_{1A}^{II}(1) + D_{2A}^{II}(1)) &= ED_{2A}^I + R_{Bs}^{II}(1)(1-\alpha)\mu \geq ED_{2A}^I \\
\sigma_{(D_{2A}^{II}(1)|D_{1A}^{II}(1))} &= [\sigma_{D_{2A}^I} - R_{Ae}^{II}(1)\theta\mu\alpha] \times \\
&\quad \sqrt{1 - Corr^2(D_{1A}^{II}(1), D_{2A}^{II}(1))} \leq \sigma_{D_{2A}^I}.
\end{aligned}$$

Then it can be shown from (2.27) that:

$$\begin{aligned}
\pi_A^{II}(1) &= (p-c)E[D_{1A}^{II}(1) + D_{2A}^{II}(1)] - (p-s)\phi(w)\sigma_{(D_{2A}^{II}(1)|D_{1A}^{II}(1))} \\
&\geq (p-c)E[D_{2A}^I] - (p-s)\phi(w)\sigma_{D_{2A}^I} \\
&= \pi_A^I.
\end{aligned}$$

■

Lemma 2 proves that if Firm B is not offering an ABD program, Firm A prefers to start an ABD program with discount coefficient 1 over not having an ABD program.[9] Having an ABD program with the optimal scenario II discount coefficient, x_A^{II}, rather than 1 only improves Firm A's situation; i.e., $\pi_A^{II} \geq \pi_A^{II}(1)$. Thus, we have shown:

$$\pi_A^{II} \geq \pi_A^I. \tag{2.33}$$

[9]Recall that we have not yet included a fixed cost for implementing an ABD program. Such a cost would clearly alter the conclusion of Lemma 2.

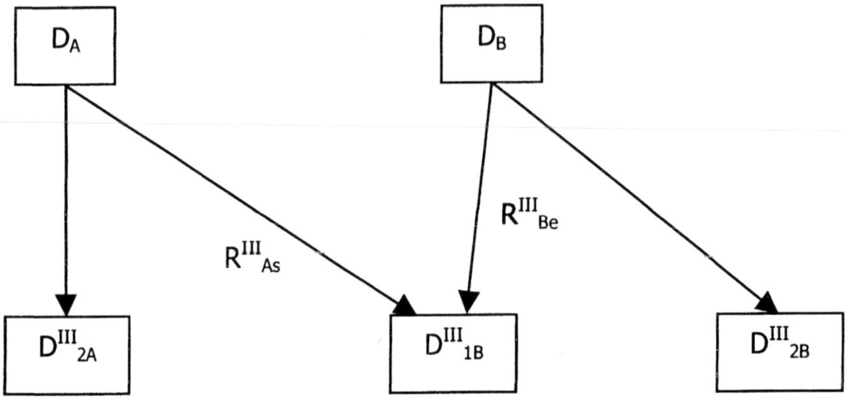

Figure 2.5. Effective Demands under Scenario III: Only Firm B Launches ABD Program

3.3 Scenario III: Only Firm B Offers ABD

The setup, and hence, the results of this section are perfectly symmetric with the previous section.

In scenario III as depicted in Figure 2.5, Firm A does not employ an ABD program and earns expected profits π_A^{III} as a function of Firm B's discount coefficient x_B:

$$\pi_A^{III} = (p-c)E(D_{2A}^{III}) - (p-s)\phi(w)\sigma_{2A}^{III}, \tag{2.34}$$

which can be simplified as:

$$\pi_A^{III} = [1 - R_{As}^{III}(x_B^{III})]\alpha\mu[(p-c) - (p-s)\phi(w)\theta]. \tag{2.35}$$

Comparing π_A^{III} with π_A^I in (2.17), we can conclude that Firm A earns lower profits in Scenario III than in Scenario I:

$$\pi_A^{III} \leq \pi_A^I. \tag{2.36}$$

We now turn our attention to Firm B which employs an ABD program with discount coefficient x_B^{III}. Substituting the definitions of the consumer response functions (1) and (2) into the equations for the parameters of the demand distributions for Firm B, and then optimizing the expected profit function yields the first order condition for Firm B's discount coefficient x_B^{III}. By simplifying terms, we have:

$$x_B^{III} = \frac{p\hat{r}(1+b) + U_3 r_{Be}(a+b) + c(a+b)\hat{r}}{2(a+b)p\hat{r}}, \tag{2.37}$$

where

$$\begin{aligned}
\hat{r} &= r_{As}\alpha + r_{Be}(1-\alpha) && (2.38)\\
U_3 &= (p-c)(1-\alpha) - (p-s)\phi(w)\theta(1-\alpha)T_3 && (2.39)\\
T_3 &= \sqrt{1 - Corr(D_{1B}^{II}, D_{2B}^{II})}\\
&= \alpha r_{As}\sqrt{1-\rho^2} \times\\
&\quad \sqrt{\frac{1}{\alpha^2 r_{As}^2 + (1-\alpha)^2 r_{Be}^2 + 2r_{As}r_{Be}\rho\alpha(1-\alpha)}}. && (2.40)
\end{aligned}$$

Parallel to Proposition 1, we get that the optimal firm-B period-1 discount coefficient in Scenario III, x_B^{III}, is increasing in the cost c, salvage value s, and the correlation ρ when $\rho > 0$, and is decreasing in the coefficient of variation θ.

Employing the optimal discount coefficient x_B^{III} given in (2.37), Firm B earns Scenario III expected profits π_B^{III}:

$$\begin{aligned}
\pi_B^{III} &= (x_B^{III}p - c)\frac{1}{2}\mu\hat{r}(1 + b - (a+b)x_B^{III}) +\\
&\quad U_3\mu\Big(1 - \frac{r_{Be}}{2}(1+b) + \frac{r_{Be}}{2}x_B^{III}(a+b)\Big). && (2.41)
\end{aligned}$$

With positions in scenario III switched relative to those in scenario II, it immediately follows from Lemma 2 that Firm B earns at least as much expected profit in scenario III as in scenario I:

$$\pi_B^{III} \geq \pi_B^{I}. \qquad (2.42)$$

3.4 Scenario IV: Both Firms Offer ABD

Scenario IV provided the impetus for our original question, to wit, what is the equilibrium behavior when two competing firms offer ABD programs. As in the analysis of scenarios I-III, we take the use of the ABD program as a given and solve for equilibrium first-period discount pricing and second-period ordering decisions; we put off until the next section the question of equilibrium behavior in the larger game. In scenario IV, both firms offer ABD. As depicted in Figure 2.6, both firms potentially poach some of their competitor's customers, as well as entice some of their own customers to purchase early. From (3) we get:

$$\begin{aligned}
D_{1A}^{IV} &= R_{Ae}^{IV}(x_A, x_B)D_A + R_{Bs}^{IV}(x_A, x_B)D_B, && (2.43)\\
D_{1B}^{IV} &= R_{Be}^{IV}(x_A, x_B)D_B + R_{As}^{IV}(x_A, x_B)D_A, && (2.44)\\
D_{2A}^{IV} &= [1 - R_{Ae}^{IV}(x_A, x_B) - R_{As}^{IV}(x_A, x_B)]D_A, && (2.45)\\
D_{2B}^{IV} &= [1 - R_{Be}^{IV}(x_A, x_B) - R_{Bs}^{IV}(x_A, x_B)]D_B. && (2.46)
\end{aligned}$$

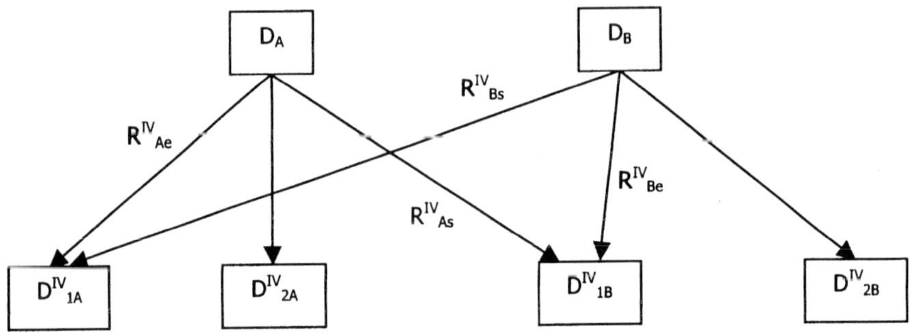

Figure 2.6. Effective Demands under Scenario IV: Both Firms Launch ABD Program

We first solve for each firm's best response function. That is, we begin by taking Firm B's discount coefficient $x_B^{IV} = y$ as given, and find Firm A's best response, $x_A^{IV}(y)$. Note that Firm A's best response is a function of Firm B's choice. Likewise we will find Firm B's best response taking Firm A's action as given. The choices will be in equilibrium if they are best responses to each other.

If Firm B uses the price discount coefficient $y < 1$, i.e., charges an ABD price of yp, then Firm A faces the maximization problem:

$$\begin{aligned} \pi_A^{IV} &= \max_{x_A} \Big\{ E_{D_{1A}^{IV}} \Big\{ (x_A \cdot p - c) D_{1A}^{IV} \\ &\quad + \max_{Q_A} E_{[D_{2A}^{IV} | D_{1A}^{IV}]} \{ p \cdot \min\{Q_A, D_{2A}^{IV}\} + \\ &\quad s[Q_A - D_{2A}^{IV}]^+ - cQ_A \} \Big\} \Big\}. \end{aligned} \tag{2.47}$$

Substituting the definitions of the consumer response functions (1) and (2) into the equations for the parameters of the demand distributions, (2.4) – (2.10), and then optimizing (2.47) yields the first-order condition for Firm A's period-one discount coefficient $x_A^{IV}(y)$ as a function of Firm B's choice y:

$$x_A^{IV}(y) = \frac{p\overline{r}(1 + by) + U_2[r_{Ae}(a + b) - r_{As}b] + c(a + b)\overline{r}}{2(a + b)p\overline{r}} \tag{2.48}$$

where $\overline{r}$, U_2, and T_2 are as given in (2.29)–(2.31).

PROPOSITION 2.3 *In Scenario IV, Firm A's best response function, denoted by $x_A^{IV}(y)$, is increasing in Firm B's strategy y. For all $y \leq 1$, $x_A^{IV}(y) \leq x_A^{II}$; i.e., Firm A has a lower ABD price when Firm B offers an ABD program than when it doesn't. Furthermore, $x_A^{IV}(y)$ is decreasing in θ and increasing in c, s, and ρ when $\rho > 0$.*

While Firm A's ABD discount in Scenario II (only Firm A offers ABD) is driven mainly by the second-period demand uncertainty reduction generated by the early orders, Firm A's ABD discount in Scenario IV (both firms offer ABD) is also driven by the competitive pressures of Firm B's ABD discount. The proposition shows that Firm A's ABD discount coefficient $x_A^{IV}(y)$ is increasing in Firm B's ABD discount coefficient y. That is, the lower the price charged by Firm B in period 1, the lower Firm A must charge in period 1 as well. If Firm A does not lower its period-1 price in response to a price decrease by Firm B, it not only loses a portion of its demand in both periods, it also loses the information content of the lost period-1 demand.

The resulting profit function for Firm A $\pi_A^{IV}(y)$ as a function of Firm B's strategy y is given by

$$\begin{aligned}\pi_A^{IV}(y) &= (x_A^{IV}p-c)\frac{1}{2}\mu\bar{r}(1+by-x_A^{IV}(a+b)) \\ &+U_2\mu\left[\left(1-\frac{r_{Ae}}{2}-\frac{r_{As}}{2}\right)+x_A^{IV}\left(\frac{r_{Ae}(a+b)}{2}-\frac{r_{As}b}{2}\right)-\right. \\ &\left. y\left(\frac{r_{Ae}b}{2}-\frac{r_{As}(a+b)}{2}\right)\right]. \end{aligned} \tag{2.49}$$

When Firm A uses strategy x, a best response function for Firm B, $x_B^{IV}(x)$, and the resulting profit function, $\pi_B^{IV}(x)$, are determined similarly. Specifically, we have

$$x_B^{IV}(x)=\frac{p\hat{r}(1+bx)+U_3[r_{Be}(a+b)-r_{Bs}b]+c(a+b)\hat{r}}{2(a+b)p\hat{r}}, \tag{2.50}$$

and

$$\begin{aligned}\pi_B^{IV}(x) &= (x_B^{IV}p-c)\frac{1}{2}\mu\hat{r}(1+bx-x_B^{IV}(a+b)) \\ &\quad +U_3\mu\left[\left(1-\frac{r_{Be}}{2}-\frac{r_{Bs}}{2}\right)+x_B^{IV}\left(\frac{r_{Be}(a+b)}{2}-\frac{r_{Bs}b}{2}\right)\right. \\ &\quad \left. -x\left(\frac{r_{Be}b}{2}-\frac{r_{Bs}(a+b)}{2}\right)\right], \end{aligned} \tag{2.51}$$

where $\hat{r}$, U_3, and T_3 are as given in (2.38) - (2.40). Parallel to Propostion 3, it follows that $x_B^{IV}(x)$ is increasing in Firm A's strategy x, decreasing in θ and increasing in c, s, and ρ when $\rho > 0$. Furthermore, Firm B has a lower ABD price when Firm A offers an ABD program than when it does not, i.e., $x_B^{IV}(x) \leq x_B^{III}$.

The strategy pair (x, y) is in equilibrium if the strategies are best responses to each other, i.e.,

$$x = x_A^{IV}(y) \tag{2.52}$$
$$y = x_B^{IV}(x). \tag{2.53}$$

PROPOSITION 2.4 *In scenario IV (both firms offer ABD), there is a uniquc equilibrium in the discount coefficients* (x_A^{IV}, x_B^{IV}). *These are given by*

$$\begin{aligned} x_A^{IV} &= [\bar{r}\hat{r}(2a+3b)(p+c(a+b)) + U_3 b\bar{r}(r_{Be}(a+b) - r_{Bs}b) + \\ &\quad 2U_2(a+b)\hat{r}(r_{Ae}(a+b) - r_{As}b)] \,/\, [p\bar{r}\hat{r}(2a+3b)(2a+b)] \\ x_B^{IV} &= [\bar{r}\hat{r}(2a+3b)(p+c(a+b)) + U_2 b\hat{r}(r_{Ae}(a+b) - r_{As}b) + \\ &\quad 2U_3(a+b)\bar{r}(r_{Be}(a+b) - r_{Bs}b)] \,/\, [p\bar{r}\hat{r}(2a+3b)(2a+b)] \end{aligned}$$

PROOF: The best response functions (2.48) and (2.50) are monotonically increasing functions (of the competitor's strategy) with slopes less than 1, i.e., they are contractions. Hence, there is a unique equilibrium [see Friedman (1986), Theorem 3.4]. Solving (2.48) and (2.50) simultaneously completes the proof. ∎

Employing the lattice-theoretic method detailed in Lippman, Mamer, and McCardle (1986), it follows that the Scenario IV equilibrium discount coefficients maintain the same comparative statics with regard to the parameters of the model as the best response functions. That is,

PROPOSITION 2.5 *The equilibrium discount coefficients,* (x_A^{IV}, x_B^{IV}), *are decreasing in the level of demand uncertainty* θ, *and are increasing in cost* c, *salvage value* s, *and correlation* ρ *when* $\rho > 0$.

COROLLARY 2.6 *Assume* $r_{Ae} = r_{Be} = r_{As} = r_{Bs}$. *Then in the unique Scenario IV equilibrium, the firm with the larger market share charges the higher discount price* xp *in period 1. That is,* $x_A^{IV} \geq x_B^{IV}$ *if and only if* $\alpha \geq 0.5$.

PROOF: By the assumption that $r_{Ae} = r_{Be} = r_{As} = r_{Bs}$, it follows that $\hat{r} = \bar{r}$. Then from Proposition 4, $x_A^{IV} \geq x_B^{IV}$ if and only if $U_2 \geq U_3$ as given in (2.30) and (2.39), respectively. But $U_2 \geq U_3$ if and only if $\alpha \geq 0.5$. ∎

Substituting the equilibrium discount coefficients given in Proposition 4 into the profit function for Firm A (2.47) yields the equilibrium expected profit $\pi_A^{IV} = \pi_A^{IV}(x_B^{IV})$: similarly for Firm B.

To summarize the results of Section 3, we have determined the within-Scenario equilibrium ABD discounts coefficients x_A^k and x_B^k, and expected profits π_A^k and π_B^k for each of the Scenarios, $k = I, II, III, IV$.

These values completely specify the payoff matrix in Table 2.1. In the next section we determine the equilbrium across Scenarios, which we refer to as the ABD Equilibrium.

4. ABD Equilibrium

To determine the equilibrium behavior in the larger game as depicted in Table 2.1, it is necessary to incorporate the decision of whether or not to have an ABD program. To evaluate if a retailer should offer an ABD program or not, we now compare the expected profit associated with different scenarios and the change in expected profit when moving from one scenario to another.

First, let us suppose that Firm B does not offer ABD. In this case, as noted from (2.33), Firm A prefers to offer ABD; i.e., Firm A prefers Scenario II (only Firm A offers ABD) to Scenario I (neither firm offers ABD). Second, let us suppose that Firm A does not offer ABD. In this case, as noted from (2.42), Firm B prefers to offer ABD; i.e., Firm B prefers Scenario III (only Firm B offers ABD) to Scenario I. It remains to examine two situations: when Firm B offers ABD, what would Firm A prefer to do; and when Firm A offers ABD, what would Firm B prefer to do? Instead of comparing the within-Scenario equilibrium expected profits generated by Firm A in Scenario III and Scenario IV (or Firm B in Scenarios II and IV), we examine a slightly different question that enables us to show that Scenario IV (both firms offer ABD) yields an unique equilibrium to the ABD game. Specifically, the question we address in this section is, for example, given that Firm B offers ABD and Firm A does not (Scenario III), would Firm A want to implement an ABD program? Lemma 7 answers this question in the affirmative.[10]

LEMMA 2.7 *If Firm B offers an ABD program with a discount coefficient $y \leq 1$, then offering an ABD program with discount coefficient 1 is at least as good for Firm A as not having an ABD program, i.e., $\pi_A^{IV}(1, y) \geq \pi_A^{III}(y)$.*

PROOF: If Firm A does not offer ABD, then Firm A will obtain the expected profit as given in Scenario III (only Firm B offers ABD at a discount coefficient $y = x_B^{III}$). From (2.34) in Section 3.3, we have:

$$\pi_A^{III}(y) = (p - c)E(D_{2A}^{III}(y)) - (p - s)\phi(w)\sigma_{D_{2A}^{III}(y)}. \tag{2.54}$$

If Firm A adds an ABD program with discount coefficient 1 and Firm B does not change what it is doing, Firm A would then earn Scenario

[10] Lemma 7 is similar in statement and proof to Lemma 2 in Section 3.2.

IV expected profits, $\pi_A^{IV}(1,y)$, (in this case, Firm A offers ABD with a discount coefficient 1 and Firm B offers ABD with discount coefficient $y = x_B^{III}$). Substituting in (2.11), solving for the optimal order quantity, and rearranging terms yields,

$$\begin{aligned}
\pi_A^{IV}(1,y) &= E_{D_{1A}^{IV}(1,y)}\Big\{(p-c)D_{1A}^{IV}(1,y) \\
&\quad + \mathrm{Max}_Q E_{[D_{2A}^{IV}(1,y)|D_{1A}^{IV}(1,y)]}\{p\ \min\{Q, D_{2A}^{IV}(1,y)\} + \\
&\quad s[Q - D_{2A}^{IV}(1,y)]^+ - cQ\}\Big\} \\
&= (p-c)E[D_{1A}^{IV}(1,y) + D_{2A}^{IV}(1,y)] - \\
&\quad (p-s)\phi(w)\sigma_{(D_{2A}^{IV}(1,y)|D_{1A}^{IV}(1,y))}. \qquad (2.55)
\end{aligned}$$

From equations (2.4)–(2.10), note that:

$$\begin{aligned}
E(D_{1A}^{IV}(1,y) + D_{2A}^{IV}(1,y)) &= ED_A^{III}(y) + \\
&\quad R_{Bs}^{IV}(1,y)(1-\alpha)\mu \geq ED_A^{III}(y) \\
\sigma_{(D_{2A}^{IV}(1,y)|D_{1A}^{IV}(1,y))} &= [\sigma_{D_A^{III}(y)} - R_{Ae}^{IV}(1,y)\theta\mu\alpha] \times \\
&\quad \sqrt{1 - Corr^2(D_{1A}^{IV}(1,y), D_{2A}^{IV}(1,y)} \\
&\quad \leq \sigma_{D_A^{III}(y)}.
\end{aligned}$$

It then follows that:

$$\begin{aligned}
\pi_A^{IV}(1,y) &= (p-c)E[D_{1A}^{IV}(1,y) + D_{2A}^{IV}(1,y)] - \\
&\quad (p-s)\phi(w)\sigma_{(D_{2A}^{IV}(1,y)|D_{1A}^{IV}(1,y))} \\
&\geq (p-c)E[D_{2A}^{III}(y)] - (p-s)\phi(w)\sigma_{D_{2A}^{III}(y)} \\
&= \pi_A^{III}(y)
\end{aligned}$$

∎

From Lemma 7 we can conclude that $\pi_A^{IV}(1, x_B^{III}) \geq \pi_A^{III}(x_B^{III})$. This implies that when Firm B offers an ABD program, Firm A prefers having an ABD program: therefore, Scenario III does not represent an equilibrium of the larger game. By symmetry, an identical argument can be used to show that $\pi_B^{IV}(x_A^{II}, 1) \geq \pi_B^{II}(x_A^{II})$. This implies that when Firm A offers an ABD program, Firm B prefers having an ABD program: therefore, Scenario II does not represent an equilibrium of the larger game.

Based on Lemma 7, Firm A reasons as follows: given that Firm B has an ABD program with discount coefficient x_B^{III}, Firm A is better off with an ABD program with a discount coefficient 1 than without an ABD program; and, Firm A is even better off if it offers an ABD program with a discount coefficient of $x_A = \text{argmax}_x \ \pi_A^{IV}(x, x_B^{III})$.[11] By symmetry, Firm B applies the same reasoning to justify the fact that Firm B is better off offering an ABD program when Firm A has an ABD program with discount coefficient x_A^{II}. Once both firms are offering ABD programs, the resulting equilibrium is as given in Scenario IV. This observation is captured in the following Proposition.

PROPOSITION 2.8 *If it is costless to impelement an ABD prgram, the ABD game has a unique equilibrium: both firms offer ABD with discount coefficients as stated in Proposition 4 for Scenario IV.*

PROOF: We have already shown that Firm A prefers to start an ABD program both when Firm B does not have an ABD program (Scenario I) and when Firm B does have an ABD program (Scenario III). By symmetry, we can show that Firm B prefers to start an ABD program both when Firm A does not have an ABD program (Scenario I) and when Firm A does have an ABD program (Scenario II). All that remains to be shown is that once both firms have an ABD program (Scenario IV), neither firm prefers to cancel its program; that is, neither firm would wish to deviate from Scenario IV.

Assume the firms are in Scenario IV with discount coefficients as given in Proposition 4. Without loss of generality, consider Firm A. Firm A prefers $\pi_A^{IV}(x_A^{IV}, x_B^{IV})$ to $\pi_A^{IV}(1, x_B^{IV})$ because the equilibrium in Scenario IV is unique. Furthermore, by Lemma 7 and the argument immediately following, Firm A prefers $\pi_A^{IV}(1, x_B^{IV})$ to $\pi_A^{III}(x_B^{IV})$. Thus, Firm A will not deviate (eliminate its ABD program) from the equilibrium in Scenario IV: similarly for Firm B. ∎

By assuming that there are no fixed costs for a firm to launch an ABD program, we have proved that the ABD game (as depicted in Table 2.1) has a unique equilibrium: both firms offer ABD programs with discount coefficients as stated in Proposition 4. The key argument of our proof hinges on the following 4 inequalities:

$$\pi_A^{II} \geq \pi_A^{I} \tag{2.56}$$

$$\pi_B^{IV}(x_A^{II}, x_B^{IV}(x_A^{II})) \geq \pi_B^{II} \tag{2.57}$$

$$\pi_B^{III} \geq \pi_B^{I} \tag{2.58}$$

$$\pi_A^{IV}(x_A^{IV}(x_B^{III}), x_B^{III}) \geq \pi_A^{III} \tag{2.59}$$

[11] We do not claim that the Scenario-IV firm-A expected payoff dominates the Scenario-III firm-A expected payoff. That is, we do not claim that π_A^{IV} as given in (2.47) is at least as great as π_A^{III} as given in (2.35) because, as opposed to (2.35), (2.47) assumes that Firm B uses the best response possible when it offers ABD.

We now utilize these 4 inequalities to evaluate the equilibrium of the ABD game for the case in which a fixed cost K is incurred when a firm offers ABD. In this case, the expected profits listed in the above inequalities remain the same except the following: Firm A's expected profits in Scenarios II and IV become $\pi_A^{II} - K$ and $\pi_A^{IV}(x_A^{IV}(x_B^{III}), x_B^{III}) - K$, respectively. Similarly, Firm B's expected profits in Scenarios III and IV become $\pi_B^{III} - K$ and $\pi_B^{IV}(x_A^{II}, x_B^{IV}(x_A^{II})) - K$, respectively. By considering these effective changes and by rearranging the terms, the above inequalities can be rewritten as:

$$\pi_A^{II} - \pi_A^{I} \geq K \tag{2.60}$$
$$\pi_B^{IV}(x_A^{II}, x_B^{IV}(x_A^{II})) - \pi_B^{II} \geq K \tag{2.61}$$
$$\pi_B^{III} - \pi_B^{I} \geq K \tag{2.62}$$
$$\pi_A^{IV}(x_A^{IV}(x_B^{III}), x_B^{III}) - \pi_A^{III} \geq K. \tag{2.63}$$

For $K = 0$, all four inequalities hold. They are proved via Lemmas 2 and 7, and the material that follows immediately. It follows that, for K small enough (near zero), all of these inequalities continue to hold. By applying Lemmas 2 and 7 and the material that follows these lemmas, we can conclude that there is a unique equilibrium in which both firms offer ABD. On the other hand, if K is large enough, none of these inequalities hold. In this case, we can utilize the same argument as presented in this section to show that there is a unique equilibrium in which neither firm offers ABD. For very large K, whatever benefit there might accrue from having an ABD program is outweighed by the implementation cost K. What will happen for moderate values of K so that some but not all inequalities hold?

Suppose for some set of parameters, inequality (2.60) holds, but (2.61) and (2.62) do not hold. Then there is a unique equilibrium in which Firm A offers ABD and Firm B does not, i.e., Scenario II. Similarly, if inequality (2.62) holds, but (2.60) and (2.63) do not hold, then there is a unique equlibrium in which Firm B offers ABD but Firm A does not, i.e., Scenario III. Finally, if both (2.60) and (2.62) hold, but (2.63) and (2.61) do not hold, there are two equilibria represented by Scenario II and Scenario III. That is, it is in equilibrium for either firm to offer ABD, but not both. These results are summarized in Figure 2.7. Numerical examples of each of these cases are provided in the next section.

To determine how parameters such as the degree of product demand uncertainty θ, demand correlation ρ, and market share α affect equilibrium behavior, it is important to recognize that any change in these parameters that would increase the difference in profits with and without the ABD program would also ensure (2.60) through (2.63) are more

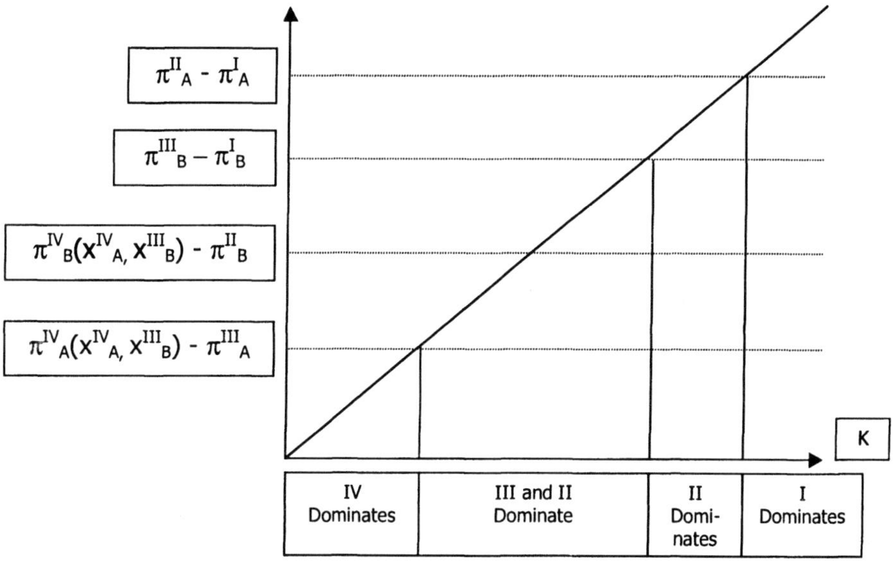

Figure 2.7. Impact of K on ABD Equilibrium

easily satisfied. This, in turn, would make sure Scenario IV is the preferred strategy for both firms. We examine this issue in greater detail in the numerical analyses described in the next section.

5. Numerical Analysis

To better illustrate the ideas in this chapter, we develop a numerical example with parameters as described in Table 2.2. We first calculate the optimal discount coefficient (where appropriate), the profits for each firm and total profits across the four scenarios described in Section 3. These results are summarized in Table 2.3. This table shows that $\pi_B^{II} = 1994 < \pi_B^I = 2091$ and that $\pi_A^{II} = 2417 > \pi_A^I = 2091$, as expected from (2.26) and (2.33) respectively. In addition, $\pi_A^{III} = 1994 < \pi_A^I = 2091$ and $\pi_B^{III} = 2417 > \pi_B^I = 2091$, as expected from (2.36) and (2.42) respectively. Consistent with Proposition 7, these results also imply that the optimal strategy for both firms is to offer the ABD program; the optimal discount and profits are given in Scenario IV of Table 2.3. Also, note from this table that total expected profits in this example are highest when both firms offer the ABD program. However, the best

Parameter	Value	Parameter	Value
Selling Price (P)	100	r_{Ae}	0.9
Cost (C)	50	r_{Be}	0.9
Salvage Value (S)	25	r_{As}	0.6
Mean (μ)	100	r_{Bs}	0.6
Coefficient of Variation (θ)	0.3	a	0.9
Correlation (ρ)	0.4	b	0.05
Brand A Market Share (α)	0.5		

Table 2.2. Parameters for Illustrative Example.

case scenario for each individual firm is to offer the ABD while the other does not.[12]

To better understand the degree to which product demand uncertainty θ affects the optimal discount coefficient and expected profits, we varied θ from 0 to 1.8 and calculated these variables across all the four scenarios. As expected from Propositions 1 and 3, the optimal discount coefficients are decreasing in the level of demand uncertainty. This is because the firm values the information content of early orders more and, thus, lowers its first-period price to increase period-1 sales. However, this increased discount and demand uncertainty contributes to a decrease in total expected profits. We also found that the decline in expected profits at a firm is much greater if it does not implement the ABD program. In addition, this decline is even worse if the competing firm offers the ABD program. This suggests that under increased demand uncertainty, it is even more critical for a firm to offer the ABD program when the competing firm has instituted this program.

To analyze the impact of demand correlation ρ between firms on the optimal discount coefficients and total expected profits, we varied ρ from 0 to 0.99 and calculated these variables across all the appropriate scenarios. As expected from Propositions 1 and 3, the optimal discount coefficients and expected profits are increasing with the level of demand

[12]In conformance with this observation, we found the equilibrium prices to be lower than the prices at the best case scenario for each firm and this trend was repeated in all of the analyses described in this section.

Scenario I: Neither firm offers ABD	**Scenario II: Firm A offers ABD and Firm B does not**
Firm A: Expected Profit (π_A^I) = 2091 Firm B: Expected Profit (π_B^I)= 2091 Total Expected Profits ($\pi_A^I + \pi_B^I$) = 4182	Firm A: Optimal Discount Coefficient (x_A^{II}) = 0.942 Firm A: Expected Profit (π_A^{II}) = 2417 Firm B: Expected Profit (π_B^{II}) = 1994 Total Expected Profits ($\pi_A^{II} + \pi_B^{II}$) = 4411
Scenario III: Firm B offers ABD and Firm A does not	**Scenario IV: Both Firms offer ABD**
Firm A: Expected Profit (π_A^{III}) = 1994 Firm B: Optimal Discount Coefficient (x_B^{III}) = 0.942 Firm B: Expected Profit (π_B^{III}) = 2417 Total Expected Profits ($\pi_A^{III} + \pi_B^{III}$) = 4411	Firm A: Optimal Discount Coefficient (x_A^{IV}) = 0.935 Firm A: Expected Profit (π_A^{IV}) = 2305 Firm B: Optimal Discount Coefficient (x_B^{IV}) = 0.935 Firm B: Expected Profit (π_B^{IV}) = 2305 Total Expected Profits ($\pi_A^{IV} + \pi_B^{IV}$) = 4610

Table 2.3. Results for the Four Scenarios.

correlation, when $\rho > 0$. This is because as the degree of demand correlation $\rho > 0$ increases, the information content of early orders increases, reducing the need for first period demand and consequently, the firm increases period 1 price. This higher price and lowered variance due to increased information content in the first period in turn increase expected profits. Recall that as π_A^I and π_B^I are independent of ρ, expected profits do not change with changing levels of demand correlation when both firms do not implement the ABD in Scenario I. However, it is important to note that expected profits of the firm not implementing the ABD program also increase with ρ in Scenarios II and III. This occurs since the increasing level of absolute demand correlation reduces the need for the firm offering the ABD to discount heavily. This in turn reduces the number of customers who switch from the firm not offering the ABD and, thus, increases their profits. Nevertheless, across this range of ρ, we found that a firm is always worse off by not offering the ABD program when the competing firm does offer this program, more than when both do not.

To evaluate the impact of market share on optimal discount coefficients and expected profits, we varied α from 0 to 1 across all the

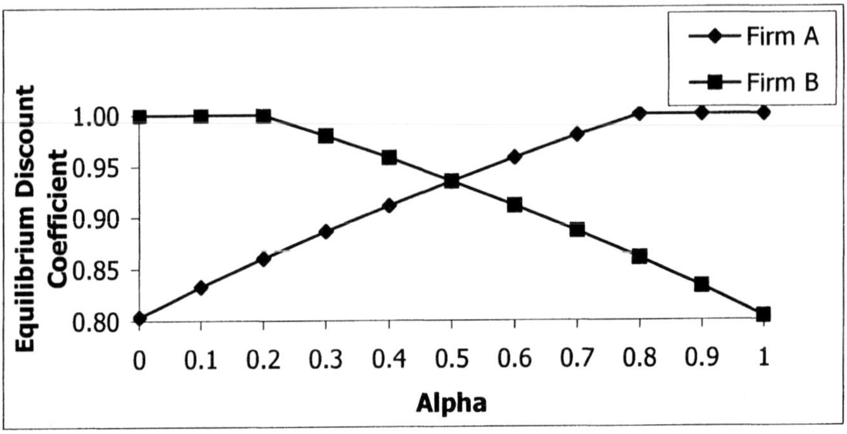

Figure 2.8. Equilibrium Discount Coefficients Versus α in Scenario IV

relevant scenarios. As expected, when the market share of a firm increases, then the optimal discount coefficient and profits increase. This is because with higher market share, there is only a small gain in additional demand that does not justify lowering the price. Such higher prices and increased demand due to higher market share in turn contribute to higher expected profits. Figure 2.8 represents the change in optimal discount coefficients with market share for both firms under Scenario IV. The corresponding changes in expected profits for each firm and total expected profit are shown in Figure 2.9. Observe from this figure that total profits across both firms are maximized when one firm dominates the other, so that their market shares are vastly dissimilar. This is because as market share becomes more asymmetric, the dominant firm charges a higher price on a larger fraction of demand, while the dominated firm charges a lower price on a smaller fraction of demand. Thus, the resulting gain in expected profit for the dominant firm offsets the loss in expected profits of the dominated firm. To assess how r_{ie}, the degree of customer risk aversion to incurring a stockout during the regular season affects the optimal discount coefficient and expected profits, we set $r_{Ae} = r_{Be} = r_{ie}$ and varied r_{ie} from 0 to 1 across all of the appropriate scenarios. We found that the optimal discount coefficient is increasing in r_{ie}, since increasing the level of risk aversion of the customers induces them to choose the ABD program, without the need for lowering the discount price. Such higher prices in turn contribute to higher expected profits. To understand how these results change when this parameter changes only at one firm, we set $r_{Be} = r_e$, varied r_{Ae}

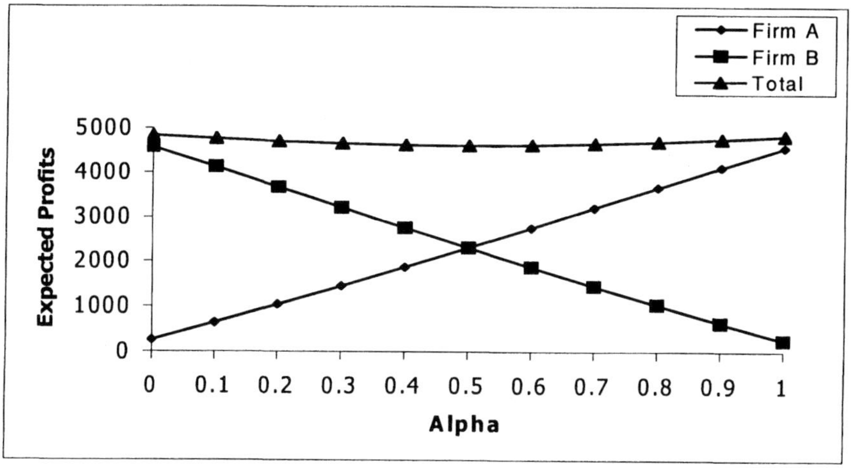

Figure 2.9. Expected Profits Versus α in Scenario IV

from 0 to 1, and considered these in the context of Scenario IV. In this case, Firm A discounts more aggressively until $r_{Ae} = r_e$, to compensate for the fact that their customers are less risk averse than customers of Firm B. However, once $r_{Ae} > r_e$, Firm A discounts less than Firm B, as the same level of early sales can now be achieved with a higher discount price since their customers are now more risk averse than customers of Firm B.

We also evaluate how r_{is}, the degree of customer loyalty, impacts the optimal discount coefficient and expected costs. To perform this analysis, we set $r_{As} = r_{Bs} = r_{is}$ and varied r_{is} from 0 to 1 across all of the appropriate scenarios. We found that the optimal discount coefficient is decreasing in r_{is}. This is because steeper discounts are required to retain increasingly disloyal customers, which in turn contribute to lower expected profits. To estimate how these results change when this parameter changes only at one firm, we set $r_{As} = r_s$, varied r_{Bs} from 0 to 1, and again considered this in the context of Scenario IV. In this case, Firm A discounts less than Firm B until $r_{Bs} = r_s$. This occurs as a steeper discount by Firm A would elicit a proportionately larger discount from Firm B and this in turn would cause a net loss of customers to Firm A since, until this threshold, Firm A's customers are less loyal than Firm B's customers. However, once $r_{Bs} > r_s$, Firm A discounts more than Firm B since Firm B's customers are now less loyal than customers of Firm A and steeper discounts now will entice Firm B's customers to defect to Firm A.

We next consider the impact of the setup cost K on the equilibrium behavior of the larger game. Note from Table 2.2 that $x_A^{II} = 0.942$ and $x_B^{III} = 0.942$. We use these results in (2.50) and (2.48) to compute $x_B^{IV}(x_A^{II}) = 0.935$ and $x_A^{IV}(x_B^{III}) = 0.935$. These in turn are used in (2.51) and (2.49) to calculate $\pi_B^{IV}(x_A^{II}, x_B^{IV}(x_A^{II})) = 2310$ and $\pi_A^{IV}(x_A^{IV}(x_B^{III}, x_B^{III}) = 2310$. Using these values along with appropriate values from Table 2.2 in (2.60) to (2.63), we find that when $K < 316$, Scenario IV is the optimal strategy so that both retailers offer the ABD program. If $K > 326$, then this results in Scenario I wherein both retailers do not offer the ABD program. Finally, when $316 \leq K \leq 326$, there are two equilibria represented by Scenarios II and III.

We also consider how demand uncertainty, demand correlation and market share affect equilibrium behavior in the larger game, when the fixed costs for implementation are given. In this regard, it is useful to recollect that increased demand uncertainty increases the benefit of implementing the ABD program and thus moves the firms toward Scenario IV. To illustrate this point, we assumed that $K = 300$ and found that when $\theta > 0.28$, (2.60) through (2.63) hold so that Scenario IV is the optimal strategy. When $\theta < 0.27$, none of the inequalities (2.60) through (2.63) hold and the optimal strategy for both firms is Scenario I. This is because the degree of uncertainty is not large enough to justify the fixed cost of implementing the ABD program. When $0.27 \leq \theta \leq 0.28$, (2.60) and (2.62) hold, but (2.63) and (2.61) do not, and there are two equilibria represented by Scenarios II and III. Similarly, when the degree of positive demand correlation increases, the benefit of implementing the ABD program increases and both firms move towards scenario IV. To demonstrate this, we conducted an equivalent analysis on ρ and found that when $\rho > 0.265$ the optimal strategy is Scenario IV; when $\rho < 0.184$, the optimal strategy is Scenario I; and when $0.184 < \rho < 0.265$, Scenarios II and III represent the two equilibria.

Lastly, as the market share of a firm increases, the firm is more inclined to implement an ABD program. This is because the benefits of implementing the ABD program are achieved on a larger portion of demand and the resulting gain in profits offsets the fixed costs of implementation. In this example, we calculate that when $0.468 \leq \alpha \leq 0.532$, (2.60) through (2.63) hold so that both firms prefer to implement the program and Scenario IV is the optimal strategy. When, $\alpha \geq 0.547$, (2.60) holds, but (2.61) and (2.62) do not, so that only Firm A prefers to implement the program resulting in Scenario II. When $\alpha \leq 0.453$, (2.62) holds, but (2.60) and (2.63) do not and, thus, only Firm B prefers to implement the program resulting in Scenario III. Finally, when $0.453 \leq \alpha \leq 0.468$ or $0.532 \leq \alpha \leq 0.547$, both (2.60) and (2.62) hold, but (2.63) and (2.61) do not. Consequently, there are two equilibria represented by Scenario II and III. These results are summarized by Table 2.4.

Alpha	Optimal Strategy
0.000 – 0.453	Scenario III
0.453 – 0.468	Scenario II & III
0.468 – 0.532	Scenario IV
0.532 – 0.547	Scenario II & III
0.547 – 1.000	Scenario II

Table 2.4. Optimal Strategies under Varying α.

6. Concluding Remarks

In this chapter we extended the single-firm Advanced Booking Discount model of Tang et al. (2004) to include competition. We showed that, in general, in a two-firm competitive model, if it is optimal for one firm to adopt an ABD program, the unique equilibrium has both firms adopting an ABD program. One of the main advantages of an ABD program is the information content in the pre-orders, which works to reduce the uncertainty regarding regular season sales. As is well known from the inventory literature, reducing the uncertainty in a newsvendor-type model reduces shortage and spoilage costs and, hence, increases profits. An additional advantage of the ABD program in a competitive model is that it allows a firm to "steal" customers away from the competition. The equilibrium discount coefficients were shown to be increasing in salvage value and production cost, and decreasing in demand uncertainty. We also constructed a numerical example to illustrate how changes in demand uncertainty, demand correlation, and market share affect discount coefficients and profits in each scenario, and how these parameters and the fixed costs of implementing the ABD program affect equilibrium behavior across scenarios.

While the model presented here entailed only two firms, we expect the results (when the cost $K = 0$) to extend to an n-firm competitive model. The difficulty is in the specification of the consumer response functions (1) and (2) when there are more than 2 firms. There are other limitations to our model, and we plan to use the current model as a basic building block to address them (much as Tang et al. (2004) served as a building block for this chapter). For example, we aim to extend the analysis to more than two periods with multiple replenishments; more than one product with joint capacity constraints; to include asymmetries in prices, costs, and salvage values across firms; and to include discounting and price competition in both the pre- and regular season.

References

Achabal, D.D., McIntyre, S.A. Smith, 1990. "Maximizing Profits from Periodic Department Store Promotions," *Journal of Retailing* 66, 383-407.

Bhardwaj, P., and Sismeiro, D., 2001. "Can Manufacturers and Retailers Agree on Store Brand Provision?" Working Paper, The Anderson School at UCLA.

Bickel, P. and K. Doksum, 1977. *Mathematical Statistics*, Holden Day Publisher, San Francisco.

Choi, S.C., 1991. "Price Competition in a Channel Structure with a Common Retailer," *Marketing Science* 10 (Fall), 271-296.

Eliashberg, J., and R. Steinberg, 1991. "Marketing-Production Joint Decision Making," in J. Eliashberg and G. Lilien (Eds), *Management Science in Marketing*, Handbooks in Operations Research and Management Science, North Holland.

Fisher, M., Rajaram, K. and Raman, A., 2001. "Optimizing Inventory Replenishment of Retail Fashion Products," *Manufacturing and Service Operations Management* Summer 2001 3(3), 230-241.

Friedman, J., *Game Theory with Applications to Economics*, Oxford University Press, New York, 1986.

Lal, R., 1990. "Price Promotions: Limiting Competitive Encroachment," *Marketing Science* 9, 247-262.

Lippman, S.A., J. Mamer, and K.F. McCardle, 1986. "Comparative Statics in Non-Cooperative Games via Transfinitely Iterated Play," *Journal of Economic Theory* 41, 288-303.

Lollar, C., 1992. "Trade Secrets of the Travel Game," *Working Woman*, 17, May, 94.

McCardle, K.F., K. Rajaram, and C.S. Tang, 2004. "Advance Booking Discount Programs under Retail Competition," *Management Science* 50, 701-708.

McVea, R., 1997. "Wild West Lives Again in Suburbs," *Chicago Tribune*, August 3, 1997.

Narasimhan, C., 1984. "A Price Discrimination Theory of Coupons," *Marketing Science* 3, 128-147.

Raju, J.S., Sethuraman, R. and Dhar, S.K., 1995. "The Introduction and Performance of Store Brands," *Management Science* 41, 957-978.

Shubik, M. and Levitan, R., 1980. *Market Structure and Behavior*, Harvard University Press, Cambridge, MA.

Silver, E.A., D.F. Pyke, and R. Peterson, 1998. *Inventory Management and Production Planning and Scheduling*, Third Edition, John Wiley.

Smith, S. A., and D.D. Achabal, 1998. "Clearance Pricing and Inventory Policies for Retail Chains," *Management Science* 44, 285-300.

Tang, C.S., Rajaram K., Alptekinoglu, A., and Ou, J., 2004. "The Benefits of Advance Booking Discount Programs: Model and Analysis," *Management Science.* 50, 465-478.

Weng, K., and Parlar, M., 1999. "Integrating Early Sales with Production Decisions: Analysis and Insights," *IIE Trans. Scheduling Logist.* 31, 1051-1060.

Xie, J., and Shugan, S., 2001. "Electronic Tickets, Smart Cards, and Online Prepayments: When and How to Advance Sell," *Marketing Science* 20, 3, 219-243.

Chapter 3

THIRD PARTY LOGISTICS PLANNING WITH ROUTING AND INVENTORY COSTS

Alexandra M. Newman
Division of Economics and Business
Colorado School of Mines
Golden, CO 80401

Candace A. Yano
Department of Industrial Engineering and Operations Research and
The Haas School of Business
University of California, Berkeley, CA 94720-1777

Philip M. Kaminsky
Department of Industrial Engineering and Operations Research
University of California, Berkeley, CA 94720-1777

Abstract We address a scheduling and routing problem faced by a third-party logistics provider in planning its day-of-week delivery schedule and routes for a set of existing and/or prospective customers who need to make shipments to their customers (whom we call "end-customers"). The goal is to minimize the total cost of transportation and inventory while satisfying a customer service requirement that stipulates a minimum number of visits to each customer each week and satisfaction of time-varying demand at the end-customers. Explicit constraints on the minimum number of visits to each customer each week give rise to interdependencies that result in a dimension of problem difficulty not commonly found in models in the literature. Our model includes two other realistic factors that the third-party logistics provider needs to consider: the cost of holding inventory borne by end-customers if deliveries are not made "just-in-time" and the possibility of multiple vehicle visits to an end-customer in the same period (day).

We develop a solution procedure based on Lagrangian relaxation in which the particular form of the relaxation provides strong bounds. One of the subproblems that arises from the relaxation serves to integrate the impact of the timing of deliveries to the various end-customers with inventory decisions, which not only contributes to the strong lower bound that the relaxation provides, but also yields a mathematical structure with some unusual characteristics; we develop an optimal polynomial-time solution procedure for this subproblem. We also consider two variants of the original problem with more restrictive assumptions that are usually imposed implicitly in many vehicle routing problems. Computational results indicate that the Lagrangian procedure performs well for both the original problem and the variants. In many realistic cases, the imposition of the additional restrictive assumptions does not significantly affect the quality of the solutions but substantially reduces computational effort.

Keywords: Third-party logistics; vehicle scheduling; period vehicle routing problem; inventory routing problem; delivery scheduling

1. Introduction

Our research was motivated by a problem faced by a California-based third-party logistics (3PL) provider that offers shipping services in the form of full-truckload (or as-if-full-truckload) moves to its customers. Most of its customers are manufacturing firms that supply components to downstream manufacturers or finished goods to distribution centers, or distribution centers that supply large retail firms. Some of the manufacturing customers use the 3PL to provide transportation to support vendor-managed inventory (VMI) programs. For clarity, we use the term *customer* to refer to a purchaser of 3PL services and *end-customer* to refer to a customer's customers. Each customer needs to supply items to its end-customers to satisfy the end-customers' daily demands on or before their respective due dates. The 3PL provider has a contract with each of its customers to transport these goods, usually with a requirement on the minimum number of deliveries per week for each end-customer. Typically, the 3PL provider's customers would view more frequent delivery as an element of better customer service. Perhaps more importantly, the end-customers prefer more frequent deliveries to reduce their inventory holding costs, and the customers who are involved in VMI programs directly benefit from reduced inventory at the end-customers (if the customers own this inventory, as is common). We explicitly consider these factors in our model.

In this paper, we address the 3PL provider's problem of selecting routes to execute on each day of the week to service its existing and/or prospective customers. We consider the problem from the viewpoint of the 3PL provider, who is seeking a cost-efficient solution while satisfying customer service requirements that stipulate conditions such as frequency of delivery. Although the solution of this problem can be used for operational purposes, the 3PL management initially approached us seeking to estimate the cost of servicing a prospective customer in the context of preparing bids and/or negotiating the terms of a contract.

It is important to emphasize that the focus of this paper is on route selection and delivery quantity decisions, and not on the construction of candidate routes, because a typical 3PL provider has little difficulty generating a set of viable routes, taking into account the structure of the road network, traffic patterns, etc. Our primary concern is to find an effective solution procedure for our problem given a practical set of candidate routes.

The remainder of this paper is organized as follows. In the next section, we describe our problem in more detail and present a mathematical formulation. Section 3 contains a review of the related research literature. In Section 4, we discuss two restricted variants of our problem, and in Section 5, we present our proposed solution approach for the original problem and the two variants. Computational results are reported in Section 6, and Section 7 concludes the paper.

2. Model Description

Our research was motivated by a 3PL provider which, for reasons of material handling efficiency, usually requires its customers to palletize the goods to be shipped. For this reason and for ease of exposition, we assume that the volume of goods can be expressed in terms of a homogeneous unit, such as a standard pallet. In the formulation that follows, we assume that all customers use the *same* standard unit, but we only require that each customer's basic unit of shipment be sufficiently standardized that we do not have to address the "bin-packing" aspect of the truck loading problem.

We assume that the 3PL provider owns, leases, or otherwise controls a fleet of trucks and that the trucks are homogeneous. We also assume that the 3PL provider has sufficient trucks to service the selected routes. 3PL firms often have standing arrangements for rental vehicles when needed, and additional drivers are available except in unusual circumstances. Thus, although limitations due to the number of vehicles or drivers may exist, they do not play a major role in a 3PL provider's route planning decisions. When using our procedure to estimate the cost of servicing a new customer, the 3PL provider recognizes that additional vehicles may be needed, and generally would not want to constrain the number of vehicles *a priori*. In the concluding section, we explain how our approach can be generalized to accommodate non-homogeneous vehicles and the cost of rental trucks when required.

Our problem framework has many customers and many end-customers. To distinguish among origin-destination pairs, we use the term *job type* to refer to the movement of goods from a single customer location to a single end-customer location. Each "route" in our problem consists of a sequence of jobs types. To service a job type, a truck must arrive empty at a *customer*, pick up its goods, and deliver all of these goods to an *end-customer*. Thus, routes consist of (customer, end-customer, customer, end-customer,...) sequences, where for every customer, the end-customer is known. There are a number of reasons for assuming that routes have this structure. The first reason is that the transport capacity is sold to customers in full-truckload (or as-if-full-truckload) increments, and the vast majority of customers specify the minimum number of visits per week such that the transport capacity is reasonably well utilized (over 50% on the average). Second, many of the parts being transported are valuable and at risk of theft, so customers may require that the truck be sealed in transit or that the truck proceed directly to the end-customer. One example of such a part is a compact disc (CD) containing software. The manufacturing cost of the CD is negligible, but the significantly higher retail price makes it a target for thieves. Third, when using our model to estimate the cost of servicing a prospective customer, the 3PL provider may not be able to estimate in advance the opportunities for consolidation of loads that would facilitate the construction of appropriate candidate routes. This is especially true because our approach *determines* the shipment quantities rather than taking them as given. In the concluding section, we discuss how one can make heuristic adjustments to take advantage of consolidation opportunities *ex post*.

We emphasize that our motivating application precludes the need for focusing on the classical routing problem. The number of possible routes is severely constrained by geographical and practical considerations. For example, due to driver workday restrictions and the time required to load and unload goods, it is unlikely that a route would contain more than four or five (customer, end-customer) pairs. Thus, the 3PL provider can easily enumerate and determine the cost of all practical routes, and use these costs to select the best route containing any given subset of (customer, end-customer) pairs.

Under mild conditions, namely (i) if route costs are linear in distance (or time), (ii) the triangle inequality for driving distance (alternatively, time) holds, and (iii) the 3PL provider's objective is to minimize total route cost, it is optimal to service each job type as few days per week as possible (i.e., one day per week if the total weekly shipment quantity is less than or equal to one truckload, or the minimum number of trucks required to transport the load otherwise). Without additional economic incentives, the selected schedule will contain the minimal number of shipments for each customer. In order to lower their inventory holding costs, the end-customers desire as many shipments as possible. Indeed, they may desire just-in-time delivery of their daily demands. Thus, the customers may be willing to pay more to provide more frequent service for their end-customers. As a proxy for the customers' willingness to pay, we include the cost of holding inventory *at the end-customer.* Thus, we implicitly assume that end-customers are willing to reward customers for decreased holding costs, and that customers in turn are willing to reward the logistics provider. Alternatively, if the customer is providing VMI services and owns the inventory at the end-customer, the customer benefits directly from reduced inventory levels. We also impose a lower bound on the number of delivery days for each job type. Both the inventory holding costs and the lower bounds on delivery days encourage better service (i.e., smaller, more frequent deliveries). The goal is to choose the routes to execute each day (and thus implicitly the job types to service) and the delivery quantity for each job type to minimize the sum of route costs and inventory holding costs over a horizon of T periods. In doing so, we must meet demands on time at the end-customers and satisfy a lower bound on the number of delivery days for each end-customer. Of course, some of these decisions may be fixed in advance to represent the unchangeable portion of the existing schedule. A formulation follows.

Indices:

- j: job type (specifies (customer, end-customer) pair)
- r: truck route (specifies a sequence of stops)
- t: time (day of week), $t = 1, ..., T$

Data:

- c_r: total cost for executing route r
- D_{jt}: demand of job type j on day t, expressed in standard units (e.g., pallets)
- h_j: one-period inventory holding cost for one unit of demand for job type j

- b_j: minimum number of days of service for job type j
- α_{jr}: 1 if route r includes job type j; 0 otherwise
- CAP: capacity of a vehicle in standard units (e.g., pallets)

Decision Variables:

- z_{rt}: number of times route r is executed on day t
- x_{jt}: number of units of job type j shipped on day t
- v_{jt}: number of vehicles servicing job type j on day $t = \sum_r \alpha_{jr} z_{rt}$
- y_{jt}: 1 if there is one or more delivery of job type j on day t; 0 otherwise (implicit decision)
- I_{jt}: inventory of job type j remaining at the end-customer at the end of period t (implicit decision)

(P)

$$minimize \sum_r \sum_t c_r z_{rt} + \sum_j \sum_t h_j I_{jt}$$

$$s.t.$$

$$\sum_t y_{jt} \geq b_j \quad \forall j \tag{3.1}$$

$$y_{jt} \leq v_{jt} \quad \forall j,t \tag{3.2}$$

$$v_{jt} \leq \sum_r \alpha_{jr} z_{rt} \quad \forall j,t \tag{3.3}$$

$$I_{jt} = I_{j,t-1} + x_{jt} - D_{jt} \quad \forall j,t \tag{3.4}$$

$$x_{jt} \leq CAP * v_{jt} \quad \forall j,t \tag{3.5}$$

$$y_{jt} \text{ binary} \quad \forall j,t$$

$$z_{rt}, v_{jt} \text{ non-negative integers} \quad \forall j,r,t$$

$$x_{jt}, I_{jt} \text{ non-negative} \quad \forall j,t$$

The first set of constraints ensures that each job type receives its minimum required days of service (or more). Without these constraints, a job type's "frequency of service" requirements may not be met, particularly if the end-customer's demands and holding costs are low, and the incremental cost of servicing the job type is high. The incremental cost of servicing a job type is high if the end-customer and/or the corresponding shipment origin is located far from the truck depot and/or from the other customers and end-customers. The second set of constraints ensures that a service day is credited only if the number of truck visits (for that job type) is one or greater. The third set of constraints defines the v_{jt} variables, and the fourth set of constraints represents the inventory balance equations, which, in conjunction with non-negativity constraints on I_{jt}, preclude shortages. The fifth set of constraints limits the shipment quantity to the corresponding shipment capacity. Observe that constraints (3.3) are expressed as inequalities rather than equalities. Due to the structure of constraint sets (3.1), (3.2) and (3.3), it is optimal for the v_{jt} values to be as large as possible; thus, constraints (3.3) are always satisfied as equalities. The inequality representation aids in our solution procedure, as we explain in more detail later.

The formulation above addresses the problem for a finite horizon. At the 3PL provider that motivated our work, the plan is expected to repeat periodically, usually weekly. For this reason, it is not essential for the system to start and end the week with zero inventory. We do require, however, that the plan be repeatable, and we therefore impose the constraints:

$$I_{jT} = I_{j0} \quad \forall j. \tag{3.6}$$

Of course, the model above does not explicitly represent all the possible complexities of real world problems. It can, however, be modified to capture at least some of these complexities, including:

- multiple truck types (the formulation is for a single truck type);
- constraints on the number of routes or the number of truck-hours available in a day (unconstrained here);
- delivery time constraints (unconstrained here; such constraints can be considered easily in the generation of routes);
- multi-day routes (single-day routes assumed here); and
- time-varying route and inventory holding costs (assumed time-invariant here).

Also, recall that we are addressing a problem in which each route consists of a series of one or more (pick-up, drop-off) operations, which reflects the usual mode of operation at the 3PL provider that motivated our research. Figure 3.1 shows an example of an allowable route. As a consequence of this assumption, truck capacity limits apply only to a single delivery. Of course, for some applications, goods from two or more job types may be loaded onto a truck simultaneously. Modifying our model to accommodate this problem variant would significantly increase the complexity of the model. (We discuss this issue further in the concluding section.)

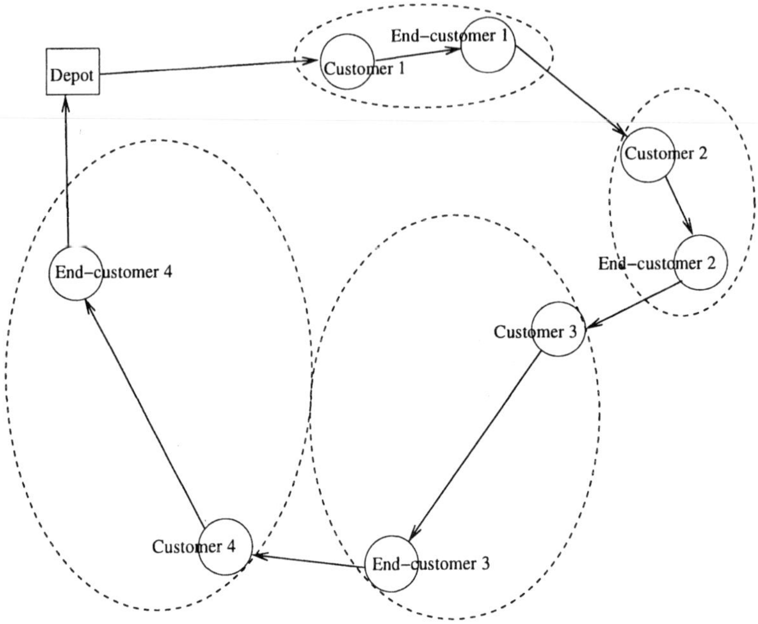

Figure 3.1. Depiction of an example route.

3. Literature Review

Our problem involves selecting routes for each day of the week and determining shipment quantities for each customer (within the capacity constraints defined by the selected routes) to satisfy demands that may vary by period. The latter decisions are similar to lot sizing decisions. The long history of research on deterministic *single-stage*, single-item lot sizing models begins with the seminal work of Wagner and Whitin (1958) for the uncapacitated model. Aggarwal and Park (1990), Federgruen and Tzur (1991), and Wagelmans et al. (1992) developed faster exact algorithms for the uncapacitated case. For the capacitated problem, Florian and Klein (1971) characterized the optimal solution for the case of constant capacity. Baker et al. (1978) developed algorithms for the case of time-varying capacity, and Love (1973) characterized optimal solutions when production and storage costs have a piecewise concave structure. Lippman (1969) analyzed the multiple setup cost case, where there is a fixed charge for each increment of capacity (such as one truckload). We discuss his results in more detail in Section 5.

Although we do not explicitly solve the routing problem, our problem contains features of both the *Period Vehicle Routing Problem (PVRP)* and the *Inventory Routing Problem (IRP).* The PVRP is a multi-period vehicle routing problem in which the decisions are the service day(s) for each customer and the vehicle routes for a service provider on each day. Very few PVRPs in the literature consider inventory or other costs associated with the selection of a particular day-of-week schedule. The emphasis is on minimizing routing costs and/or the number of required vehicles. The most common assumptions are that the number of visits during the horizon is fixed and that the delivery or pick-up quantity is the same for each visit. A few authors do allow for different numbers of visits and/or different delivery quantities. Russell and Gribbin (1991) allow an arbitrary allocation of the week's goods among a specified number of deliveries for each customer. Gaudioso and Paletta (1992) assume constant demand and equal delivery quantities, and impose spacing constraints between deliveries, without accounting for the cost of holding the inventory required to support such a delivery pattern while avoiding shortages. Similarly, Chao et al. (1995) explicitly account for the effect of time-varying demand and the delivery patterns on the quantities to be delivered, but do not consider inventory holding costs.

In contrast to the PVRP, IRPs more strongly emphasize the tradeoff between delivery and inventory-related costs. Typical objective functions include vehicle routing costs, inventory holding costs, and shortage costs. For articles on continuous-time problems with constant demand, see Dror and Trudeau (1996), Herer and Roundy (1997), Federgruen and Van Ryzin (1997), Viswanathan and Mathur (1997) and Chan et al. (1998), and references therein. For single-period problems with stochastic demands, see Federgruen and Zipkin (1984) and Federgruen et al. (1986). For multi-period problems with stochastic demand, see Webb and Larson (1995), Herer and Levy (1997) and Bard et al. (1998) and the references therein. Finally, for continuous-time problems with stochastic demand, see Larson (1988), Dror and Trudeau (1996), and Qu et al. (1999). It should be noted that the vast majority of the multi-period problems and continuous time problems with stochastic demand assume that demand is stationary.

Several IRP papers address problems that are closely related to ours. Chien et al. (1989) consider a single-period model with deterministic demand in which the supplier has a limited quantity of the product, and the goal is to maximize revenue less transportation and shortage costs subject to supply and demand availability, and vehicle capacity constraints. Our problem is essentially a multi-period generalization of the Chien et al. model with additional costs for inventory and constraints on the delivery patterns. Dror and Levy (1986) consider a multi-period model in which the demand is constant but the required shipment quantity depends on the delivery day, and each customer is serviced no more than once during the time horizon. Bell et al. (1983) develop a model to minimize the cost of distributing industrial gases considering the forecasted inventory levels of the customers. Dror et al. (1985) study a finite-horizon problem in which each customer receives at most one delivery during the horizon and deliveries must occur before the customer is projected to deplete his supply. Dror and Levy (1986), and Dror and Trudeau (1996) consider generalizations and additional solution approaches for these models. Chandra (1993) addresses the joint problem of warehouse procurement decisions and delivery (routing) to retailers for multiple products over a finite horizon, and develops a heuristic for the problem. Chandra and Fisher (1994) examine a similar problem in which a production schedule, rather than a procurement schedule, must be decided. Metters (1996) examines the problem of coordinating delivery and sortation of mail when there are deadlines for the completion of sortation, and solves the problem using commercial optimization software. Carter et al. (1996) consider the problem of planning the delivery of multiple grocery items during multiple periods over a finite, repeating horizon. A delivery pattern must be selected for each customer, and inventory allocations and vehicle routes must be chosen on each day. Vehicle capacity, vehicle availability, route duration and delivery time window restrictions apply. They develop a heuristic procedure for solving this problem. As the size of the fleet is an important constraint in their motivating application, their procedure emphasizes smoothing vehicle use. A variant of our problem without constraints on the number of service days per week and with deliveries only (i.e., no intermediate stops for pickups) is addressed by Lee et al. (2003), who construct annealing heuristics and derive certain properties of the optimal solution for their problem.

In contrast to the vast majority of PVRP models, our model specifically accounts for effects of different delivery patterns on the inventory that must be held by the end-customer. In contrast to many IRP models, our model directly addresses a multi-period problem with time-varying demand that may need to be satisfied by more than one shipment during the horizon. Equally important is that our formulation of the problem permits an exact representation of route costs (versus a fixed cost per delivery, cf. Carter et al.), an exact representation of inventory costs incurred by the end-customer as a consequence of the delivery schedule, as well as constraints on the number of deliveries per week for each job type.

None of the articles cited above accounts for all of the factors and constraints that we consider. For this much more general and accurate representation of real-world, multi-period shipment problems involving both pickup and delivery with days-of-service constraints that 3PL providers are facing, our major contribution is developing an approach that produces near-optimal solutions relatively quickly.

It is important to highlight the complications introduced by the minimum-days-of-service constraints. Our problem exhibits strong links across multiple periods, not only because of the inventory costs induced by the day-of-week delivery pattern for an individual customer, but because the incremental cost of servicing a customer and the difficulty of satisfying that customer's days-of-service constraint depend upon the delivery patterns of the other customers. Although the *economic* interactions noted above arise in the IRP and some versions of the PVRP, in those contexts, they tend to induce "soft" constraints that can often be negotiated via earlier shipments (and the associated inventory costs). On the other hand, the "hard" days-of-service constraints in our problem create structural linkages that cause the (general) integral route selection decisions to play a stronger role in the solution of our problem.

4. Problem Variants

In addition to (P) formulated in Section 2, we examine two restricted versions that may be applicable in many problem environments. These restricted problems are not only realistic but can be solved using the same solution framework as that described in the next section and with less computational effort. As our discussion proceeds, we will explain why the problems are easier to solve. Here, we present the motivation for the restrictions and the related changes in the problem formulations.

Variant 1:

In the first problem variant, we impose the constraint that each job type receives at most one visit per day. In this case, at most a partial truckload could be shipped ahead of schedule on a given day. Such a constraint would be imposed in practice if the customer insists on a low-inventory, almost-just-in-time solution.

The formulation changes as follows:

- The z_{rt} and v_{jt} variables are now binary.
- The y_{jt} variables are now equivalent to the v_{jt} variables and can be removed from the formulation by substituting v_{jt} wherever y_{jt} appears and removing redundant constraints (e.g., (3.2)).
- We add the constraint

$$\sum_r \alpha_{jr} z_{rt} \leq 1 \quad \forall j, t.$$

If the demand on a single day exceeds one truckload, then the job type can be replaced by multiple ("dummy") job types with the same physical origin and destination, where each of these job types has demand of up to one truckload per day. (Of course, it is most economical to subdivide the goods into as few truckloads as possible.) In this case, for each job type j whose demand exceeds a truckload on day t, we define J_{jt} as the set of corresponding "dummy" job types and rewrite the constraint that defines visits as:

$$y_{jt} \leq \sum_{i \in J_{jt}} \sum_{r} \alpha_{ir} z_{rt}$$

Although this constraint has a different form than the constraints in the original formulation, it has the same structure as certain constraints in the formulation on which our solution procedure is based. Consequently, we can easily handle demands exceeding a truckload for any job type on any day.

Variant 2:

The second problem variant does not restrict the number of visits for each job type; it simply permits us to execute each route at most once on each day. The motivation for this constraint is the very small likelihood of needing, much less choosing, the same route more than once on the same day. Such a need would arise only if several job types that could comprise a relatively efficient route could all benefit from receiving more than one truckload of goods on the same day.

The only required change in the formulation is to make the z_{rt} variables binary.

5. Solution Approach

The set partitioning problem, an NP-hard problem (Garfinkel and Nemhauser 1969), is a special case of (P), which implies that (P) is NP-hard. To see this, consider the special case of our problem in which inventory costs are ignored and inventory non-negativity constraints are not enforced except at the end of the horizon. In this case, it is optimal to service each job type with as few vehicles as possible, and without regard to the day of the week, so each "week" can be regarded as a single time period. Each job type requiring more than one truckload in a week is replaced by an appropriate number of "dummy" jobs, in the same way as in Variant 1. The z_{rt} become binary rather than general integer variables. The routes are defined for the set of "dummy" job types and the α_{jr} values are defined accordingly. With these redefinitions and appropriate simplifications of the objective function and constraints, the problem reduces to the set partitioning representation of the single-period (standard) vehicle routing problem.

The inventory non-negativity constraints and inventory costs link the periods, thereby creating a problem that is more difficult than a set partitioning problem. Indeed, as we report in more detail later, even small instances of (P) are impractical to solve using commercial software. This necessitates the development of a solution procedure that takes advantage of the structure of our problem. We first present a solution approach for the general problem, and then explain how the procedure should be modified for the two problem variants.

We propose a Lagrangian approach to the problem in which constraints (3.2) are relaxed using multipliers μ_{jt} (≥ 0) and constraints (3.3) are relaxed using multipliers λ_{jt} (≥ 0). Observe that expressing constraints (3.3) as inequalities allows us to dualize them using non-negative Lagrange multipliers. This, in turn, provides for a more meaningful interpretation of the λ_{jt} values, and, as we observed in preliminary computational studies, a more stable solution procedure.

Because we have relaxed constraints (3.2), the introduction of constraints (3.7) and (3.8), shown below, which are redundant in the original problem, provides a stronger formulation for the Lagrangian procedure. Constraints (3.7) ensure that a service day is credited only if at least one appropriate route is selected, while constraints (3.8) ensure that the total number of truck visits is large enough to satisfy each job type's demand.

$$y_{jt} \leq \sum_{r} \alpha_{jr} z_{rt} \qquad \forall j, t \tag{3.7}$$

$$\sum_{r} \sum_{t} \alpha_{jr} z_{rt} \geq \lceil CAP^{-1} \sum_{t} D_{jt} \rceil \quad \forall j \tag{3.8}$$

Relaxing constraints (3.2) and (3.3) and adding constraints (3.7) and (3.8) yields two subproblems for fixed λ_{jt} and μ_{jt} values:

(P1)

$$minimize \sum_r \sum_t (c_r - \sum_j \lambda_{jt} \alpha_{jr}) z_{rt} + \sum_j \sum_t \mu_{jt} y_{jt}$$

$$s.t.$$

$$\sum_t y_{jt} \geq b_j \quad \forall j$$

$$y_{jt} \leq \sum_r \alpha_{jr} z_{rt} \quad \forall j,t$$

$$\sum_r \sum_t \alpha_{jr} z_{rt} \geq \lceil CAP^{-1} \sum_t D_{jt} \rceil \quad \forall j$$

$$y_{jt} \text{ binary} \quad \forall j,t$$

$$z_{rt} \text{ non-negative integers} \quad \forall r,t$$

and (P2)

$$minimize \sum_j \sum_t h_j I_{jt} + \sum_j \sum_t (\lambda_{jt} - \mu_{jt}) v_{jt}$$

$$s.t.$$

$$I_{jt} = I_{j,t-1} + x_{jt} - D_{jt} \quad \forall j,t$$

$$x_{jt} \leq CAP * v_{jt} \quad \forall j,t$$

$$v_{jt} \text{ non-negative integers} \quad \forall j,t$$

$$x_{jt}, I_{jt} \text{ non-negative} \quad \forall j,t$$

Subproblem (P1), the routing subproblem, is a variant of the PVRP with lower bounds on the number of service days and on the number of truck visits over the horizon for each job type. In the objective function, there is an adjusted cost for each route that accounts for the value of that route's shipping capacity in meeting the needs of customers on that route on that day, as well as additional "costs" corresponding to satisfying the service-day requirements of the customers. The "standard" PVRP includes only terms containing the z_{rt} variables, and consequently, is much simpler. Problem (P1) is NP-hard for the same reasons as (P), via the same reduction.

Observe that without the added valid inequalities (3.7) and (3.8), the optimal solution to (P1) would be to identify, for each j, the b_j smallest values of μ_{jt} and to set the corresponding values of y_{jt} to 1 (otherwise set y_{jt} to zero) and to set z_{rt} to 1 if its coefficient (reduced cost) in the objective function is negative (otherwise set z_{rt} to zero). Such a solution does not ensure consistency between the selected routes and the days of service, and provides no guarantee that the number of routes is sufficient to handle each job type's weekly demand. Consequently, the "solution" is far from being feasible and its cost is a very loose bound on the actual cost. The inclusion of constraints (3.7) and (3.8) provides substantially better bounds, primarily because the resultant routing solution is feasible: the constraints on number of service days are satisfied and there are sufficient routes to accommodate the required freight flows.

Subproblem (P2), the shipment scheduling subproblem, is a capacitated lot sizing problem with multiple setup costs, one for each capacitated vehicle. Because it is a subproblem in a relaxation with a form that allows some of the (adjusted) setup costs to be negative, when viewed as a stand-alone problem, (P2) could have an unbounded objective. However, in our problem context, despite these negative setup costs, we must devise a method that provides a strong bound in order to solve the original problem. It is from this vantage point that we analyze (P2) and develop an optimal polynomial-time solution procedure for it.

5.1 Analysis of (P2)

The second subproblem is separable by job type. Thus, in the remainder of this section, we consider the problem for a single job type and omit the job type subscript. Lippman (1969) studies a class of multiple setup cost problems that includes ours as a special case. He shows that there exists an optimal solution consisting of regeneration intervals. (A regeneration interval is a set of consecutive periods with zero initial and terminal inventory and with all intermediate periods having positive inventory.) Thus, the strategy is to find the optimal solution for each potential regeneration interval, then to find the best combination of regeneration intervals using a shortest path algorithm.

Lippman also shows that there exists an optimal solution such that:

$$I_{t-1}(x_t \quad mod \quad CAP) = 0.$$

In other words, in each period, either entering inventory is zero or the shipment quantity is a multiple of a full truckload (or both). Although not explicitly stated in his paper, a further implication of this result is that *within a regeneration interval*, only the first period can have a partial-truckload shipment, as all of the remaining periods must have $I_{t-1} > 0$. As such, if all trucks have the same capacity (as in our problem), we know exactly how many vehicles are sent within the regeneration interval.

Lippman's result is based on the assumption that the setup costs are non-negative, and his result characterizes the shipment quantities but does not explicitly state how many trucks should be sent. Of course, when setup costs are positive, it is optimal to send as few trucks as possible to accommodate the shipment quantity in each period. Lee (1989) and Anily and Tzur (2002), among others, have studied variants of this lot-sizing problem in which multiple capacitated shipments (of arbitrary quantities) are allowed in each period, but all of these models have the implicit assumption of positive setup costs. Pochet and Wolsey (1993) study the special (restrictive) case in which the batch size must be some integer multiple of some basic batch size, but they, too, assume that setup costs must be positive.

Our second subproblem has the unusual and distinctive characteristic that some of the (adjusted) setup costs may be negative, and it is this characteristic that necessitates a different solution approach. The approaches in the literature cannot be applied directly to our problem because they do not allow for the combination of negative and time-varying setup costs. Both of these aspects arise in our second subproblem.

In our problem, it may be optimal to send extra trucks, including some that are completely empty. To avoid an unbounded solution, we impose the constraint

$$v_t \leq \lceil \sum_t D_t/CAP \rceil, \quad \forall t$$

which simply limits the number of trucks in any period to the number that would be required to service all of the demand in a single period. Let $\bar{v} = \lceil \sum_t D_t/CAP \rceil$. (We later obtain stronger bounds on v, but for the purposes of our present analysis, this particular upper bound is useful.) It is clear that $v_t^* = \bar{v}$ for periods in which the corresponding coefficients are negative. The problem is now to determine how to use this "free capacity" and how to make shipments in the remaining periods. We show that this modified problem (with constraints on v_t) has the same property as that derived by Lippman. We then show how to construct an optimal solution (both the truck schedule and the shipment quantities) for this problem. For ease of exposition, let K_t denote the coefficient associated with v_t.

Proposition 1: For a setup cost structure of the form $K_t v_t$ where some of the K_t values may be negative, the optimal solution satisfies:

$$I_{t-1}(x_t \quad mod \quad CAP) = 0.$$

Proof: Suppose, to the contrary, that we have an optimal solution, x_t^*, in which $I_{t-1} > 0$ and $x_t^* \quad mod \quad CAP > 0$. The latter condition implies there is excess transportation capacity in period t. There exists some

$\epsilon = min\{CAP - (x_t^* \mod CAP), I_{t-1}\} > 0$ such that we can ship, in period t, ϵ units that had been shipped in some prior period, at a minimum savings of $h\epsilon$. This contradicts the optimality of the original solution. □

Observe that the possibility of trucks being sent empty does not change the implication of Proposition 1 with respect to the timing of a partially-filled truck. Thus, only the first period in a regeneration interval can have a partial truck shipment. The complication in our problem is that we do not know in advance how many trucks will be sent in a regeneration interval. However, we do know how many non-empty trucks will be sent.

Consider a solution for a regeneration interval consisting of periods a through b, constructed as follows:

Algorithm A2:

Step 1. For $t = a, ..., b$, set

$$x_t = \sum_{k=a}^{b} D_k - CAP\lfloor(\sum_{k=t+1}^{b} D_k/CAP)\rfloor - \sum_{k=a}^{t-1} x_k$$

Step 2. For $t = a + 1, ..., b$,

$n^* = \arg\max_{a \le n \le t}\{K_t - CAP * h * (t - n) - K_n\}$

$S = \max_{a \le n < t}\{K_t - CAP * h * (t - n) - K_n\}$

If $S > 0$

set $x_{n^*} = x_{n^*} + x_t$

set $x_t = 0$

Step 3. For $t = a, ..., b$, set

$v_t = \lceil x_t/CAP \rceil$

Step 4. For $t = a, ..., b$,

if $K_t < 0$ and $v_t < \bar{v}$, set $v_t = \bar{v}$.

In Step 1, the shipment quantity is set so that the fractional truckload is shipped in the first period and beyond this, just enough full truckloads are shipped in each period so that demand is satisfied on time. This tentative solution can be viewed as the schedule that is as close to just-in-time as possible while retaining properties of the optimal solution.

In Step 2, we determine, for each period, the best earlier period into which we could shift full truckloads. If the savings is positive, we shift all relevant truckloads. In Step 3, we set the truck variable v_t equal to the minimum number of trucks necessary to ship the quantity x_t. Finally, in Step 4, we identify periods with $K_t < 0$ in which the maximum number of trucks is not yet assigned, and set the corresponding v_t values to their upper bounds. Note that Steps 2, 3, and 4 retain properties of the optimal solution.

Proposition 2: Algorithm A2 produces an optimal solution (x_t^* and v_t^*) for a given regeneration interval.

Proof: We first note that by Proposition 1, in some optimal solution for the regeneration interval, all shipments except possibly that in the first period of the interval are integer multiples of truck capacity. In this proof, we restrict our attention to schedules for which the property in Proposition 1 holds. Thus, Step 1 of the algorithm assigns shipments so that the minimum number of non-empty trucks, and indeed, the *only possible number of non-empty trucks* under the property in Proposition 1, are scheduled. Moreover, the schedule constructed in Step 1 is such that each full truckload is scheduled as late as possible. Thus, the only feasible changes entail moving full truckloads to earlier periods. Recall that the maximum number of trucks in each period ($\bar{v}$) is sufficient to ship the entire demand during the regeneration interval. Thus, for any moves of full truckloads to earlier periods, we can consider the best time to dispatch each individual non-empty truckload independently. Now, by construction of the algorithm, we cannot move an entire truckload from one period to another while reducing costs (cf. Step 2). Had it been possible to reduce costs by removing a truck with $K_t > 0$ and shifting the load into a period with $K_t \leq 0$, that shift would have been implemented in Step 2. Furthermore, by Step 4 of the algorithm, we cannot reduce the total cost by adding empty trucks to the solution. Thus, Algorithm A2 produces an optimal solution for the regeneration interval. □

We note that regeneration intervals can be considered independently. Thus, the optimal solution for each possible regeneration interval can be determined using this algorithm, and a shortest path algorithm can be used to select the optimal set of regeneration intervals.

Observe that Steps 1, 3 and 4 have linear time complexity and Step 2 has $O(T^2)$ complexity. Thus the computation of the costs for the $O(T^2)$ regeneration intervals has $O(T^4)$ complexity. The shortest path problem (to find the best combination of regeneration intervals) has $O(T^2)$ complexity. Consequently, the overall procedure has $O(T^4)$ complexity. Observe, however, that the computations and comparisons are extremely simple.

We note that for any time interval that could be covered by a single regeneration interval, there may be an alternate dominant solution consisting of a set of shorter regeneration intervals. Such dominant solutions are identified by the shortest path procedure. Thus, it is only necessary for our procedure to find an optimal solution that satisfies the regeneration interval property for the time interval under consideration. Among all such optimal solutions, we restrict our search to those satisfying an established property of the optimal solution within a regeneration interval. By doing so, we are able to construct a very efficient solution procedure for (P2).

Because of its structure, (P2) is easy to solve and produces a much stronger bound than its LP relaxation, and thereby provides a significant contribution to strengthening the overall lower bound. In addition, as we will see later, including the fixed charge costs (via the dual variables) and the vehicle capacity constraints in this subproblem also provides relatively fast convergence of the Lagrangian procedure because the solutions of (P1) and (P2) are more consistent than they would be without these considerations in (P2).

5.2 Solution Procedure for (P)

Recall that in order to solve (P), we propose a Lagrangian approach in which we relax two sets of constraints – those ensuring that each service day is correctly accounted for, and those defining the v_{jt} variables. Correspondingly, we associate a set of multipliers with each set of relaxed constraints. We employ variants of the subgradient optimization method to update the multipliers at each iteration. (See the Appendix for details.) For each set of multipliers, we first solve (P1) and (P2) optimally to provide a lower bound, which we update if it has improved. We then construct a feasible solution by using the z^*_{rt} values from (P1), computing

$$v_{jt} = \sum_r \alpha_{jr} z^*_{rt} \quad \forall j, t$$

and substituting the values of v_{jt} (as fixed quantities) in (P2). We then solve (P2), which is a linear program when the v_{jt} values are fixed.

We observed that solutions constructed by the method described above often result in excess truck movements, i.e., the v_{jt} values are larger than necessary to handle the resulting shipment quantities. Therefore, we also construct another feasible solution by taking the solution for (P2) and checking its feasibility with respect to the customer service constraints. If the solution satisfies these constraints, for each day of the week, we solve the associated problem (P1). Of course, if the solution for (P1) exactly satisfies the shipping requirements from (P2), the solution is optimal and there is no need to re-solve the routing subproblem. Although we could make incremental changes to the routes from

(P1) (i.e., eliminate excess stops), in exploratory tests, such a method did not consistently produce good solutions.

We compute the objective value of the feasible solution and update the upper bound if the newly-constructed solution has a better objective than the current upper bound. The multipliers are then updated and the process is repeated until optimality is achieved, or the best feasible solution is within some tolerance of the lower bound, or until the step size reaches virtually zero (thereby precluding any significant improvement in the objective function value).

Because we allow a repeating schedule rather than restricting the solution to one with zero initial and ending inventories, constraints (3.8), which are retained in (P1), are sufficient to ensure that the z_{rt} values from (P1) will yield a feasible solution for (P2). If one is solving a finite horizon problem, then it may be necessary to impose lower bound constraints on $\sum_{k=1}^{t}\sum_{r}\alpha_{jr}z_{rk}$ for $t = 1, ..., T$ and for all j to ensure that the timing of trucks allows for a feasible solution of (P2) with the v_{jt} values implied by (P1).

5.3 Modification for Problem Variants

Variant 1:

In this case, we only need to impose the tighter of the customer service (number of visit days) constraint or the constraint related to total delivery capacity for each job type j. Noting that there exists an optimal solution such that $v_{jt} = \sum\limits_{r}\alpha_{jr}z_{rt}$, we can eliminate the v_{jt} variables entirely. With these simplifications, (P1) becomes

(P1′)

$$minimize \sum_{r}\sum_{t}(c_r - \sum_{j}\lambda_{jt}\alpha_{jr})z_{rt}$$

s.t.

$$\sum_{r}\sum_{t}\alpha_{jr}z_{rt} \geq \max\{b_j, \lceil CAP^{-1}\sum_{t}D_{jt}\rceil\} \quad \forall j$$

$$\sum_{r}\alpha_{jr}z_{rt} \leq 1 \quad \forall j, t$$

$$z_{rt} \text{ binary} \quad \forall r, t$$

The only change in (P2) is that the v_{jt} variables are now binary, so the problem becomes a lot sizing problem with standard (binary) setups. Recall, however, that because (P2) is derived from a relaxation, the setup costs may be negative.

Variant 2:

In this case, the only changes are that the z_{rt} variables are binary. The same solution procedure can be used, but the constraint space is much smaller. Next, we discuss ways to further limit the search space.

5.4 Bounding the v_{jt} Values

In this subsection, we describe methods to obtain upper and lower bounds on the values of v_{jt} and consequently also on $\sum_r \alpha_{jr} z_{rt}$ for the original problem and for Variant 2. The upper bounds on $\sum_r \alpha_{jr} z_{rt}$ also have obvious implications for the individual z_{rt} values in the general model where the z_{rt} values may be greater than 1. These bounds and their derivations are intuitive and we state them without proof.

Simple Bounds

The simple bounds are based on the observation that for customer j, service must occur on at least b_j days. Thus, the most that one would ship on a single day is the sum of demands on the $T-(b_j-1)$ days with the greatest demands, as the demands on all of the other days would be shipped "just-in-time." This bound may be quite loose, but it is easy to compute and does not differ by day of week.

Cost-Based Bounds

The cost-based bounds recognize the economic tradeoffs between the "setup" (transportation) and holding costs. We can derive an upper bound on the number of times job type j is serviced on day t as follows: First, we let the setup cost on day t be equal to a lower bound on the smallest incremental cost of servicing job j, which can be determined by finding the least expensive way to insert job type j into any (already-generated) executable route. Then, we let the setup cost for all other days be equal to an upper bound, for example, that derived from serving job type j alone. (Note that this route is always feasible.) With these bounds on setup costs, we solve the associated lot sizing problem with multiple setups. The number of trucks on day t in the solution of this problem is a tentative upper bound on $v_{jt} = \sum_r \alpha_{jr} z_{rt}$. In other words, we would not service job type j on day t any more times than if transportation costs were as cheap as possible on day t and as expensive as possible on the other days. This would be a valid bound if we did not have a constraint on the number of service days. To account for this constraint, we note that if the initial upper bound is equal to zero, it is economically unfavorable to service that job type on that day. To incur the minimum penalty while contributing to the days of service, we would service that job type at most once on that day. Thus, we can, without loss of optimality, set the upper bound on v_{jt} equal to 1 in this case.

Similarly, we can find lower bounds on the v_{jt} values by setting the setup cost on day t equal to an upper bound and the cost on the other days equal to a lower bound and solving the same lot sizing problem. In this case, the days-of-service constraint does not necessitate any adjustment.

Other Bounds

Observe that an upper bound on v_{jt} can be used as an upper bound on all z_{rt} such that $\alpha_{jr} = 1$. Such bounds may be useful when the upper bound on v_{jt} is small (e.g., 1) and the corresponding z_{rt} values would otherwise be constrained only by much larger values obtained from the simple bounds described above.

The analysis can be taken a step further by noting that another upper bound on z_{rt} is:

$$\min_j \{\text{upper bounds on } v_{jt} \text{ such that } \alpha_{jr} = 1\}.$$

In other words, the maximum number of times we would select a route is the minimum among the upper bounds on v_{jt} for the locations on the route. Bounds of this type may be useful when many locations have small upper bounds on v_{jt}. We did not implement this type of bound because of the computational effort required for the large number of routes in our problems.

6. Computational Results

We perform a series of computational tests in order to evaluate the effectiveness of our algorithm on the original problem and on the two problem variants. Before describing our computational study, it is important to point out that preliminary computational tests showed that both (i) the bounds described in the previous section and (ii) constraint sets (3.7) and (3.8) that are redundant in (P) but not redundant in (P1) are critical in finding solutions quickly. Without them, our procedure is not efficient, and the standard implementation of CPLEX applied to (P) is rarely able to find feasible solutions, even for problems of modest size. We report results in which both solution approaches, i.e., our Lagrangian approach and applying CPLEX to (P), are afforded the benefits from these additional valid inequalities. We next detail problem generation, and then discuss results.

6.1 Problem Generation and Execution of Algorithms

We generate a variety of problem instances for our computational study. Each problem instance has a five day time horizon. For each customer, the minimum number of days of service in a week is randomly generated from a discrete uniform distribution on [1,3]. We chose this range for the minimum days of service because problems with a four-day minimum service requirement are easier to solve and a five-day delivery requirement eliminates day-of-week decisions altogether.

The depot is located at the center of a 100 "mile" x 100 "mile" area. The location of each customer and end-customer is determined by randomly generating (i.i.d.) horizontal and vertical coordinates from a U[0, 100] distribution. All customer locations are thus distinct, so there is a one-to-one relationship between customers and job types. Once customer and end-customer locations are generated, we use Euclidean distances and assume that transportation costs are linear in the travel distance (normalized to $1 per "mile," which is roughly equal to the true variable cost for many 3PL providers). All trucks are assumed to have a capacity of 20 units (e.g., pallets).

Each customer's daily inventory holding cost per unit is generated from a U[0.5,5] distribution. This range of inventory holding costs corresponds to goods whose value may be as much as approximately $20,000 per (full) truckload. For such goods, less-than-daily delivery may be warranted, necessitating day-of-week decisions.

To generate candidate routes, we initially generated all combinations of 1, 2, 3, or 4 job types. In typical applications involving the transport of components to manufacturers and finished goods from distribution centers to retailers, the combination of transit times between customers and end-customers and enroute loading and unloading time limits the number of customers that a single vehicle can service to about 4 job types in a typical work shift, especially in congested urban areas.

Before solving the traveling salesman problem (TSP) for each combination, we apply a filter which eliminates those combinations for which a very loose lower bound on total route time exceeds a 7.5 hour workday. We assume an average driving speed of 40 miles per hour (which is similar to the value used by regional delivery companies in major metropolitan areas), and a total of 30 minutes for loading, unloading and waiting time associated with one delivery. For those combinations that pass the filter, we solve the TSP (by enumeration) and eliminate the combination if the route time for the optimal TSP solution exceeds the 7.5 hour threshold, or retain the best TSP routing if the route time is below the threshold. However, we retain all single-job-type routes, even if they exceed the route time threshold, to ensure that a feasible solution exists.

We solve one set of ("small") problems with 25 customers and their respective end-customers. The problem sizes in this set are limited in order to allow us to compare our heuristic solutions with those obtained from commercial software. The second set contains ("large") problems with 50 customers and their respective end-customers. All computations are performed on a Sunblade 1000 with 1 GB RAM. CPLEX 7.0 is utilized for (P) and its variants, as well as for the subproblems in the Lagrangian procedure. We use AMPL as the matrix generator for the solver, and as the scripting language for the Lagrangian procedure. In all instances, we record only the *solve* time required.

For each problem instance, we solve the original version of the problem (P), as well as Variants 1 and 2. We first generate bounds on the v_{jt} values, as described in Section 5.4, to be applied in the original version and in Variant 2. We execute our Lagrangian procedure as well as CPLEX applied to each variant of (P), utilizing all relevant bounds on the v_{jt} values. We employ an optimality tolerance of 2% for both procedures, and, because preliminary results indicate that the quality of the CPLEX solutions for the original problem and for Variant 2 do not improve significantly after several hours, we impose a time limit of 4 hours on both the 25-customer and the 50-customer problems. The four hour time limit allows for a reasonable tradeoff in both solution procedures between optimality and solution time. In executing the Lagrangian procedure, we terminate it when either the optimality tolerance or the time limit is reached, or, additionally, when the step size becomes zero to within the precision of the computer (precluding significant improvement in the objective function value), whichever comes first. (See the Appendix for details of CPLEX parameter settings and numerical implementation issues.) We discuss parameters specific to the problem sets, along with computational results, below.

6.2 25-Customer Problems

For this set of 10 problems, we generate demand for each customer and each day from a truncated Normal ($\mu = 10, \sigma = 3$) distribution, rounded to the nearest integer. For these problems, it is unlikely that more than one vehicle will visit a customer on a given day in a good solution (i.e., it is unlikely that $v_{jt} > 1$). Consequently, even Variant 1 would not be overly constraining for these problems. For this set of problems, after applying our filter, the number of remaining routes is about 7,000, corresponding to between 8,000 and 34,000 binary variables for Variants 1 and 2 (and a few hundred integer variables for Variant 2), and a corresponding number of *integer* variables for the original problem. The problems contain about 500 constraints.

In Table 3.1, we report the objective values for the problem variants and for the two solution procedures, along with the corresponding optimality gaps (best objective from procedure/lower bound from procedure - 1). Where optimality gaps are not reported, the gap is less than 2%. For Variant 1, both procedures solve all 10 problems to within approximately 2% of their respective lower bounds fairly quickly.

Variant 2 and the original problem are more difficult to solve. For Variant 2, the Lagrangian procedure identifies solutions within 2% of their respective lower bounds for all 10 scenarios in less than 25 minutes of CPU time, whereas after 4 hours of computing, CPLEX applied to (P) identifies solutions with optimality gaps of between 5% and 15%. For the original problem, the Lagrangian procedure identifies solutions within 5% of their respective lower bounds for 3 of the problems and within 8.3% for all 10 problems before the step size becomes virtually zero. CPLEX applied to (P) identifies a solution within 5% of its lower bound for only one problem, within 10% of its lower bound for 7 of the 10 problems, and within 14.3% for the remaining problems.

Although the Lagrangian solutions have optimality gaps of up to 2% for Variant 2 and up to 8.3% for the original version of the problem, for each of the 10 problems, the Lagrangian procedure finds a solution that ranges from one-half of one percent to more than 7% percent better than the solution identified by CPLEX applied to (P), with an average improvement of approximately 2-3%. In addition to providing better solutions, the Lagrangian procedure consumes, on average, less than 5% of the CPU time for our (fine-tuned) CPLEX implementation on the corresponding instances of Variant 2 and the original problem.

For the distribution from which we generated demands, the $v_{jt} \leq 1$ and $z_{rt} \leq 1$ constraints present in Variant 1 are unlikely to affect the optimal solution. The imposition of these constraints reduces the search space so the Variant 1 problems require considerably less CPU time than the other problem variants; thus, Variant 1 can be solved with a straightforward implementation of CPLEX applied to (P). Thus, where it is reasonable to assume that $v_{jt} \leq 1$ in an optimal solution, the application of Variant 1 appears to be a practical alternative.

Prob.	Variant 1		Variant 2		Original Problem (P)		Benchmark
No.	CPLEX	Lagrangian	CPLEX	Lagrangian	CPLEX	Lagrangian	
			Objective Function Values (% Gap)				
1	8302 †	8353	8483 (5.0%)	8351	8390 (3.9%)	8298 (4.0%)	10701
2	7778	7816	7842 (5.6%)	7782	8001 (7.4%)	7793 (8.3%)	8962
3	7581	7574	7889 (10.2%)	7588	7838 (9.6%)	7617 (7.8%)	9282
4	9015	9054	9757 (14.7%)	9074	9710 (14.3%)	9054 (6.2%)	11621
5	7535	7587	7672 (7.3%)	7582	7782 (8.7%)	7577 (7.7%)	9263
6	7091	7126	7298 (9.3%)	7149	7366 (10.1%)	7160 (7.4%)	8727
7	7011	7049	7172 (8.1%)	6979	7138 (7.6%)	6979 (4.5%)	8864
8	8244	8336 (2.3%)	8509 (10.4%)	8274	8422 (9.5%)	8287 (7.6%)	10348
9	8288	8291	8587 (11.4%)	8303	8825 (13.7%)	8264 (5.3%)	9869
10	6792	6829	6860 (5.8%)	6827	6878 (6.0%)	6806 (5.0%)	8526
			Solution Times (seconds)				
1	130	312	*	164	*	421	
2	1300	599	*	925	*	1021	
3	860	176	*	225	*	660	
4	47	21	*	103	*	205	
5	1500	570	*	294	*	571	
6	2600	324	*	1371	*	1033	
7	1000	825	*	406	*	1183	
8	980	874	*	301	*	472	
9	310	189	*	79	*	284	
10	3100	2164	*	604	*	1381	

† If gap percentage is not reported, gap is less than 2%.
* Indicates that CPLEX reached the time limit of 4 hours.

Table 3.1. Objective values, corresponding gaps and solution times for 25-customer problems

A typical 3PL provider would not explicitly consider the consequences of its selected route schedule on the cost of holding inventory at the end-customer. But the 3PL provider has the opportunity to offer tangible value to the party who bears the cost of holding inventory by offering more frequent service, and this value can translate into additional revenue for the 3PL provider. For this reason, we are interested in the consequences of ignoring inventory holding costs. To make this assessment, we use CPLEX applied to (P) (although, in principle, we could have used either procedure) to solve Variant 1 with the inventory holding costs set to zero. This version of the model is identical to the standard PVRP with the usual assumptions that (i) at most one visit is made to each end-customer (drop-off location) each day and thus also (ii) each route is executed at most once each day. This models the situation faced by the "traditional" 3PL provider, who optimizes his own costs (while ignoring those of his customers). To tabulate the full cost of this solution, we add the consequent inventory costs. We treat this total cost as a benchmark which we then compare to optimal or near-optimal solutions to estimate the system-wide benefit of explicitly considering customer inventory costs. Observe that whether or not the 3PL provider explicitly considers end-customer inventory holding costs when determining a delivery schedule, the end-customers *will* incur these costs directly, in addition to indirectly incurring the cost of transportation from the 3PL provider.

The benchmark objective values are about 15% to 30% greater than the corresponding values from Variant 1 (the most constrained version of the problem), suggesting that accounting for inventory costs leads to significantly better solutions if the 3PL provider is currently ignoring inventory costs. Even if the 3PL provider considers inventory costs indirectly by offering customers the possibility of better service at a higher price, there may still be opportunity from using more accurate "value pricing," particularly for customers with expensive goods. In view of the thin profit margins in the trucking industry, even a portion of a 15% to 30% gap is likely to be large enough to have a significant effect on the bottom line.

6.3 50-Customer Problems

Our primary reason for solving 50-customer problems is to demonstrate that problems of the sizes observed in some practical applications can be solved by our procedure. Large logistics providers typically subdivide their customers into geographical districts, and/or according to the type of vehicle required, e.g., standard, refrigerated, extra shock protection (used for sensitive electronic goods), small vehicles (for narrow roads or hilly terrains). The corresponding problems would then also be separable by these job type categories.

Recall that one of our goals in constructing the Lagrangian procedure is to develop a viable method of solving problems for which the usual PVRP assumption of $v_{jt} \leq 1$ might be unnecessarily restrictive. Because only highly correlated demands among customers on relatively "efficient" routes (with little deadheading) would lead to $z_{rt} > 1$, we generate customer demands in such a way that we could test Variant 2 and our original problem for cases with some individual demands exceeding a truckload. We solve 10 problems with 50 (customer, end-customer) pairs. Demands are generated from a truncated Normal ($\mu = 20, \sigma = 5$) distribution, rounded to the nearest integer.

After applying the route filter, these problem instances contain about 60,000 routes. In general, these problems contain between 250,000 and 600,000 binary variables (and a few hundred integer variables) for Variant 2, and a corresponding number of *integer* variables for the original version of the problem. The problems contain between 1000 and 1500 constraints, on average.

Results for the 50-customer problems appear in Table 3.2. Where optimality gaps are not reported, the gap is less than 2%. CPLEX applied to (P) fails to find a feasible solution within 4 hours of CPU time for 9 out of the 10 problem instances (for both Variant 2 and the original problem). On the other hand, the Lagrangian procedure identifies a solution within 2% of optimality in 9 of the 10 cases for Variant 2, and in the tenth case, the optimality gap is only 2.3%.

The Lagrangian procedure also identifies solutions within 2% of optimality for 3 of the 10 cases of the original problem. In the remaining 7 cases of the original problem, the Lagrangian procedure provides solutions that are generally within 8% of the corresponding lower bound, but the gaps range up to 18%. We allowed the Lagrangian procedure to run to termination (i.e., until the step size equals virtually zero) for the two problems that have large (> 10%) gaps at the 4 hour time limit, and found that at termination, solutions within 7% of the respective lower bounds were achieved.

	Variant 2				Original Problem P			
	CPLEX		Lagrangian		CPLEX		Lagrangian	
Prob. No.	obj. value (% gap)	time (sec.)	obj. value (% gap)	time (sec.)	obj. value (% gap)	time (sec.)	obj. value (% gap)	time (sec.)
1	—	*	21106†	10700	—	*	21145 (7.9%)	*
2	23876 (5.7%)	*	23673	4760	24121 (6.7%)	*	23612 (5.5%)	*
3	—	*	23770	13800	—	*	26640(18.0%)	*
4	—	*	22300	12400	—	*	21038	9660
5	—	*	21455 (2.3%)	*	—	*	22480 (14.7%)	*
6	—	*	20323	11900	—	*	20327 (7.2%)	*
7	—	*	22526	7410	—	*	21499	7540
8	—	*	22535	13400	—	*	22129 (6.1%)	*
9	—	*	18432	2000	—	*	18432	3220
10	—	*	23443	5910	—	*	23397 (6.0%)	*

* Indicates that the procedure reached the time limit of 4 hours.
† If gap percentage is not reported, gap is less than 2%.

Table 3.2. Objective values, corresponding gaps and solution times for 50-customer problems

Solving the original problem with 50 customers is evidently quite difficult, but the Lagrangian procedure reliably finds what appear to be good feasible solutions, while, for the vast majority of problem instances, CPLEX applied to (P) is unable to find any feasible solution.

A portion of the difficulty of solving 50-customer problems is due to the large number of routes. In practical applications with several dozen job types, a 3PL provider would rarely consider including 50,000+ routes as we have done in our computational study. In practice, routes would be eliminated due to factors other than route time, so the usable set would be much smaller. Consequently, problems with more job types and proportionally fewer routes are within the range of what could be solved in practice.

Although the Lagrangian procedure produces excellent solutions in most cases, it could also be used to provide strong bounds in a branch-and-bound framework if one desired to use an enumerative procedure to find better solutions. The Lagrangian procedure could be executed differentially at various nodes in the branch-and-bound tree to take best advantage of its flexibility.

Overall, the Lagrangian procedure appears to be a promising approach, especially for solving these difficult problems in which v_{jt} may exceed 1, where there are strong interactions among the decisions across both locations and time periods.

7. Summary and Conclusions

We have modeled a multi-customer, multi-period delivery scheduling problem faced by a third-party logistics provider in which routes must be selected, and delivery quantities must be decided while satisfying constraints on the number of customer visits during a specified horizon. We have developed a solution procedure based on Lagrangian relaxation to minimize the total cost of transportation and inventory.

In this paper, we focus on the route selection and delivery quantity decisions because a typical 3PL provider has little difficulty generating a practical set of candidate routes, taking into account the structure of the road network, traffic patterns, etc. Our primary concern was to find an effective solution procedure given a good set of candidate routes. We constructed a relaxation that provides strong bounds, owing largely to the combination of the following: (i) the identification of additional valid inequalities that "tighten" the formulation and the relaxation, (ii) the economic structure of the relaxation in which one of the subproblems integrates the impact of the *timing* of deliveries to the various end-customers with the inventory decisions, and (iii) upper bounds on the values of specific decision variables that are derived from the solutions to variants of certain subproblems. The subproblem mentioned in (ii) above has the unusual feature of potentially negative setup costs (for some values of the Lagrange multipliers) and we develop an optimal polynomial-time solution procedure for it.

We also consider two variants of the problem in which we impose one or both of the constraints implicitly assumed in much of the literature. The weaker of the two constraints permits a route to be used at most once each day, and the stronger constraint limits the number of routes servicing each customer each day to at most one. Computational results indicate that the Lagrangian procedure performs well on difficult problem instances for which it is ineffective to simply apply CPLEX to (P). The results also suggest that the imposition of the additional simplifying constraints does not significantly affect the quality of the solutions when it is unlikely that two trucks will be sent to a single customer on the same day in an optimal solution, and that the resulting problems require much less computational effort to solve. When demand is such that more than one stop per day is required at a customer (i.e., Variant 2 or (P) is appropriate), the Lagrangian procedure obtains very good solutions fairly quickly. More notably, the Lagrangian procedure produces very strong bounds, and thus may be valuable within a branch-and-bound procedure.

Several generalizations can be handled with no modification or only minor modifications to our approach. Time varying costs require no change in the solution procedure. Constraints on route duration and delivery and pick-up time windows can be considered in the route generation routine. Heterogeneous truck types can be handled by generating routes applicable to each truck type. If a job type can be serviced by more than one type of vehicle, then our algorithm for (P2) cannot be used directly, but because this subproblem is separable by job type, it can be solved using commercial software with a concomitant increase in the CPU time. If truck availability imposes practical limitations and rental vehicles are available, it would be possible to add the cost of a rental vehicle to each route and solve the problem in the usual way. The ability to avoid rental costs for the routes covered by the 3PL provider's own vehicles would create a "sunk" benefit in the model (i.e., it would appear as a non-controllable "cost" in the objective function that would not actually need to be paid), and the rental costs for all additional vehicles would be properly accounted for.

Multi-day routes can be handled with a modification to the formulation to account for the actual day of delivery and the extra cost of inventory due to goods in transit. Also, allowing multiple shipments to be loaded onto a truck at the same time can be handled, in principle, for small, pre-defined sets of job types that typically have small shipments. Considerable "bookkeeping" effort in the route generation scheme may be required, as well as a change in the solution method for P2. (The revised version of P2 could still be solved easily using commercial software.) Of course, if the solution generated by our procedure allows for consolidation of loads for customers that happen to be on the same route, then any such route can be modified to take advantage of such opportunities if they reduce costs.

The Lagrangian procedure may be enhanced by devising more effective multiplier adjustment methods especially designed for this problem, or other methods for constructing feasible solutions from the Lagrangian solutions. Also, if the problem contains a particularly large number of potential routes, it may be possible to generate only a subset of the routes *a priori* and to utilize a column generation-based approach to construct other economically viable routes. Further research is needed to explore the implications of such a strategy. Recall that subproblem (P2) already is separable by customer, so it is easily solved for large numbers of customers. It may be necessary, however, to devise a more efficient solution method for subproblem (P1).

Further research is also needed to consider more rigid constraints on allowable day-of-week combinations (e.g., MWF or Tu-Th) and the possibility of backorders, and to handle uncertainty in demand and transit times.

Appendix: Computational Implementation Issues

CPLEX

For all executions of the CPLEX software on the original problem and its variants, we use strong branching, i.e., the branching variable is selected whose resolution is most likely to yield the greatest improvement in the objective function value. This setting provides the best overall performance.

Lagrangian Procedure

Within the Lagrangian procedure, we employ variants of the standard subgradient optimization method to update the multipliers. For Assumption 1, we use the variant of the subgradient procedure (Held et al. 1974) described in Camerini et al. (1975). For Variant 2 and the original problem, we use a version in which the scale factor is halved if the lower bound has not improved after 5 iterations. We also update the multipliers using a weight of 0.35 on the slack from the prior iteration and a weight of 1 on the slack from the current iteration.

To solve subproblem (P2), which can be solved easily using CPLEX (thus obviating the need for the special-purpose algorithm developed in Section 5.1), we use the default branch-and-bound algorithmic settings, including the default optimality tolerance of 0.0001.

Subproblem (P1) is more difficult to solve than (P2). For the original formulation and for Variant 2, we use the solution from the prior iteration as a "warm start" for the next iteration. We do not utilize the CPLEX-generated cuts because we observed that they do not provide much benefit relative to the CPU effort. We also select the CPLEX parameter setting that emphasizes optimality over feasibility. We solve each subproblem to within 5% of optimality or stop after 1000 seconds, whichever occurs sooner. For Variant 1, we similarly use the solution from the prior iteration as a "warm start" for the next iteration and we turn off the CPLEX-generated cuts. We use a search strategy in which the branching variable is selected based on "pseudo-reduced" costs, i.e., estimates of the change in the objective from rounding a fractional variable to the nearest integer; the branching node is selected based on the best integer objective that can be achieved from solving the subproblem corresponding to all nodes eligible for selection. We solve each subproblem to within 1% of optimality. The ease with which these subproblems are solved in contrast to those in Variant 2 and the original problem obviates the need for a time limit.

References

Aggarwal, A. and J.K. Park, 1990. Improved Algorithms for Economic Lot-Size Problems. Working Paper, Laboratory for Computer Science, MIT, Cambridge, MA.

Anily, S. and M. Tzur, 2002. Shipping Multiple Items by Capacitated Vehicles–An Optimal Dynamic Programming Approach. Working paper, Faculty of Management, Tel Aviv University, Tel Aviv, Israel. To appear in *Transportation Science.*

Baker, K.R., P. Dixon, M.J. Magazine and E.A. Silver, 1978. An Algorithm for the Dynamic Lot-Size Problem with Time-Varying Production Capacity Constraints. *Management Science* **24** (16), 1710-1720.

Bard, J.F., L. Huang, P. Jaillet and M.A. Dror, 1998. A Decomposition Approach to the Inventory Routing Problem with Satellite Facilities. *Transportation Science* **32** (2), 189-203.

Bell, W.J., L.M. Dalberto, M.L. Fisher, A.J. Greenfield, R. Jaikumar, P. Kedia, R.G. Mack and P.J. Prutzman, 1983. Improving the Distribution of Industrial Gases with an On-Line Computerized Routing and Scheduling Optimizer. *Interfaces* **13** (6), 4-23.

Camerini, P., L. Fratta and F. Maffioli, 1975. On Improving Relaxation Methods by Modified Gradient Techniques. *Mathematical Programming Study* **3**, 26-34.

Carter, M.W., J.M. Farvolden, G. Laporte and J. Xu, 1996. Solving an Integrated Logistics Problem Arising in Grocery Distribution. *INFOR* **34** (4), 290-306.

Chan, L.M., A. Federgruen and D. Simchi-Levi, 1998. Probabilistic Analyses and Practical Algorithms for Inventory-Routing Models. *Operations Research* **46** (1), 96-106.

Chandra, P., 1993. A Dynamic Distribution Model with Warehouse and Customer Replenishment Requirements. *Journal of the Operational Research Society* **44** (7), 681-692.

Chandra, P. and M.L. Fisher, 1994. Coordination of Production and Distribution Planning. *European Journal of Operational Research* **72** (3), 503-517.

Chao, I.M., B.L. Golden and E.A. Wasil, 1995. A New Heuristic for the Period Traveling Salesman Problem. *Computers and Operations Research* **22** (5), 553-565.

Chien, T.W., A. Balakrishnan and R.T. Wong, 1989. An Integrated Inventory Allocation and Vehicle Routing Problem. *Transportation Science* **23** (2), 67-76.

Dror, M., M. Ball and B. Golden, 1985. A Computational Comparison of Algorithms for the Inventory Routing Problem. *Annals of Operations Research* **4**, 3-23.

Dror, M. and L. Levy, 1986. A Vehicle Routing Improvement Algorithm Comparison of a "Greedy" and a Matching Implementation for Inventory Routing. *Computers and Operations Research* **13** (1), 33-45.

Dror, M. and P. Trudeau, 1996. Cash Flow Optimization in Delivery Scheduling. *European Journal of Operational Research* **88**, 504-515.

Federgruen, A. and G. van Ryzin, 1997. Probabilistic Analysis of a Combined Aggregation and Math Programming Heuristic for a General Class of Vehicle Routing and Scheduling Problems. *Management Science* **43** (8), 1060-1078.

Federgruen, A. and P. Zipkin, 1984. A Combined Vehicle Routing and Inventory Allocation Problem. *Operations Research* **32** (5), 1019-1037.

Federgruen, A., G. Prastacos and P.H. Zipkin, 1986. An Allocation and Distribution Model for Perishable Products. *Operations Research* **34** (1), 75-82.

Federgruen, A. and M. Tzur, 1991. A Simple Forward Algorithm to Solve General Dynamic Lot Sizing Models with n periods in $O(n \log n)$ or $O(n)$ Time. *Management Science* **37** (8), 909-925.

Florian, M. and M. Klein, 1971. Deterministic Production Planning with Concave Costs and Capacity Constraints. *Management Science* **18** (1), 12-20.

Garfinkel, R.S. and G.L. Nemhauser, 1969. The Set Partitioning Problem: Set Covering with Equality Constraints. *Operations Research* **17**, 848-856.

Gaudioso, M. and G. Paletta, 1992. A Heuristic for the Periodic Vehicle-Routing Problem. *Transportation Science* **26** (2), 86-92.

Held, M., P. Wolfe and H. Crowder, 1974. Validation of Subgradient Optimization. *Mathematical Programming* **6**, 62-88.

Herer, Y.T. and R. Levy, 1997. The Metered Inventory Routing Problem, an Integrative Heuristic Algorithm. *International Journal of Production Economics* **51** (1, 2), 69-81.

Herer, Y. and R. Roundy, 1997. Heuristics for a One-Warehouse Multi-retailer Distribution Problem with Performance Bounds. *Operations Research* **45** (1), 102-115.

Larson, R.C., 1988. Transporting Sludge to the 106-mile Site: An Inventory/Routing Model for Fleet Sizing and Logistics System Design. *Transportation Science* **22** (3), 186-198.

Lee, C.-G., Y.A. Bozer and C.C. White, III, 2003. A Heuristic Approach and Properties of Optimal Solutions to the Dynamic Inventory Routing Problem. Working paper, Department of Mechanical and Industrial Engineering, University of Toronto, Toronto, Ontario, Canada.

Lee, C.-Y., 1989. A Solution to the Multiple Set-Up Problem with Dynamic Demand. *IIE Transactions* **21** (3), 266-270.

Lippman, S., 1969. Optimal Inventory Policy with Multiple Set-up Costs. *Management Science* **16** (1), 118-138.

Love, S.F., 1973. Bounded Production and Inventory Models with Piecewise Concave Costs. *Management Science* **20** (3), 313-318.

Metters, R.D., 1996. Interdependent Transportation and Production Activity at the United States Postal Service. *Journal of the Operational Research Society* **47** (1), 27-37.

Pochet, Y. and L.A. Wolsey, 1993. Lot-Sizing with Constant Batches: Formulation and Valid Inequalities. *Mathematics of Operations Research* **18** (4), 767-785.

Qu, W.W., J.H. Bookbinder, and P. Iyogun, 1999. An Integrated Inventory-Transportation System with Modified Periodic Policy for Multiple Products. *European Journal of Operational Research* **115**, 254-269.

Russell, R.A. and D. Gribbin, 1991. A Multiphase Approach to the Period Vehicle Routing Problem. *Networks* **21** (7), 747-765.

Viswanathan, S. and K. Mathur, 1997. Integrating Routing and Inventory Decisions in One-Warehouse Multiretailer Multiproduct Distribution Systems. *Management Science* **43** (3), 294-312.

Wagelmans, A., S. Van Hoesel and A. Kolen, 1992. Economic Lot Sizing: An $O(n \log n)$ Algorithm that Runs in Linear Time in the Wagner Whitin Case. *Operations Research* **40**, Suppl. No. 1, S145-S156.

Wagner, H.M. and T.M. Whitin, 1958. Dynamic Version of the Economic Lot Size Model. *Management Science* **5** (1), 89-96.

Webb, I.R. and R.C. Larson, 1995. Period and Phase of Customer Replenishment–A New Approach to the Strategic Inventory/Routing Problem. *European Journal of Operational Research* **85** (1), 132-148.

Chapter 4

OPTIMAL INVESTMENT STRATEGIES FOR FLEXIBLE RESOURCES, CONSIDERING PRICING*

Ebru K. Bish
Grado Department of Industrial and Systems Engineering
Virginia Polytechnic Institute and State University
Blacksburg, VA, 24061-0118

1. Introduction

The resource (capacity) investment decision is one of the determining factors of a firm's profitability. The resource investment process in many industries is characterized by long lead-times and economies of scale in investment costs. As a result, this decision needs to be made early, using highly uncertain long-term demand forecasts, and is costly and difficult to change later on. An example is the automotive industry, where the resource investment decision needs to be made 3-5 years before production starts; the mean demand forecast in this stage deviates from the actual sales by 40% on average [Biller, Bish, and Muriel (2002); Jordan and Graves (1995)]. Similar examples can be found in other manufacturing and service industries. Under such high uncertainty, manufacturers need to be *flexible* so that they can effectively match their supply with demand.

Investing in *resource flexibility* is one strategy that is gaining importance in today's competitive environment. The term "flexible resource" refers to a resource (such as a plant or an assembly line) with the ability to produce multiple products (or satisfy multiple service types); this is also referred to as "process flexibility" or "manufacturing flexibility" in the literature [Sethi and Sethi (1990)]. Although flexible resources are generally more expensive to acquire than "dedicated resources," which can only produce a single product, their benefits can be significant. For

*This research has been supported in part by NSF Grant # DMI-0010032.

example, "Chrysler experienced a staggering loss of more than $2 billion on its Town&Country and Voyager minivans in 2000 due to overestimating the market demand" [Goyal and Netessine (2004); Greenberg (2001)]. Resource flexibility allows the firm to "adjust to [such] market twists and turns" [BusinessWeek, December 22, 2003], providing a *risk-pooling* effect. As Corrington, manufacturing vice president at Daimler Chrysler, says, "With so much competition, the days of one product, one plant are starting to diminish" [The Detroit News, August 5, 2003]. Indeed, as the marketplace is becoming more competitive, more and more manufacturers are investing in flexible resources. Sony can quickly shift from one model of camcorder to another [Goyal and Netessine (2004)]. The automotive industry is well-known for utilizing manufacturing flexibility, with Japanese auto-makers leading [BusinessWeek, December 22, 2003]. For example, "Nissan's new Canton, Mississippi assembly plant can send a minivan, pickup truck, and sport-utility vehicle down the same assembly line, one after the other, without interruption. ... [As a result,] the Japanese can keep their plants busy pretty much no matter how the market shifts. ... Nissan, Toyota, and Honda all run their plants at 100% capacity, while the American Big Three use about 85%" [BusinessWeek, December 22, 2003]. However, the Big Three are catching up. "For the first time ever, a Chrysler Group manufacturing facility is able to produce two entirely different products on the same line. ... This will result in a $100 million cost savings in production launch for the all-new 2004 Chrysler Pacifica while simultaneously reducing tooling expenditures by approximately 40 percent" [DaimlerChrysler News, 2002]. Half of GM's thirty-five North American assembly lines can now make multiple vehicles [BusinessWeek, December 22, 2003]. "By the end of the decade Ford expects 75% of its vehicle assembly operations to be changed over to the flexible process. ... The savings over the next decade are expected to be in the $1.5- to $2-billion range" [Goyal and Netessine (2004)]. For resource flexibility to be beneficial, however, the key factors that affect its value need to be understood so that it can be incorporated into the resource investment decision.

Resource flexibility has received much attention in the economics and manufacturing literature [see the references in Beach et al. (2000); De Groote (1994); De Toni and Tonchia (1998); Jones and Ostroy (1984); Kouvelis (1992); Sethi and Sethi (1990)]. In addition, several other strategies for risk-pooling, such as centralization of inventory, delayed product differentiation, component commonality, and lateral transshipments, have been proposed and analyzed, and a large number of companies have reported success using these strategies [see Tayur, Ganeshan, and Magazine (2000) and the references therein]. However, it is only recently that flexible resource investment and management issues have been incorporated into operations management models; see Van Mieghem (2003) for an excellent review of research in this area. Most researchers model this decision problem as a two-stage stochastic programming problem. In the first stage (the planning stage), a resource investment decision is made under demand uncertainty; and in the second stage (the production stage), uncertainty is resolved and resources are allocated to product demands [see, for instance, Bish and Suwandechochai (2003); Bish and Wang (2004); Chod and Rudi (2002); Eppen, Martin, and Schrage (1989); Fine and Freund (1990); Gupta, Gerchack and Buzacott (1992); Netessine, Dobson, and Shumsky (2002); Van Mieghem (1998)]. Using a similar framework, Van Mieghem and Rudi (2002) introduce a class of models, called *newsvendor networks*, that can be used to study stochastic capacity investment decisions, including resource flexibility, and derive some nice properties of newsvendor networks. Researchers analyzing more complex stochastic programming problems had to resort to numerical methods [see, for instance, Chen, Li, and Tirupati (2002); Eppen, Martin, and Schrage (1989)]. See also Caulkins and Fine (1990), Eberly and Van Mieghem (1997), and Harrison and Van Mieghem (1999) for multi-period extensions; Jordan and Graves (1995) and Graves and Tomlin (2000) for multi-product multi-plant systems; Goyal and Netessine (2004) for including competition and Van Mieghem (2004) for considering risk aversion in the investment decision; and Cattani, Dahan, and Schmidt (2003) for analysis of resource flexibility under a consumer choice model. Recognizing that the flexible resource can also be seen as a financial option that can be exercised after demand uncertainty is resolved, several researchers have used financial options theory to study this problem [see, for instance, Andreou (1990); Birge (2000); Dangl (1999); Triantis and Hodder (1990)].

In addition to investing in resource flexibility, the firm might be able to utilize pricing control to better match its supply and demand. Although the extent to which the firm can set and adjust prices depends on the market structure rather than being the firm's choice, if the firm can exercise pricing power, then this will affect its resource investment strategy. "As the manufacturing gets more flexible, it is not far off where – between Internet and DirectTV – one will be able to do one-on-one marketing" [Brandweek, April 9, 2001]. In such an environment, using pricing in conjunction with resource flexibility can be quite a powerful tool, as it allows the firm to adjust prices, while quickly changing the "mix" of products it sells over time. For example, "Ford managed to earn \$7.2 billion [in 1999], more than any auto maker in history," in part due to a new pricing strategy that helped change the mix of vehicles it sells [BusinessWeek, April 10, 2000].

The firm's optimal "resource investment portfolio" (capacities and mix of flexible and dedicated resources) depends on several factors, including the firm's *cost structure* (i.e., investment and operating costs), *demand characteristics* (i.e., demand uncertainty, variability, correlations, consumers' willingness to substitute the products with each other), *market characteristics* (i.e., competition, the firm's ability to set and adjust prices), and *the firm's risk management strategy*, among others. Researchers have started analyzing the impact of these factors on the optimal investment portfolio. In particular, the effects of investment costs and demand correlations on the firm's optimal investment portfolio have been studied when the firm does not have pricing power [Van Mieghem (1998)], and the effect of the firm's pricing strategy on the optimal capacity investment has been studied considering a single-product firm that invests in only one dedicated resource [Van Mieghem and Dada (1999)]. Two interesting questions then arise, which are the focus of this chapter:

(a) How does the opportunity to set prices affect the value that resource flexibility has for the firm?

(b) How does such a possibility alter the effects of investment costs and demand correlations on the firm's optimal investment portfolio?

The analysis in this chapter is based on the recent works by Bish and Suwandechochai (2003) and Bish and Wang (2004). We start, in the following section, by presenting the firm's optimal investment decision problem under a "postponed pricing" scheme, where the firm has pricing power in a monopolistic situation, and can delay its pricing decision to a time when demand uncertainty is resolved. We then discuss the structure of the firm's optimal resource investment portfolio under the price postponement scheme, and analyze how this portfolio is impacted by investment costs, demand patterns, and correlations. Then, in Section 3, we discuss the impact of the firm's pricing scheme on the capacity investment of the flexible resource. Finally, in Section 4, we suggest directions for further research.

2. The Firm's Optimal Investment Portfolio under a Postponed Pricing Scheme

This section is based on the analysis in Bish and Wang (2004). We refer the interested reader to Bish and Wang (2004) for proofs of the subsequent results.

The work in Bish and Wang (2004) builds upon the seminal works by Fine and Freund (1990) and Van Mieghem (1998). In particular, Fine and Freund (1990) consider the problem of determining the firm's optimal investment portfolio so as to maximize the firm's expected profit over a set of possible scenarios. The pricing decision is implicitly considered through a concave revenue function. Although some of the subsequent results in this section are similar to theirs, by modeling demand uncertainty through continuous random variables rather than a set of possible scenarios, the model in this section attempts to obtain a new characterization of the optimal investment portfolio. This characterization then allows the analytical study of the impact of demand parameters, correlations, and investment costs on the optimal investment portfolio under a postponed pricing scheme. On the other hand, Van Mieghem (1998) studies the optimal investment portfolio for a two-product firm, assuming that prices are exogenously determined, and models this decision problem as a multi-dimensional newsvendor model. In the first stage, the resource investment decision is made under demand uncertainty, given exogenously determined prices; and in the second stage, demands are realized and resources are allocated to demands. In that sense, the subsequent work can be seen as an extension of the work in Van Mieghem (1998) to incorporate the ex-post pricing decision into the capacity planning framework. Van Mieghem (1998) shows that when demands are perfectly positively correlated, it might still be optimal to invest in the flexible resource, but only when the prices are different. His numerical results suggest that as demand correlation between the products increases, "the optimal levels of dedicated resources increase in a concave manner, while the optimal level of the flexible resource decreases in a convex manner." Recently, Van Mieghem (2004) shows the equivalence between flexible resource strategies and component commonality strategies under exogenously set prices, and finds that "while the value of commonality strategy decreases in the correlation between product demands, commonality is optimal even when the product demands move in lockstep (perfectly positively correlated) if there is a sufficient profit differential between the two products," similar to his earlier result on flexible capacity.

A recent paper by Chod and Rudi (2002) also focuses on the interaction between resource flexibility and postponed pricing, as is done here. Chod and Rudi consider a two-product firm and a linear demand curve for each product, which is a function of its own price as well as the price of the other product. They model this decision problem as a two-stage stochastic programming problem. The resource investment decision is made in the first stage, under uncertainty on demand intercepts, and pricing and resource allocation decisions are made in the second stage, when demand uncertainty is resolved. In that sense, there are similarities between their model and the model presented below. However, there are also significant differences in the assumptions and objectives of the two models. While the model presented here does not include cross-price effects in the demand curves, Chod and Rudi do not consider dedicated resources in the investment decision. Hence, the firm's only investment decision in their model is the capacity of the flexible resource, whereas in our model it is the firm's investment portfolio, which consists of flexible as well as dedicated resources. Thus, while our focus is on how the trade-offs between dedicated and flexible resources affect the firm's optimal resource investment portfolio, the focus of Chod and Rudi is on how the value of resource flexibility under ex-post pricing depends on demand variability and correlation. Most recently, Goyal and Netessine [2004] introduce competition between two firms into this framework, and study the firm's flexible resource investment decision considering a model similar to Chod and Rudi.

2.1 Model and Assumptions

We consider a firm that produces two products, since this case is analytically tractable, while being sufficient to capture the important elements of the problem. The firm has the option to invest in two dedicated resources, each of which can satisfy only one product, and/or in a more expensive, flexible resource that can satisfy both products. The firm employs a postponed pricing strategy and seeks a resource investment portfolio that maximizes its expected profit. We model this decision problem as a two-stage stochastic programming problem. In the first stage, the resource investment decision is made under highly uncertain demand patterns. In this stage, the shape (i.e., the slope) of each demand curve is known. However, its location (i.e., the intercept) is uncertain. This represents the uncertainty in the market size. Then, in the second stage, demand curves are realized, and the firm jointly determines its pricing and resource allocation decisions, constrained by its earlier resource investment.

We model the demand for each product i (d_i) as a downward-sloping linear function of its own price (or contribution margin), denoted as p_i. That is, for $i = 1, 2$,

$$d_i = \xi_i - \alpha_i p_i,$$

where α_i (> 0) is the slope of the demand-curve and ξ_i is its intercept. Let $\vec{\xi} = (\xi_1, \xi_2)$.

In the first stage of our stochastic program, we model each $\xi_i, i = 1, 2$, as a continuous random variable with positive support. At this time, the firm makes its resource investment decision, $\vec{K} = (K_1, K_2, K_f)$, so as to maximize its expected profit, where K_i corresponds to the capacity investment for dedicated resource i, $i = 1, 2$, and K_f that for the flexible resource. Let $V(\vec{K})$ denote the expected profit in Stage 1, which equals the expected revenue ($E[\Pi^*(\vec{K}, \vec{\xi})]$) less the investment costs. Then, in Stage 2 uncertainty is resolved (i.e., the realization ϵ_i of random variable ξ_i is observed for $i = 1, 2$) and the firm maximizes its revenue through pricing and resource allocation decisions, constrained by its earlier investment decision. Let $\vec{x} = (y_1, y_2, z_1, z_2)$ denote the resource allocation vector in stage 2, where y_i and z_i respectively correspond to the amount of product i produced using the dedicated resource and the flexible resource, for $i = 1, 2$. As in the earlier literature, we assume that investment costs are linear and that the variable cost of production is the same for the dedicated and the flexible resource. Let c_i denote the unit cost of investing in resource i, $i = 1, 2, f$, where $c_1, c_2 < c_f$. In addition, we consider that $c_f < c_1 + c_2$; otherwise the problem becomes trivial (i.e., the firm would never invest in the flexible resource). Throughout the paper, we do not make any distributional assumptions on $\xi_i, i = 1, 2$. All of the following results hold for any continuous distribution of $\xi_i, i = 1, 2$.

This decision problem can be formulated as the following stochastic program:

$$\text{(Stage 1)}\quad P_1: \quad \max_{\vec{K}} V(\vec{K}) = E[\Pi^*(\vec{K},\ \vec{\xi})] - \sum_{i=1,2,f} c_i K_i \tag{4.1}$$

$$\textit{subject to}\ \ K_i \geq 0, i = 1, 2, f. \tag{4.2}$$

$$\text{(Stage 2)}\quad P_2: \quad \Pi^*(\vec{K}, \vec{\epsilon}) = \max_{\vec{x},\ \vec{p}} \sum_{i=1}^{2} p_i(y_i + z_i) \tag{4.3}$$

subject to

$$y_i \leq K_i, \quad i = 1, 2 \tag{4.4}$$

$$z_1 + z_2 \leq K_f \tag{4.5}$$

$$y_i + z_i \leq \epsilon_i - \alpha_i p_i, i = 1, 2 \tag{4.6}$$

$$p_i \leq \frac{\epsilon_i}{\alpha_i}, i = 1, 2 \tag{4.7}$$

$$y_i, z_i, p_i \geq 0, i = 1, 2, \tag{4.8}$$

where $E(.)$ denotes the expected value operator. In the above formulation, constraints (4.4) and (4.5) are the capacity constraints for the dedicated and flexible resources, respectively; constraints (4.6) ensure that the total production of each product does not exceed its demand, induced by the firm's pricing decision; and constraints (4.7) and (4.8) are the nonnegativity constraints for demands, allocation quantities, and prices, respectively. We note that demand nonnegativity constraints in (4.7) are redundant and are included in the formulation for the sake of completeness.

In our formulation, we do not consider any penalty cost for a lost sale other than a forfeited profit. This is because the firm is a price-setter and determines how much demand to satisfy through pricing. In fact, it is easy to show that any solution with excess demand will be sub-optimal in the second stage. In addition, a salvage value for unused capacity can be included in the model without changing the structure of the results.

Observe that when the flexible resource is not available or is not considered in the investment decision, the optimal dedicated resource capacity for each product can be obtained independently. We will refer to this case as the "dedicated system." Based on properties of the optimal solution, the decision problem for product $i, i = 1, 2$, in the dedicated system can be written as follows:

$$\text{(Stage 1 Product } i) \quad P_1(i): \qquad \max_{K_i \geq 0} V_i(K_i) \equiv E[\Pi_i^*(K_i, \xi_i)] - c_i K_i. \tag{4.9}$$

$$\text{(Stage 2 Product } i) \quad P_2(i): \qquad \Pi_i^*(K_i, \epsilon_i) = \max_{p_i} \; p_i(\epsilon_i - \alpha_i p_i) \tag{4.10}$$

$$\textit{subject to } p_i \geq \frac{\epsilon_i - K_i}{\alpha_i}. \tag{4.11}$$

Let $\vec{K}^I = (K_1^I, K_2^I)$ denote the optimal investment vector in the dedicated system.

Consider an *unconstrained* version of Problem $P_2(i)$, $i = 1, 2$, in which the resource capacity constraint in (4.11) is relaxed. Let p_i^U and d_i^U denote the optimal price and the corresponding demand for product $i, i = 1, 2$, in the unconstrained Problem $P_2(i)$. Then, one can show that $p_i^U = \frac{\epsilon_i}{2\alpha_i}$, or equivalently, $d_i^U = \frac{\epsilon_i}{2}, i = 1, 2$. This result will be used subsequently in our analysis.

2.2 Impact of Investment Costs and Demand Parameters on the Optimal Investment Portfolio

The following lemma establishes the necessary and sufficient conditions for investing in the flexible resource.

LEMMA 4.1 *The optimal investment strategy is such that $K_f > 0$ only if $c_f < \underline{c}_f$, where the threshold value $\underline{c}_f$ can be computed by solving two independent optimization problems, one for each dedicated resource.*

Thus, the flexible resource will be beneficial only when its unit investment cost is *not too expensive.* Lemma 4.1 extends Theorem 1 of Fine and Freund (1990) to our multi-dimensional newsvendor model with ex-post pricing and has a similar interpretation: The firm invests in the flexible resource "only when the expected value of its best usage exceeds its cost."

We next study the structure of the firm's optimal investment strategy when it consists of an investment in the flexible resource. We first note that when $K_f > 0$ in the optimal solution, the solution must be one of the following forms, each of which corresponds to a boundary solution of the feasible region for the Stage 1 Problem:

$$\begin{aligned}
\vec{K}^F &= (K_1^F = 0, \quad K_2^F = 0, \quad K_f^F > 0), \\
\vec{K}^{1F} &= (K_1^{1F} > 0, \quad K_2^{1F} = 0, \quad K_f^{1F} > 0), \\
\vec{K}^{2F} &= (K_1^{2F} = 0, \quad K_2^{2F} > 0, \quad K_f^{2F} > 0), \\
\vec{K}^A &= (K_1^A > 0, \quad K_2^A > 0, \quad K_f^A > 0).
\end{aligned}$$

The following theorem analyzes the structure of the firm's optimal investment strategy, considering the case where one product would be priced higher than the other one if resource capacities were not constraining.

THEOREM 4.2 *Consider the case where* $\Pr(\frac{\xi_1}{\alpha_1} < \frac{\xi_2}{\alpha_2}) = 1$, *that is,* $\alpha_2\xi_1 < \alpha_1\xi_2$ *with probability 1. Then, if* $c_f < \underline{c}_f$, *the optimal strategy must be one of the following forms:*

1 *Invest in dedicated resource 2 and the flexible resource only (solution* $\vec{K}^{2F}$*). In this case,* $K_2^{2F} > \frac{\alpha_2}{\alpha_1} K_f^{2F}$*; or*

2 *Invest in all three resources (solution* $\vec{K}^A$*). In this case,* $K_2^A > \frac{\alpha_2}{\alpha_1}(K_1^A + K_f^A)$.

Thus, if the firm prices product 2 higher than product 1 under unlimited resource capacities (i.e., $p_1^U = \frac{\epsilon_1}{2\alpha_1} < p_2^U = \frac{\epsilon_2}{2\alpha_2}$), then the optimal investment strategy will be to always invest in dedicated resource 2, which can satisfy the more desirable product at a lower investment cost. If a flexible resource investment is made, then the firm should invest either in dedicated resource 2 and the flexible resource only (solution $\vec{K}^{2F}$); or in all three resources (solution $\vec{K}^A$). No other solution can be optimal. Thus, Theorem 4.2 highlights the effect of the relationship between the demand functions on the firm's optimal resource investment portfolio. This result, together with Lemma 4.1, extends Proposition 2 in Van Mieghem (1998), who shows that the optimal investment strategy, when prices are exogenously determined, is to invest either in dedicated resources only, or in the dedicated resource for the higher priced product and the flexible resource, or in all three resources. When prices are decision variables, however, the relationship between the demand functions impacts the optimal investment strategy. Furthermore, Theorem 4.2 shows that when product 2 is more desirable, then the optimal capacity of dedicated resource 2 is greater than the combined optimal capacities of the other resources (flexible resource plus the dedicated resource of the other product) times the relative slope of product 2 demand (α_2/α_1).

2.3 Impact of Demand Correlation on the Optimal Investment Strategy

Let ρ denote the correlation coefficient between ξ_1 and ξ_2. In the following theorem, we first analyze the case where the demand patterns are *perfectly positively correlated.*

THEOREM 4.3 *Consider the case where $Pr(\xi_1 = a\xi_2) = 1$ for some constant $a > 0$; that is, the two demand patterns are perfectly positively correlated ($\rho = +1$). Then the optimal capacity investment decision has the following structure:*

1 If $\{\frac{c_2}{c_1} < \frac{\alpha_1}{a\alpha_2} \leq 1\}$ or $\{\frac{\alpha_1}{a\alpha_2} = \frac{c_2}{c_1}\}$ or $\{1 \leq \frac{\alpha_1}{a\alpha_2} < \frac{c_2}{c_1}\}$, then the optimal strategy is to invest in dedicated resources only ($\vec{K}^D$).

2 If $\frac{\alpha_1}{a\alpha_2} > \max\{\frac{c_2}{c_1}, 1\}$, then:

- *if $c_f > \underline{c}_f$, then the optimal strategy is to invest in dedicated resources only ($\vec{K}^D$);*
- *if $c_f < \underline{c}_f$, then the optimal strategy is either to (1) invest only in dedicated resource 2 and the flexible resource ($\vec{K}^{2F}$); or (2) invest in all three resources ($\vec{K}^A$). Furthermore,*
    - *(a) if strategy $\vec{K}^{2F}$ is optimal, then $\frac{K_1^I}{a} < \frac{K_f^{2F}}{a} < K_2^{2F} < K_2^I$;*
    - *(b) if strategy $\vec{K}^A$ is optimal, then $\frac{K_1^A}{a} < \frac{K_1^I}{a} < \frac{K_1^A + K_f^A}{a} < K_2^A < K_2^I < K_2^A + K_f^A$, where $\vec{K}^I$ denotes the optimal solution to the dedicated system.*

3 Symmetrically, if $\frac{\alpha_1}{a\alpha_2} < \min\{\frac{c_2}{c_1}, 1\}$, then:

- *if $c_f > \underline{c}_f$, then the optimal strategy is to invest in dedicated resources only ($\vec{K}^D$);*

- *if $c_f < \underline{c}_f$, then the optimal strategy is either to (1) invest only in dedicated resource 1 and the flexible resource ($\vec{K}^{1F}$); or (2) invest in all three resources ($\vec{K}^A$). Furthermore,*
  (a) if strategy $\vec{K}^{1F}$ is optimal, then $K_2^I < K_f^{1F} < \frac{K_1^{1F}}{a} < \frac{K_1^I}{a}$;
  (b) if strategy K^A is optimal, then $K_2^A < K_2^I < K_2^A + K_f^A < \frac{K_1^A}{a} < \frac{K_1^I}{a} < \frac{K_1^A + K_f^A}{a}$.

Theorem 4.3 states that if $c_f < \underline{c}_f$, then the firm should invest in the flexible resource only if (1) $\frac{\alpha_1}{a\alpha_2} > \max\{\frac{c_2}{c_1}, 1\}$; or (2) $\frac{\alpha_1}{a\alpha_2} < \min\{\frac{c_2}{c_1}, 1\}$.

The condition, $\frac{\alpha_1}{a\alpha_2} > \max\{\frac{c_2}{c_1}, 1\}$, ensures that when the two products compete for the flexible resource, the priority with which the flexible resource is allocated to products changes, depending on how demand uncertainty is resolved. This is a requirement for investing in the flexible resource. Otherwise, the firm would be better off investing in more dedicated capacity for the higher priority product instead of the flexible capacity. Thus, financial factors, rather than risk pooling, determine whether or not the firm should invest in the flexible resource in this case.

In order to interpret these conditions in financial terms, consider again the condition, $\frac{\alpha_1}{a\alpha_2} > \max\{\frac{c_2}{c_1}, 1\}$. This condition implies that (1) the firm would price product 2 higher if resource capacities were not constraining; that is, $p_2^U > p_1^U$; and (2) the dedicated resource investment cost per unconstrained price ratio for product 2 is lower; that is, $\frac{c_2}{p_2^U} < \frac{c_1}{p_1^U}$. The second condition, $\frac{\alpha_1}{a\alpha_2} < \min\{\frac{c_2}{c_1}, 1\}$, is a symmetric condition for product 1. We note here that these conditions are somewhat similar to those of Van Mieghem (1998), who shows that under exogenously fixed prices, a necessary condition for flexible resource investment in the case of perfect positive correlation is a price differential between the products (in our case, it is the unconstrained optimal price differential). However, our optimal resource allocation strategy is quite different from Van Mieghem's. This is because when prices are exogenously fixed and are independent of the demands (such as the case in Van Mieghem), the optimal allocation strategy will be to always allocate the flexible resource to the higher priced product, if needed. This is not true in our case, where an inverse relationship exists between prices offered and demands generated.

Theorem 4.3 also provides insights on the *substitution effect* of the flexible resource. Consider again the condition that $\frac{\alpha_1}{a\alpha_2} > \max\{\frac{c_2}{c_1}, 1\}$. Theorem 4.3 shows that in this case, any flexible resource investment will be accompanied by investment in dedicated resource 2, which can satisfy product 2 at a cheaper cost: If a flexible resource investment is made, then the firm invests either in dedicated resource 2 and the flexible resource only (solution $\vec{K}^{2F}$) or in all three resources (solution $\vec{K}^A$); no other solution can be optimal. Furthermore, if it is optimal to invest in dedicated resource 2 and the flexible resource only (solution $\vec{K}^{2F}$), then the structure of the optimal strategy is such that $K_1^I < K_f^{2F}$ and $K_2^{2F} < K_2^I$; that is, the flexible resource *substitutes* dedicated resource 1 in the dedicated system (solution $\vec{K}^I$) and is acquired more, while less of dedicated resource 2 is acquired, compared to solution $\vec{K}^I$ (part 2(a) of the theorem). Similarly, if it is optimal to invest in all resources (solution $\vec{K}^A$), then the firm invests less in each dedicated resource, while the total resource capacity that can produce each product is higher ($K_i^A < K_i^I < K_i^A + K_f^A, i = 1, 2$), compared to the optimal solution to the dedicated system (part 2(b) of the theorem).

Consider next the condition that $\frac{\alpha_1}{a\alpha_2} > \max\{\frac{c_2}{c_1}, 1\}$ for the special case where $a = 1$, that is, $Pr(\xi_1 = \xi_2) = 1$ and $\frac{\alpha_1}{\alpha_2} > \max\{\frac{c_2}{c_1}, 1\}$. In this case, product 2's demand curve is *dominating* over all possible prices, since the same demand corresponds to a higher price for product 2 than product 1. In addition, the dedicated resource cost per unconstrained price ratio for product 2 is lower. Then, the firm's optimal investment strategy always consists of the flexible resource, with the capacity of dedicated resource 2 being higher than the combined capacities of the other resources (i.e., if solution $\vec{K}^{2F}$ is optimal, then $K_f^{2F} < K_2^{2F}$; if solution $\vec{K}^A$ is optimal, then $K_1^A + K_f^A < K_2^A$). Thus, both the unconstrained price and the total capacity that can satisfy it are higher for product 2.

Next we analyze the case where demand patterns are *perfectly negatively correlated* (i.e., the sum of the random variables ξ_1 and ξ_2 is known with certainty, but the split between the two products is uncertain in Stage 1). The following theorem characterizes the structure of the firm's optimal investment strategy in this case.

THEOREM 4.4 *Consider the case where $Pr(\xi_1 + \xi_2 = a) = 1$ for some constant $a > 0$; that is, the two demand patterns are perfectly negatively correlated ($\rho = -1$). Then, the structure of the firm's optimal resource investment strategy is as follows:*

1. *If $K_1^I + K_2^I \geq \frac{a}{2}$, then $K_f > 0$ in the optimal solution regardless of the value of c_f (in the range $0 < c_1, c_2 < c_f < c_1 + c_2$).*

2. *If $K_1^I + K_2^I < \frac{a}{2}$, then:*
   - *if $c_f < \underline{c}_f$, then $K_f > 0$ in the optimal solution;*
   - *if $c_f > \underline{c}_f$, then $K_f = 0$ in the optimal solution and the optimal solution is given by $\vec{K}^D$.*

Theorem 4.4 shows that if the total capacity investment in the dedicated system is high (i.e., $K_1^I + K_2^I \geq \frac{a}{2}$), then the firm's optimal strategy is to invest in the flexible resource, regardless of the investment cost of the flexible resource, in the range considered. However, when the total capacity investment in the dedicated system is lower (i.e., $K_1^I + K_2^I < \frac{a}{2}$), then it is *not* always optimal to invest in the flexible resource. This depends on the investment cost of the flexible resource.

In order to understand the implications of Theorem 4.4, consider the first case (i.e., $K_1^I + K_2^I \geq \frac{a}{2}$). In this case, the total expected unused capacity in the dedicated system is greater than or equal to the total expected lost demand (considering unconstrained optimal demands given by $d_i^U = \frac{\epsilon_i}{2}, i = 1, 2$). Thus, the two products are profitable enough for the firm to invest in high capacity levels in the dedicated system such that, on expectation, the investment decision favors unused capacity over lost demand (considering unconstrained optimal demands). In this case, it will always be optimal for the firm to invest in the more expensive, flexible resource. Similarly, when capacity levels in the dedicated system are not high (i.e., $K_1^I + K_2^I < \frac{a}{2}$), then the investment decision in the dedicated system favors lost demand over unused capacity. In this case, the products are not as profitable. As a result, it may not always be optimal for the firm to invest in the flexible resource.

In the next section, we study the impact of the postponed pricing strategy on the value of the flexible resource.

3. Impact of the Postponed Pricing Strategy on the Optimal Investment Decision

The model presented above considers that the firm uses a price postponement strategy, together with resource flexibility. The next question, then, is how the firm's ability to postpone the prices affects the optimal capacity investment of the flexible resource. Van Mieghem and Dada (1999) study the effect of a postponed pricing strategy on the firm's optimal capacity investment, but considering a single product, and hence, a dedicated resource only. They show that the firm's optimal capacity investment increases as it has more ex-post flexibility in the production decision (i.e., it can postpone its production decision to a later time, when demand uncertainty is resolved). Our objective in this section is to study how the firm's flexible resource capacity investment changes as it has more ex-post flexibility in the pricing decision. This section is based on the work by Bish and Suwandechochai (2003).

We consider the two-product model of the previous section. In order to simplify the analysis, however, we consider that the firm only invests in one flexible resource that can produce both products. All other assumptions of the previous model still hold. We let $\vec{q} = (q_1, q_2)$ denote the production vector in the second stage. Overall, the firm makes three sets of decisions: (1) flexible resource investment, (2) pricing, and (3) production. As in the previous model, we model the firm's different pricing strategies using two-stage stochastic programming formulations. In the first stage, the firm needs to determine K_f, the capacity of the flexible resource; and in the second stage, the firm allocates its resource capacity to the two products. We construct two different models in order to analyze the impact of the price postponement strategy on the firm's resource investment decision. The two models differ depending on when the price vector, $\vec{p} = (p_1, p_2)$, is determined. In our first model, we consider that the firm implements a price postponement strategy and postpones its pricing decision to the second stage. Therefore, the firm determines the flexible resource capacity, K_f, in Stage 1; and the price vector, $\vec{p} = (p_1, p_2)$, and the production vector, $\vec{q} = (q_1, q_2)$, in Stage 2. We refer to this model as PP, *the price postponement strategy.* We note that this model is similar to the one studied in Chod and Rudi (2002), but without cross-price effects. In our second model, the firm determines the flexible capacity investment, K_f, and the price vector, $\vec{p} = (p_1, p_2)$, under demand uncertainty in Stage 1, before the values of random variables ξ_i, $i = 1, 2$, are observed; and the production vector, $\vec{q} = (q_1, q_2)$, in Stage 2. We refer to this model as NP, *the no-price postponement strategy.*

Specifically, Problem PP can be formulated as the following two-stage stochastic problem:

$$\begin{aligned}
\text{PP}_1 \text{ (Stage 1)} : \max_{K_f} \ V^{PP} &\equiv E[\Pi^{*PP}(K_f)] - c_f K_f \\
\text{subject to} \quad & K_f \geq 0. && (4.12) \\
\text{PP}_2 \text{ (Stage 2)} : \Pi^{*PP}(K_f) &\equiv \max_{\vec{q},\vec{p}} \sum_{i=1}^{2} p_i q_i \\
\text{subject to} \quad & \sum_{i=1}^{2} q_i \leq K_f && (4.13) \\
& q_i \leq \epsilon_i - \alpha_i p_i, \quad i = 1,2 && (4.14) \\
& p_i \leq \frac{\epsilon_i}{\alpha_i}, \quad i = 1,2 && (4.15) \\
& q_i, p_i \geq 0, \quad i = 1,2, && (4.16)
\end{aligned}$$

where V^{PP} denotes the expectation of the profit function in Stage 1, given by the expected revenue, $E[\Pi^{PP}(K_f)]$, less the investment cost, $c_f K_f$. Constraint (4.13) ensures that the total production quantity does not exceed the resource capacity determined in the first stage; constraints (4.14) imply that production of each product does not exceed its demand, induced by the firm's pricing decision; constraints (4.15) are the nonnegativity constraints for demands; and constraints (4.16) are the nonnegativity constraints for the production quantities and prices. We note that demand nonnegativity constraints in (4.15) are redundant and are included in the formulation for the sake of completeness.

In a similar way, Problem NP can be formulated as follows:

$$\begin{aligned}
\text{NP}_1 \text{ (Stage 1)} : \max_{K_f, \vec{p}} \ V^{NP} &\equiv E[\Pi^{*NP}(K_f, \vec{p})] - c_f K_f \\
\text{subject to} \quad & K_f \geq 0 && (4.17) \\
& p_i \geq 0, \quad i = 1, 2. && (4.18) \\
\text{NP}_2 \text{ (Stage 2)} : \Pi^{*NP}(K_f, \vec{p}) &\equiv \max_{\vec{q}} \sum_{i=1}^{2} p_i q_i \\
\text{subject to} \quad & \sum_{i=1}^{2} q_i \leq K_f, && (4.19) \\
& q_i \leq (\epsilon_i - \alpha_i p_i)^+, i = 1, 2 && (4.20) \\
& q_i \geq 0, \qquad i = 1, 2. && (4.21)
\end{aligned}$$

Observe that the constraints of Problem NP are similar to those of Problem PP, except that the pricing decision is now made in Stage 1 instead of Stage 2, prior to the realization (ϵ_1, ϵ_2) of random variables (ξ_1, ξ_2). Consequently, the induced demands in the second stage corresponding to a price vector $\vec{p}$ and a realization (ϵ_1, ϵ_2), given by $\epsilon_i - \alpha_i p_i$, $i = 1, 2$, can be negative. Therefore, we need to consider $(\epsilon_i - \alpha_i p_i)^+$, $i = 1, 2$, as the effective demand for product i in Stage 2, where $x^+ = \max\{0, x\}$. That is, if the price for a product has been set too high for a realization (ϵ_1, ϵ_2), then the effective demand of that product will be zero.

In the following, we present results of our numerical study that compares the optimal capacity investments and expected profits under the price postponement (model PP) and the no-postponement (model NP) strategies.

3.1 Impact of the Pricing Strategy on the Investment Decision

In this section, we consider several scenarios, each characterized by the demand curve parameters for the two products (α_1, α_2, and distribution and parameters for ξ_1 and ξ_2), and study how the optimal capacity investments and expected profits compare under the price postponement (PP) and no-postponement (NP) strategies.

For each set of parameters, $(\alpha_1, \alpha_2, E[\xi_1], E[\xi_2])$, we consider that $\xi_i, i = 1, 2$, are independently distributed with either exponential or uniform distributions, and determine the parameter values of the corresponding distribution so as to match the first moment of $\xi_i, i = 1, 2$. Table 4.1 presents $(\alpha_1, \alpha_2, E[\xi_1], E[\xi_2])$ for each scenario as well as the corresponding parameter values, (a, b), for the uniform distribution.

Scenario	α_1	α_2	$E[\xi_1]$	$E[\xi_2]$	**Uniform distribution parameters** (a, b)	
					for ξ_1	**for** ξ_2
1	2	2	70	70	(20, 120)	(20, 120)
2	3	3	100	100	(50, 150)	(50, 150)
3	2	2	20	20	(5, 35)	(5, 35)
4	1	1	10	10	(5, 15)	(5, 15)
5	0.5	0.5	10	10	(5, 15)	(5, 15)
6	2	1	25	15	(5, 45)	(5, 25)
7	1	1	10	5	(5, 15)	(3, 7)
8	1	0.5	10	5	(5, 15)	(3, 7)
9	0.5	0.5	10	5	(5, 15)	(3, 7)
10	1	0.5	10	10	(5, 15)	(5, 15)

Table 4.1. Parameter values for the different scenarios.

Table 4.2 presents the optimal values of K_f and the optimal expected profits, obtained analytically for Problems PP and NP considering scenarios 1-10. For each scenario, the higher capacity investment and the higher expected profit under strategies PP or NP are presented in bold. Please see Bish and Suwandechochai (2003) for details on the numerical procedure.

Our results indicate that whether the firm invests in more flexible capacity under strategy PP or NP depends on the unit investment cost, c_f. For high costs of investment, the firm acquires more flexible capacity under the postponed pricing strategy, PP, and for low costs of investment, more capacity under the no-postponement strategy, NP. This is mainly due to the trade-off between unused capacity (overage) and unsatisfied demands (underage). When the unit investment cost

Scenario	Exponential					Uniform				
		K_f^*		Profit			K_f^*		Profit	
	c_f	PP	NP	PP	NP	c_f	PP	NP	PP	NP
1	15	**61**	0	934.76	0	15	**44**	0	612.217	0
	3	126	**199**	1962.45	936.87	3	84	**108**	1339.52	1045.74
	2	141	**237**	2095.37	1154.01	2	91	**124**	1426.82	1161.40
2	18	**72**	0	978.80	0	18	**52**	38	619.85	147.68
	10	**109**	70	1678.58	86.27	10	78	**84**	1134.54	628.14
	5	149	**203**	2320.82	759.17	5	103	**120**	1584.26	1132.40
3	3.5	**20**	4	90.87	0.49	4	**13**	12	53.65	22.94
	2	26	**31**	125.19	27.73	2	19	**21**	85.21	56.72
	0.5	42	**71**	173.97	99.17	0.5	28	**36**	184.56	97.36
4	5	**7**	2	32.16	1.34	5	**6**	5	20.66	6.05
	2	12	**14**	61.33	23.01	2	9	**11**	42.57	28.35
	0.5	22	**39**	86.26	52.68	0.5	13	**18**	59.17	48.68
5	10	**8**	0	64.57	0	10	**6**	5	41.32	12.10
	5	**12**	11	112.63	14.34	5	**9**	**9**	76.13	46.59
	2	17	**19**	155.17	64.39	2	12	**15**	105.92	81.05
6	4	**22**	12	134.54	6.39	4	**16**	**16**	90.96	49.13
	2	30	**40**	185.09	59.23	2	21	**26**	127.53	89.26
	1	38	**62**	218.67	109.45	1	25	**34**	150.41	118.61
7	3	**6**	0	19.06	0	4	**4**	3	13.47	3.92
	1	**13**	7	44.85	18.94	1	8	**11**	31.67	23.92
	0.5	16	**24**	51.98	28.90	0.5	10	**12**	36.15	29.68
8	5	**6**	0	19.06	0	5	**4**	3	14.81	4.24
	3	**8**	5	36.31	1.98	5	6	**6**	24.92	13.22
	1	13	**20**	56.80	25.16	5	9	**12**	39.56	30.07
9	8	**6**	0	38.12	0	8	**4**	3	26.93	7.91
	5	**8**	6	59.10	6.20	5	**6**	**6**	42.40	22.67
	2	13	**17**	89.70	37.89	2	8	**11**	63.33	47.83
10	4	**11**	10	82.30	24.16	4	8	**9**	56.40	30.58
	2	15	**23**	108.65	35.42	2	10	**13**	74.44	52.66
	0.5	23	**43**	136.05	82.65	0.5	14	**21**	92.28	77.53

Table 4.2. The optimal values of K_f under strategies PP and NP for different values of c_f.

is low, the overage cost associated with having unused capacity is low. Hence, the firm invests in more capacity in Model NP to account for the higher variability in production levels. On the other hand, when the unit investment cost is high, the firm opts for a higher investment level under the postponement strategy, which can hedge against the overage risk by setting the prices under no uncertainty. Thus, Van Mieghem and Dada's (1999) earlier results, which show that the optimal capacity investment for a single-product single-resource system increases under postponed pricing, do not extend to our two-product model with a flexible resource. In addition, we find, not surprisingly, that the optimal capacity investments and expected profits under both strategies are decreasing in c_f, and the optimal expected profit under strategy PP is higher than that under strategy NP: The additional marketing power, obtained by the ability to set prices in the second stage, helps the firm realize higher profits.

4. Future Research Directions

Resource flexibility can provide a competitive advantage to the firm by hedging against demand uncertainty, but at the expense of a higher investment cost. Considering simple two-product models that are amenable to analytical analysis, we study the firm's optimal resource investment strategy, while incorporating resource flexibility and ex-post pricing into the investment decision. We characterize the structure of the firm's optimal resource investment portfolio as a function of demand parameters and investment costs. We then use this characterization to understand the conditions under which the firm invests in the flexible resource, and study the impact of investment costs, demand parameters and correlations, and the firm's pricing strategy on the value of the flexible resource. Specifically, we show that it can be optimal for the firm to invest in the flexible resource even when demand patterns are perfectly positively correlated. The reason for flexible capacity investment in this case is financial rather than risk pooling. On the other hand, we show that it can be optimal for the firm not to invest in the flexible resource even when demand patterns are perfectly negatively correlated. The flexible resource investment decision in this case depends on the profitability of the two products.

Numerous extensions to our models deserve further analysis so that a complete characterization of the effect of the important factors on the firm's optimal investment portfolio can be obtained. These directions include extending the models discussed in this chapter to consider different variable production costs for flexible and dedicated resources; different investment cost structures (including concave investment costs); different demand functions, including those in which consumers *substitute* the products with each other based on prices; different market characteristics (such as competition); and different attitudes towards risk (most research assumes that the firm is *risk neutral*, i.e., it is an expected profit maximizer. In reality, firms would have different attitudes towards risk and a firm might choose to be risk averse or risk seeking). There are very recent works that focus on the last two of these directions: Goyal and Netessine (2004) study the firm's optimal technology choice (flexible versus dedicated capacity investment) considering two firms that compete with each other for the stochastic market demand, under the assumption that each firm invests in either pure dedicated or pure flexible resources; and Van Mieghem (2004) studies the firm's optimal investment portfolio for a risk averse firm, under the assumption of exogenously fixed prices. We believe further work in these directions will be a promising avenue for future research.

References

Andreou, S.A., 1990. A Capital Budgeting Model for Product-Mix Flexibility. *Journal of Manufacturing Operations Management* **3**, 5–23.

Beach, R., A.P. Muhlemann, D.H.R. Price, A. Paterson, and J.A. Sharp, 2000. A Review of Manufacturing Flexibility. *European Journal of Operational Research* **122**, 41–57.

Biller, S., E.K. Bish, and A. Muriel, 2002. Impact of Manufacturing Flexibility on Supply Chain Performance, in *Supply Chain Structures: Coordination, Information, and Optimization.* Eds. J. Song and D. D. Yao, Kluwer Academic Publishers, Boston/Dordrecht/London, 73–118.

Birge, J.R., 2000. Option Methods for Incorporating Risk into Linear Capacity Planning Models. *Manufacturing and Service Operations Management*, **2**, 19–31.

Bish, E.K. and R. Suwandechochai, 2003. The Interaction between Resource Flexibility and Price Postponement Strategies. Technical Report. Department of Industrial and Systems Engineering, Virginia Polytechnic Institute and State University, Blacksburg, VA.

Bish, E.K., and Q. Wang, 2004. Optimal Investment Strategies for Flexible Resources, Considering Pricing and Correlated Demands. *Operations Research* **52**, No. 6, 954-964.

Caulkins, J.P. and C.H. Fine, 1990. Seasonal Inventories and the Use of Product-Flexible Manufacturing Technology. *Annals of Operations Research* **26**, 351–375.

Cattani, K., E. Dahan, and G. Schmidt, 2003. Spackling: Smoothing Make-to-order Production of Custom Products with Make-to-stock Production of Standard Items. Technical Report. University of North Carolina, Chapel Hill.

Chen, Z.L., S. Li, and D. Tirupati, 2002. A Scenario Based Stochastic Programming Approach for Technology and Capacity Planning. *Computers and Operations Research*, **29**(7), 781–806.

Chod, J. and N. Rudi, 2002. Resource Flexibility with Responsive Pricing. *Operations Research*. Forthcoming.

Coy, P., 2000. The Power of Smart Pricing: Companies are Fine-tuning Their Price Strategies and it is Paying off. *BusinessWeek*, April 10, 2000.

DaimlerChrysler News, 2002. Chrysler Group's Windsor Assembly Plant Launches Next Phase of Flexible Manufacturing. http://www.daimlerchrysler.com/news/top/2002/t21211a.htm.

Dangl, T., 1999. Investment and Capacity Choice under Uncertain Demand. *European Journal of Operational Research* **117**, 415–428.

De Groote, X., 1994. The Flexibility of Production Processes: A General Framework. *Management Science* **40**, 933–945.

De Toni, A. and S. Tonchia, 1998. Manufacturing Flexibility: A Literature Review. *International Journal of Production Research*, **36**, 1587–1617.

Eberly, J.C. and J.A. Van Mieghem, 1997. Multi-factor Dynamic Investment Under Uncertainty. *Journal of Economic Theory* **75**, 345–387.

Eppen, G.D., R.K. Martin and L. Schrage, 1989. A Scenario Approach to Capacity Planning. *Operations Research* **37**, No. 4, 517–527.

Fine, C.H. and R.M. Freund, 1990. Optimal Investment in Product-Flexible Manufacturing Capacity. *Management Science* **36**, No. 4, 449–466.

Goyal, M. and S. Netessine, 2004. Strategic Technology Choice and Capacity Investment under Demand Uncertainty. Technical Report. University of Pennsylvania, PA.

Graves, S.C. and B.T. Tomlin, 2000. Process Flexibility in Supply Chains. *Management Science*, **49**, No. 7, pp. 907–919.

Greenberg, K., 2001. Much Ado about Chrysler. *Brandweek*, 32-40, April 9, 2001.

Gupta, D., Y. Gerchak, and J.A. Buzacott, 1992. The Optimal Mix of Flexible and Dedicated Manufacturing Capacities: Hedging Against Demand Uncertainty. *International Journal of Production Economics* **28**, 309–319.

Harrison, J.M. and J.A. Van Mieghem, 1999. Multi-resource Investment Strategies: Operational Hedging under Demand Uncertainty. *European Journal of Operational Research* **113**, 17–29.

Jones, R.A. and J.M. Ostroy, 1984. Flexibility and Uncertainty. *Rev. Economic Studies* **51**, 13–32.

Jordan, W.C. and S.C. Graves, 1995. Principles and Benefits of Manufacturing Process Flexibility. *Management Science* **41**, No. 4, 577–594.

Kouvelis, P., 1992. Design and Planning Problems in Flexible Manufacturing Systems: A Critical Review. *Journal of Intelligent Manufacturing*, **3**, 75–99.

Li, S. and D. Tirupati, 1995. Technology Choice with Stochastic Demands and Dynamic Capacity Allocation: A Two Product Analysis. *Journal of Operations Management*, **12**, 239–258.

Netessine, S., G. Dobson, and R.A. Shumsky, 2002. Flexible Service Capacity: Optimal Investment and the Impact of Demand Correlation. *Operations Research* **50**, 375–388.

Sethi, A.K. and S.P. Sethi, 1990. Flexibility in Manufacturing: A Survey. *The International Journal of Flexible Manufacturing Systems* **2**, 289–328.

Tayur, S., R. Ganeshan, and M. Magazine (eds.), 2000. *Quantitative Models for Supply Chain Management.* Kluwer Academic Publishers, Boston.

The Detroit News, 2003. Flexible Factories will Help Dull Japanese Edge: Automakers Explore Turning Overcapacity into Multiuse Plants. *The Detroit News*, August 5, 2003.

Triantis, A.J. and J.E. Hodder, 1990. Valuing Flexibility as a Complex Option. *The Journal of Finance* **45**, No. 2, 549–565.

Van Mieghem, J.A., 2004. Risk-averse Newsvendor Networks: Resource Sharing, Substitution, and Operational Hedging. Technical Report. Northwestern University, Evanston, IL.

Van Mieghem, J.A., 2004. Commonality Strategies: Value Drivers and Equivalence with Flexible Capacity and Inventory Substitution. *Management Science* **50**, pp. 419–424.

Van Mieghem, J.A., 2003. Capacity Management, Investment and Hedging: Review and Recent Developments. *Manufacturing & Service Operations Management* **5**, 269–302.

Van Mieghem, J.A., 1998. Investment Strategies for Flexible Resources. *Management Science* **44**, 1071–1078.

Van Mieghem, J.A. and M. Dada, 1999. Price versus Production Postponement: Capacity and Competition. *Management Science* **45**, 1631–1649.

Van Mieghem, J.A. and N. Rudi, 2002. Newsvendor Networks: Inventory Management and Capacity Investment with Discretionary Activities. *Manufacturing & Service Operations Management* **4**, 313–335.

Welch, D., 2003. How Nissan Laps Detroit. *BusinessWeek*, December 22, 2003.

Chapter 5

MULTI-CHANNEL SUPPLY CHAIN DESIGN IN B2C ELECTRONIC COMMERCE

Wei-yu Kevin Chiang
Department of Information Systems
University of Maryland, Baltimore County
1000 Hilltop Circle, Baltimore, MD 21250, USA

Dilip Chhajed
Department of Business Administration
University of Illinois at Urbana-Champaign
350 Wohlers Hall, 1206 S. Sixth Street, Champaign, IL 61820, USA

Abstract The trend of engaging in the Internet-based direct sales has raised serious awareness and attention, both in industry and in academia, to the opportunities and challenges of using both integrated and non-integrated distribution channels simultaneously. In this chapter, we investigate the impact of the interplay between customers' channel preference and distribution costs on the supply chain channel design for a manufacturer that can sell through a retailer and directly to consumers. We develop economic/game-theoretical models to obtain insights and implications for the channel design problem. By comparing the profitability of three types of channel distribution strategies (retail-only distribution, dual-channel distribution, and direct-only distribution) under different scenarios, we disclose the optimal supply-chain channel design from the manufacturer's perspective. The analytical results are presented for both centralized and decentralized supply chains.

1. Introduction

While the burst of Internet bubble in 2002 was accompanied by the collapse of hundreds of the Internet companies, sales over the Internet have continued to increase. E-Commerce has continued to grow rapidly since early 2000. Just in the third quarter of 2003, on-line sales grew 51% to approximately $26 billion according to Forrester Research (Carrie and Walker 2003). Between November 1 and December 31st of 2003, on-line sales rose 30% to $12.5 billion (excluding travel and auctions)[1]. These increases are much higher than the overall increase in retail sales during the same period. The overall Internet sales have increased from $51 billion in 2001 to $73 billion in 2002, to $93 billion in 2003.

There are several factors behind this explosive growth including broader base of buyers and higher broadband penetration. According to a comScore Networks report of October 2003, "As consumers gain experience shopping online they spend more often and in greater amounts. The base of experienced shoppers grows every year, fueling a shift in spending from offline to online channels." The growth in on-line sales is fueled by a growth in new categories of products; while the traditionally popular "books" category grew 5% in 2002, "home & garden" and "fitness" categories experienced an increase in sales of over 60%[2]. Not surprisingly, the number of traditional brick-and-mortar companies selling on the Internet has grown. Many of them have opened direct stores while others have partnered with web merchants such as Amazon.com to participate in e-commerce.

The advent of e-commerce has facilitated the adoption of multi-channel distribution. Multi-channel distribution occurs when a single firm uses two or more distribution channels to reach one or more customer segments (Kotler 1997). It may take many forms, one of which is when a manufacturer both sells through intermediaries and directly to consumers (Preston and Schramm 1965). Sony, Estēe Lauder, Holmes, and Ethan Allen are a few examples of manufacturers in different industries who have adopted multi-channel distribution. Managing a multi-channel business can be a tremendous challenge because adding new distribution channels requires companies to think through several crucial distribution functions including production planning and materials procurement, warehousing and inventory management, order processing, shipping and transportation, pricing, and retail-outlet operations. Moreover, multi-channel distribution may result in "channel conflict" when the channels end up competing for the same customers. The growing popularity of Internet-based direct sales has raised serious awareness and attention, both in industry and in academia, to the opportunities

[1]Press Release, ComScore Networks, Inc., "Weekly Online Retail Sales Break Through $2 Billion Mark," December 18, 2003.

[2]Press Release, ComScore Networks, Inc., "comScore E-Commerce Sales Trends Accurately Predict US Department of Commerce Data," February 28, 2003.

Supply-Chain Channel Design Problem

	Direct channel has *lower* customer acceptance than the retail channel	Direct channel has *higher* customer acceptance than the retail channel
Direct channel is logistically *less efficient*	Covered by CCH	**Primary focus of this chapter**
Direct channel is logistically *more efficient*	Covered by CCH	Trivial: drop the retail store

Table 5.1. Supply-Chain Channel Design Problem.

and challenges of using both integrated (or centralized, where the manufacturer owns the retail channel) and non-integrated (or decentralized, where the manufacturer and retailer are independent) distribution channels simultaneously.

Research on strategic multi-channel supply chain management in the setting where the upstream echelon is both a supplier to and a competitor of the downstream echelon has emerged only recently. In this stream of literature, one class of papers (e.g., Rhee and Park 2000, Rhee 2001, Kumar and Ruan 2002, Tsay and Agrawal 2003, Chiang, Chhajed, and Hess 2003) is focused on modeling the price and/or service interactions between upstream and downstream echelons to address channel competition and coordination issues. From a logistics perspective, another class of papers models multi-channel inventory problems in single echelon (e.g., Boyaci 2003) and in multi-echelon settings (e.g., Chiang 2004, Chiang and Monahan 2005). Although there are several other papers (e.g., Bell, Wang, and Padmanabhan 2002, Peleg and Lee 2002, Yao and Liu 2002) that also address related issues regarding multi-channel supply chain, their foci are different. We refer the readers to Tsay and Agrawal (2004) for a recent detailed review of this stream of literature.

The research presented in this chapter extends and generalizes the work in Chiang, Chhajed and Hess (2003), which we will refer to as

CCH henceforth. They develop a model to conceptualize the impact of customers' attitudes toward direct versus traditional shopping on supply chain design. CCH argue that rather than avoiding channel conflict, manufacturers may want to use a direct channel (even if it is unlikely to produce sales) to motivate retailers to perform more effectively by reducing the degree of *double-marginalization*[3] (Spengler 1950). They show that the direct channel may not always be detrimental to the retailer since it will be accompanied by a wholesale price reduction. One of the main assumptions in CCH is that customers prefer the traditional bricks-and-mortar retail store to the Internet-based direct channel as their purchasing channel. While CCH help us understand the strategic use of a direct channel to increase channel efficiency, due to this one-sided preference assumption their result does not apply to every empirical case presented in the survey by Kacen et al. (2003) and Liang and Huang (1998).

In order to generate comprehensive insights and implications for the multi-channel supply-chain design problem, in this chapter we extend the problem considered by CCH as noted in Table 5.1. Specifically, we develop a more generalized model by also considering the case when the customer acceptance of the direct channel is higher than the retail store. We derive conditions under which a multi-channel strategy is preferable and investigate the effect of interplay between customers' channel preference and manufacturers' distribution costs on appropriate distribution strategy for both centralized and decentralized supply chains. The rest of this chapter is organized as follows. In section 2, we introduce the model and develop the demand functions. The vertically integrated system is analyzed in section 3. We then consider the best response of the manufacturer and the retailer when the independent retailer is participating in a price setting game. Concluding remarks are provided in section 5.

2. Customer Segmentation

In this section, we introduce demand functions when a traditional bricks-and-mortar retail store and a direct channel coexist.

We assume that consumers want to buy exactly one unit of a product, and the population of N consumers is heterogeneous in the valuation of the product. Without loss of generality, the market potential N can be normalized to one to eliminate an unnecessary parameter in the analy-

[3] *Double marginalization* occurs when the manufacturer, as a result of selling at a wholesale price above the marginal cost, induces its retailer to set a retail price above what it would be if it faced the marginal cost instead of the wholesale price.

sis. Let v be the consumption value (willingness-to-pay) of the product obtained from the retail store, where v is assumed to be uniformly distributed within the consumer population from zero to one, i.e., $v \in [0, 1]$. Note that a uniform demand distribution, which is fairly widely used in marketing literature, allows us to capture the heterogeneity in customer preference while at the same time it keeps the model tractable. Suppose that the product that is worth v in the retail store has a worth θv when it is obtained from a direct channel, where the value of the parameter θ is called the *customer acceptance of the direct channel.* Accordingly, when $\theta < 1$, the direct channel is not as well accepted as the traditional retail channel. On the other hand, when $\theta > 1$, customers' valuation of a product is higher in the direct channel than in the retail channel.

If the retailer offers the product at price p_r, a consumer who buys the product would derive a net consumer surplus $s^r = v - p_r$. Conversely, if the product is sold through direct channel at a price p_d, then the resulting consumer surplus s^d would be $\theta v - p_d$. All consumers whose valuation satisfies $s^r \geq 0$ would consider buying from the retailer. The marginal consumer whose valuation equals p_r is indifferent to buying from the retailer or not at all. Equivalently, all consumers whose valuations satisfy $s^d \geq 0$ would consider buying from the direct channel. The marginal consumer whose valuation equals p_d/θ is indifferent to buying from the direct marketer. If consumers can buy from either channel, their decisions depend on the comparison of the consumer surplus derived from the retailer and direct channel, i.e., s^r versus s^d. Rationally, consumers will buy from the channel that gives them a higher consumer surplus.

When $\theta < 1$, the consumers whose valuation equals $(p_r - p_d)/(1 - \theta)$ are indifferent between the two channels, and if the valuation exceeds this, they prefer the retailer. We can show that in the case where $p_d/\theta < p_r$, then $p_d/\theta < p_r < (p_r - p_d)/(1 - \theta)$ (see Figure 5.1(a)) and in the case where $p_d/\theta > p_r$, then $(p_r - p_d)/(1 - \theta) < p_r < p_d/\theta$ (see Figure 5.1(b)). In the former, all consumers with valuation in the interval $[p_d/\theta,\ (p_r-p_d)/(1-\theta)]$ prefer to buy from the direct channel and all those in the interval $[(p_r-p_d)/(1-\theta),\ 1]$[4] prefer to buy from the retailer. Those shoppers whose valuations are in $[0,\ p_d/\theta]$ decline to buy the product from either channel. In the latter case of $p_d/\theta > p_r$, no customers want to buy from the direct channel and all those consumers whose valuations are in the interval $[p_r,\ 1]$ buy from the retailer.

[4]Clearly, when $(p_d - p_r)/(1 - \theta) \geq 1$, the demand in the retail channel is zero. We exclude this extreme case in the analysis, as it appears to be trivial.

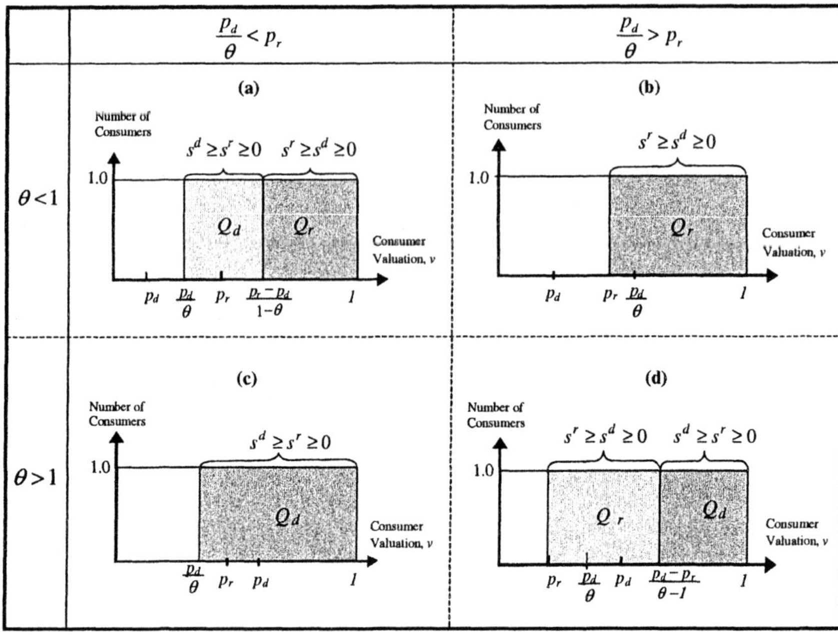

Figure 5.1. Segmentation.

When $\theta > 1$, conversely, the consumers whose valuation equals $(p_d - p_r)/(\theta - 1)$ are indifferent between the two channels, and if the valuation exceeds this they prefer the direct channel. We can show that in the case where $p_d/\theta < p_r$, no customers want to buy from the retailer and all those consumers whose valuations are in the interval $[p_d/\theta, 1]$ buy from the retailer (see Figure 5.1(c)). On the other hand, in the case where $p_d/\theta > p_r$, all consumers with valuation in the interval $[(p_d - p_r)/(\theta - 1),\ 1]$[5] prefer to buy from direct channel and all those in the interval $[p_r, (p_d - p_r)/(\theta - 1)]$ prefer to buy from the retailer (see Figure 5.1(d)). Those shoppers whose valuations are in $[0, p_r]$ decline to buy the product from either channel.

In summary, since the valuation of the consumers is uniformly distributed, demands for the retailer and direct channel (denoted by Q_r and

[5]Similarly, the extreme case when $(p_d - p_r)/(\theta - 1) \geq 1$, i.e., no demand occurs in the direct channel, is excluded in the analysis.

Q_d, respectively) correspond to piecewise linear demand functions[6]:

$$Q_r = \begin{cases} 1 - \dfrac{p_r - p_d}{1-\theta} & \text{if } \theta < 1 \text{ and } \dfrac{p_d}{\theta} < p_r \\ 1 - p_r & \text{if } \theta < 1 \text{ and } \dfrac{p_d}{\theta} > p_r \\ 0 & \text{if } \theta > 1 \text{ and } \dfrac{p_d}{\theta} < p_r \\ \dfrac{p_d - \theta p_r}{\theta - 1} & \text{if } \theta > 1 \text{ and } \dfrac{p_d}{\theta} > p_r \end{cases}, \tag{5.1}$$

$$Q_d = \begin{cases} \dfrac{\theta p_r - p_d}{\theta(1-\theta)} & \text{if } \theta < 1 \text{ and } \dfrac{p_d}{\theta} < p_r \\ 0 & \text{if } \theta < 1 \text{ and } \dfrac{p_d}{\theta} > p_r \\ 1 - \dfrac{p_d}{\theta} & \text{if } \theta > 1 \text{ and } \dfrac{p_d}{\theta} < p_r \\ 1 - \dfrac{p_d - p_r}{\theta - 1} & \text{if } \theta > 1 \text{ and } \dfrac{p_d}{\theta} > p_r \end{cases}. \tag{5.2}$$

The demand in the retail store is nonincreasing in the retail price p_r and in the customer acceptance of the direct channel θ, but is nondecreasing in the direct channel price p_d. On the other hand, the demand in the direct channel is nondecreasing in p_r and in θ, but is nonincreasing in p_d. Figure 5.2 illustrates the demand functions. Note that due to cannibalization in the channels, the retailer's demand becomes more price elastic if the retail price p_r exceeds p_d/θ when $\theta < 1$, as seen in Figure 5.2(a), while on the contrary, the direct channel's demand becomes more price elastic if the direct channel price p_d exceeds θp_r when $\theta > 1$, as seen in Figure 5.2(d).

3. Vertically-Integrated Supply Chain

When the supply chain is vertically-integrated (i.e., when the supply chain is under centralized control), given the demand functions, equations (5.1) and (5.2), the problem is to set the retail price p_r and the direct market price p_d to maximize the total channel profit

$$\pi_{vi}(p_r, p_d) = (p_r - c_r)Q_r + (p_d - c_d)Q_d, \tag{5.3}$$

where c_r and c_d are the marginal costs incurred by the manufacturer for the product sold through the retailer and the direct channel, respectively. Maximizing $\pi_{vi}(p_r, p_d)$ requires that we take into account the piecewise linear nature of the demand curves which we operationalize next.

[6]Following a conventional assumption for the analysis of linear demand functions in economics and marketing literature, we implicitly assume that the demand may not go beyond the limits, [0, 1].

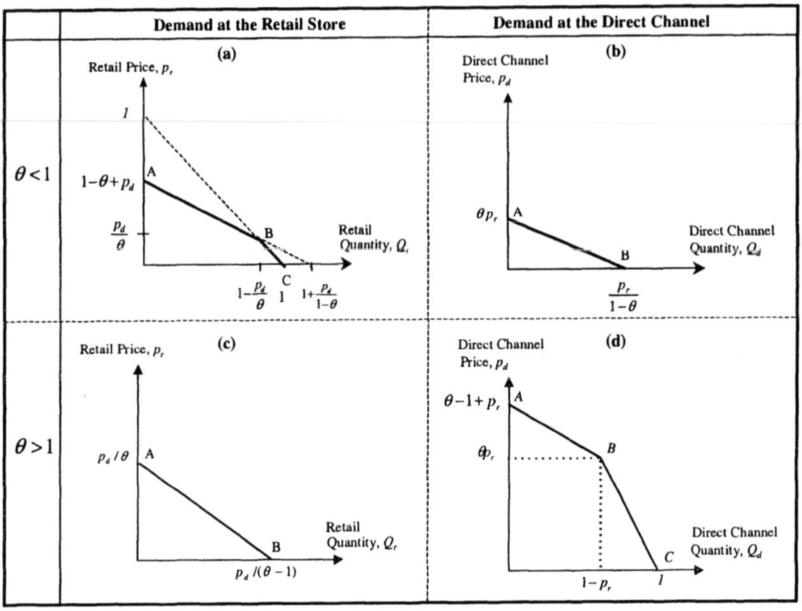

Figure 5.2. Demand Functions.

3.1 Centralized Solution

To derive the optimal pricing rule when $\theta < 1$, we first concentrate on the top lines of demand equations (5.1) and (5.2) by assuming that $p_d/\theta \leq p_r$:

$$Q_r = 1 - \frac{p_r - p_d}{1-\theta} \text{ and} \tag{5.4}$$

$$Q_d = \frac{\theta p_r - p_d}{\theta(1-\theta)}. \tag{5.5}$$

Substituting in equation (5.3), it follows that

$$\pi_{vi}(p_r, p_d) = (p_r - c_r)(1 - \frac{p_r - p_d}{1-\theta}) + (p_d - c_d)\frac{\theta p_r - p_d}{\theta(1-\theta)}. \tag{5.6}$$

It can be verified that the quadratic function $\pi_{vi}(p_r, p_d)$ in equation (5.6) is concave, and first order conditions yield

$$p_r = \frac{1+c_r}{2} \text{ and } p_d = \frac{\theta + c_d}{2}. \tag{5.7}$$

This solution satisfies $p_d/\theta \leq p_r$ only when $\theta \geq c_d/c_r$. If θ is smaller than c_d/c_r, we must have zero demand for the direct channel (the second

(a) When $\theta < 1$

Best Channel Strategy:	$0 < \theta \le \frac{c_d}{c_r}$ Retailer Only	$\frac{c_d}{c_r} \le \theta \le 1-(c_r-c_d)$ Dual-Channel Approach	$1-(c_r-c_d) \le \theta < 1$ Direct Channel Only
Price			
Direct Channel, p_d :	N/A	$\frac{\theta+c_d}{2}$	$\frac{\theta+c_d}{2}$
Retail Store, p_r :	$\frac{1+c_r}{2}$	$\frac{1+c_r}{2}$	N/A
Sales Volume			
Direct Channel, Q_d :	N/A	$\frac{c_r\theta-c_d}{2\theta(1-\theta)}$	$\frac{\theta-c_d}{2\theta}$
Retail Store, Q_r :	$\frac{1-c_r}{2}$	$\frac{1}{2}-\frac{c_r-c_d}{2(1-\theta)}$	N/A
Total, Q_d+Q_r :	$\frac{1-c_r}{2}$	$\frac{1}{2}-\frac{c_d}{2\theta}$	$\frac{\theta-c_d}{2\theta}$
Profit			
Direct Channel, π_d :	N/A	$\frac{(\theta-c_d)(c_r\theta-c_d)}{4\theta(1-\theta)}$	$\frac{(\theta-c_d)^2}{4\theta}$
Retail Store, π_r :	$\frac{(1-c_r)^2}{4}$	$\frac{(1+c_r)[(1-\theta)-(c_r-c_d)]}{4(1-\theta)}$	N/A
Total, $\pi_d+\pi_r$:	$\frac{(1-c_r)^2}{4}$	$\frac{(1-c_r)^2}{4}+\frac{(c_r-c_d)^2}{4(1-\theta)}+\frac{c_d^2-\theta c_r^2}{4\theta}$	$\frac{(\theta-c_d)^2}{4\theta}$

Note. Only the 'retailer only' option is valid when $c_d > c_r$.

(b) When $\theta > 1$

Best Channel Strategy:	$1 < \theta \le 1-(c_r-c_d)$ Retailer Only	$1-(c_r-c_d) \le \theta \le \frac{c_d}{c_r}$ Dual-Channel Approach	$\theta \ge \frac{c_d}{c_r}$ Direct Channel Only
Price			
Direct Channel, p_d :	N/A	$\frac{\theta+c_d}{2}$	$\frac{\theta+c_d}{2}$
Retail Store, p_r :	$\frac{1+c_r}{2}$	$\frac{1+c_r}{2}$	N/A
Sales Volume			
Direct Channel, Q_d :	N/A	$\frac{1}{2}-\frac{c_r-c_d}{2(1-\theta)}$	$\frac{\theta-c_d}{2\theta}$
Retail Store, Q_r :	$\frac{1-c_r}{2}$	$\frac{\theta c_r-c_d}{2(1-\theta)}$	N/A
Total, Q_d+Q_r :	$\frac{1-c_r}{2}$	$\frac{1-c_r}{2}$	$\frac{\theta-c_d}{2\theta}$
Profit			
Direct Channel, π_d :	N/A	$\frac{(\theta-c_d)[(1-\theta)-(c_r-c_d)]}{4(1-\theta)}$	$\frac{(\theta-c_d)^2}{4\theta}$
Retail Store, π_r :	$\frac{(1-c_r)^2}{4}$	$\frac{(1-c_r)(\theta c_r-c_d)}{4(1-\theta)}$	N/A
Total, $\pi_d+\pi_r$:	$\frac{(1-c_r)^2}{4}$	$\frac{-(\theta-c_d)^2+\theta(1-c_r^2)-2c_d(1-c_r)}{4(1-\theta)}$	$\frac{(\theta-c_d)^2}{4\theta}$

Note. Only the 'direct channel only' option is valid when $c_d < c_r$.

$\theta =$ customer acceptance of direct channel.
$c_d =$ marginal cost incurred by the manufacturer for the product distributed through the direct channel.
$c_r =$ marginal cost incurred by the manufacturer for the product distributed through the retailer.

Table 5.2. Channel Strategies of a Vertically Integrated Firm.

line of equation 5.2). On the other hand, if θ is too large, the demand for the retailer will drop to zero. From equation (5.4), this occurs when $p_r \geq 1 - \theta + p_d$ or $(1 + c_r)/2 \geq 1 - \theta + (\theta + c_d)/2$. In other words, when $\theta \geq 1 - (c_r - c_d)$, the demand in the retail store $Q_r = 0$ and it is optimal to use direct channel only. Table 5.2(a), which can also be found in CCH, gives a complete characterization of the optimal decisions for a centralized supply chain that could distribute through a retail or direct channel when $\theta < 1$.

When $\theta > 1$, from the demand functions in equations (5.1) and (5.2), we know that the problem corresponds to the solution of the following quadratic program:

$$\text{maximize } \pi_{vi}(p_r, p_d) = (p_r - c_r)Q_r + (p_d - c_d)Q_d, \tag{5.8}$$

$$\text{subject to} \quad Q_r = \begin{cases} 0 & \text{if } \frac{p_d}{\theta} < p_r \\ \frac{p_d - \theta p_r}{\theta - 1} & \text{otherwise} \end{cases}, \tag{5.9}$$

$$Q_d = \begin{cases} 1 - \frac{p_d}{\theta} & \text{if } \frac{p_d}{\theta} < p_r \\ 1 - \frac{p_d - p_r}{\theta - 1} & \text{otherwise} \end{cases}, \tag{5.10}$$

$$Q_r \geq 0 \text{ and } Q_d \geq 0. \tag{5.11}$$

The objective function in (5.8) is concave and, therefore, following the logic and procedure used for $\theta < 1$, we can obtain a complete characterization of the optimal decisions for a centralized supply chain when $\theta > 1$. The results are summarized in Table 5.2(b).

3.2 Channel Design for the Centralized Supply Chain

Based on the results obtained in the previous section, Figure 5.3 illustrates the best channel strategy for the centralized supply chain on the $(\theta, c_d/c_r)$ plane.

For an integrated firm, the optimal behavior of the firm as θ increases for the cases $\theta < 1$ and $\theta > 1$ is identical. So our discussion here will only focus on the case $\theta > 1$. For low values of θ $(1 < \theta \leq 1 - (c_r - c_d))$, only the retail channel is optimal; and the price, quantity and profits of the retail store do not change with θ. In this range, the higher willingness to pay for the direct channel, $\theta - 1$, does not cover the higher price of operating the direct channel, $(c_d - c_r)$.

In the middle range of θ $(1 - (c_r - c_d) \leq \theta \leq c_d/c_r)$, both channels operate with positive sales and profits. The price in the retail channel stays constant while the direct channel price keeps increasing in θ. With dual channels, an increase in θ shifts some of the sales from the retail

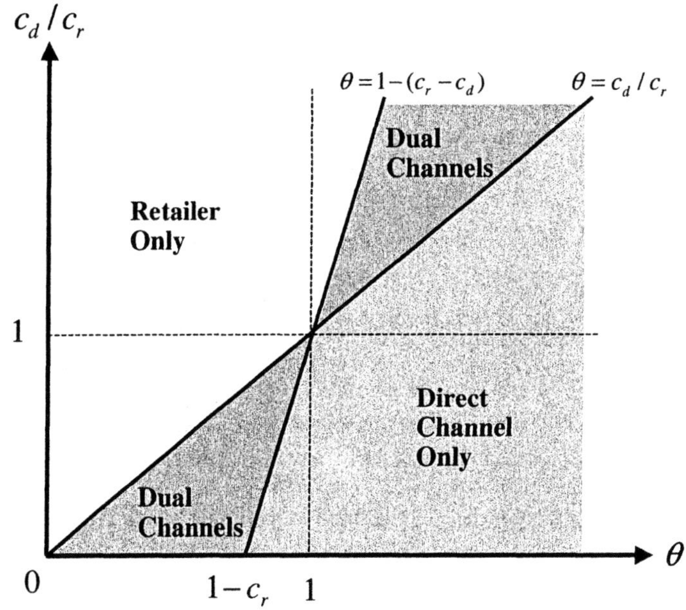

Figure 5.3. Channel Design for the Centralized Supply Chain.

store to the direct channel. The popularity of the direct channel helps in attracting an increasing number of customers to it, but these are not necessarily new customers since the total sales volume in the two channels remains constant. The optimal retail price does not vary as θ changes and this causes a sharp drop in the portion of the total profit derived by the integrated firm through the retail channel. However, the sale price in the direct channel continues to increase as demand keeps going up and the total profit of the integrated firm increases at an increasing rate.

Once θ increases beyond the threshold c_d/c_r, the sales in the retail store drop to zero and only the direct channel is used. Sales in the direct channel continue to increase with higher θ. Because of customers' greater willingness to pay, an integrated firm is able to raise prices and increase its profit at an increasing rate with θ.

4. Price-Setting Game

We have discussed the optimal channel strategy in the supply chain that is centrally managed. Would the same strategy apply if the supply chain consists of different agents who act in their own best interest? In this section, we discuss the channel design problem in the situation where the manufacturer and the retailer are independent decision-makers and each seeks its own welfare when setting prices, ignoring the collective impact of the prices on the channel as a whole. We construct a Stackelberg game to examine the interaction between the manufacturer and the retailer with the following sequence of moves. In the first stage, the manufacturer decides whether to engage in direct sales, and act as Stackelberg leader in setting the wholesale price w, and the direct channel price p_d (if the direct channel is open). In the second stage, the retailer, as a follower, chooses the retail price p_r to maximize its profit given the manufacturer's decision.

To find the equilibrium of the Stackelberg game, we start with the second stage retailer's decision followed by the first stage manufacturer's decisions.

4.1 Best Response of the Retailer

Since the retailer is the follower in the Stackelberg game, the optimal retail price hinges on the manufacturer's price decisions. Note that to keep the retailer from buying through the direct channel or other arbitrators with a lower price, the wholesale price should not be higher than the direct channel price; that is $w \leq p_d$. After the direct channel price and the wholesale price are determined by the manufacturer, the only decision variable that the retailer has control over to maximize its profit is the retail price p_r. Specifically, given p_d and w, the retailer's profit is a function of the retail price p_r :

$$\pi_r(p_r) = (p_r - w)Q_r, \tag{5.12}$$

where Q_r is the demand function given in equation (5.1). The retailer must take into account the piecewise linear demand, Q_r, when deciding the optimal retail price p_r^*.

We first focus on the case when $\theta < 1$. If $p_d/\theta < p_r$, the demand in the retail store would be $Q_r = 1-(p_r-p_d)/(1-\theta)$ (the top line in equation (5.1) or the upper line segment $\overline{AB}$ in Figure 5.2(a)). Substituting in equation (5.12), it is easy to verify that $p_r^* = (1-\theta+p_d+w)/2$. Explicitly, by backward substitution, this retail price is optimal only when the manufacturer sets the direct channel price p_d and the wholesale price w in the price region $R_1^{\theta<1}$, where

$$R_1^{\theta<1} = \{(p_d,\ w)|\frac{1-\theta+p_d+w}{2} \geq \frac{p_d}{\theta},\ w \leq p_d\}. \tag{5.13}$$

If $\theta < 1$ and $p_d/\theta > p_r$, the optimal retail price would be $p_r^* = (1+w)/2$ since the demand for the retailer would be $Q_r = 1-p_r$ (the second line

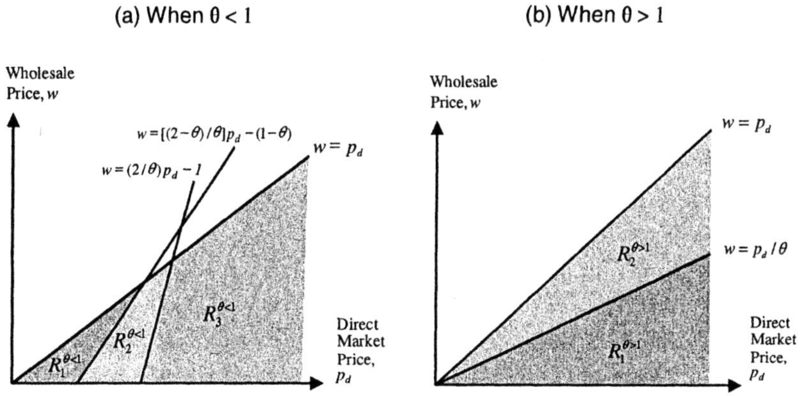

Figure 5.4. Feasible Regions for p_d and w.

in equation (5.1) or the line segment $\overline{BC}$ in Figure 5.2(a)). This optimal retail price is only valid in region $R_3^{\theta<1}$, where

$$R_3^{\theta<1} = \{(p_d,\ w)|\frac{1+w}{2} \leq \frac{p_d}{\theta},\ w \leq p_d\}. \tag{5.14}$$

If the manufacturer sets $(p_d,\ w)$ in neither region $R_1^{\theta<1}$ nor region $R_3^{\theta<1}$, then the optimal retail price is located at the kink point B in Figure 5.2(a) in the demand where the retail price is $p_r^* = p_d/\theta$. This is in response to the prices in region $R_2^{\theta<1}$, where

$$R_2^{\theta<1} = \{(p_d,\ w)|\frac{1-\theta+p_d+w}{2} \leq \frac{p_d}{\theta},\ \frac{1+w}{2} \geq \frac{p_d}{\theta},\ w \leq p_d\}. \tag{5.15}$$

When $\theta > 1$, we can show that if the manufacturer sets (p_d, w) in region $R_1^{\theta>1}$, where

$$R_1^{\theta>1} = \{(p_d,\ w)|0 \leq w \leq \frac{p_d}{\theta}\}, \tag{5.16}$$

the retail demand would be $Q_r = (p_d - \theta p_r)/(\theta - 1)$ (the bottom line in equation (5.1) or the line segment $\overline{AB}$ in Figure 5.2(c)), and therefore, the optimal retail price would be $p_r^* = (p_d/\theta + w)/2$. Finally, it is optimal for the retailer to have $p_r^* = p_d/\theta$ so that there is no demand in the retail store (the third line in equation (5.1) or the point A in Figure 5.2(c)) if the manufacturer sets (p_d, w) in region $R_2^{\theta>1}$, where

$$R_2^{\theta>1} = \{(p_d,\ w)|w \geq \frac{p_d}{\theta},\ 0 \leq w \leq p_d\}. \tag{5.17}$$

The price regions $R_i^{\theta<1}, i = 1, 2, 3$, and $R_j^{\theta>1}$, $j = 1, 2$, are illustrated in Figure 5.4. We summarize the retailer's best pricing strategy in the following theorem.

THEOREM 5.1 (BEST RESPONSE OF THE RETAILER) *Given the manufacturer's decision of wholesale price w and direct market price p_d, the optimal price for the retailer is,*

$$p_r^* = \begin{cases} \frac{1-\theta+p_d+w}{2} & \text{if } (p_d,\ w) \in R_1^{\theta<1} \\ \frac{p_d}{\theta} & \text{if } (p_d,\ w) \in R_2^{\theta<1} \\ \frac{1+w}{2} & \text{if } (p_d,\ w) \in R_3^{\theta<1} \\ \frac{p_d/\theta+w}{2} & \text{if } (p_d,\ w) \in R_1^{\theta>1} \\ \frac{p_d}{\theta} & \text{if } (p_d,\ w) \in R_2^{\theta>1} \end{cases} \tag{5.18}$$

With the retail prices determined by equation (5.18), we shift our attention to the manufacturer's pricing problem.

4.2 Channel Equilibrium

The subgame perfect equilibrium of the Stackelberg pricing game corresponds to the solution to the manufacturer's profit maximization problem given below:

$$\underset{p_d,\ w \geq 0}{\text{maximize}}\ \pi_m(p_d,\ w) = (w - c_r)Q_r + (p_d - c_d)Q_d, \tag{5.19}$$

$$\text{subject to} \quad p_r = \text{argmax}\ \pi_r(p_r), \text{ and} \tag{5.20}$$

$$w \leq p_d, \tag{5.21}$$

where Q_r and Q_d are the demand functions given in equations (5.1) and (5.2). Specifically, taking the retailer's pricing behavior into account (equation (5.20)), the manufacturer maximizes its total profit by choosing the wholesale price w and the direct market price p_d subject to constraint (5.21) which assures that the manufacturer cannot charge a higher wholesale price than direct price because the retailer would costlessly switch its purchases to the direct channel and refuse to pay the higher wholesale price. Note that the retailer's best response to the retail price in (5.20) can be replaced by the result given in (5.18).

4.2.1 When Consumers Prefer the Retail Channel to the Direct Channel. When customer acceptance level of the direct channel θ is smaller than 1, that is, when consumers prefer the retail channel to the direct channel, solving the manufacturer's profit maximization problem yields the equilibrium prices of the price-setting game

(a) Outcomes of Price-Setting Game When $\theta < 1$			(b) Comparative Statics	
			$\hat{\theta} < \theta < 1$	
	$0 < \theta < \hat{\theta}$	$\hat{\theta} < \theta < 1$	Derivative w.r.t. θ	Sign
Price				
Wholesale, w :	$\frac{1+c_r}{2}$	$\frac{\theta+c_r}{2}$	$\frac{1}{2}$	$+$
Direct Channel, p_d :	$\frac{1+c_r}{2}$	$\frac{\theta+c_r}{2}$	$\frac{1}{2}$	$+$
Retail Store, p_r :	$\frac{3+c_r}{4}$	$\frac{\theta+c_r}{2\theta}$	$-\frac{c_r}{2\theta^2}$	$-$
Sales Volume				
Direct Channel, Q_d :	0	0	0	0
Retail Store, Q_r :	$\frac{1-c_r}{4}$	$\frac{\theta-c_r}{2\theta}$	$\frac{c_r}{2\theta^2}$	$+$
Total, $Q_d + Q_r$:	$\frac{1-c_r}{4}$	$\frac{\theta-c_r}{2\theta}$	$\frac{c_r}{2\theta^2}$	$+$
Profit				
Manufacturer, π_m :	$\frac{(1-c_r)^2}{8}$	$\frac{(\theta-c_r)^2}{4\theta}$	$\frac{\theta^2-c_r^2}{4\theta^2}$	$+$
Retail Store, π_r :	$\frac{(1-c_r)^2}{16}$	$\frac{(1-\theta)(\theta^2-c_r^2)}{4\theta^2}$	$\frac{-\theta^3+(2-\theta)c_r^2}{4\theta^3}$	$-$
Total, $\pi_m + \pi_r$:	$\frac{3(1-c_r)^2}{16}$	$\frac{(\theta-c_r)(\theta-2\theta c_r+c_r)}{4\theta^2}$	$\frac{(1-\theta)c_r^2}{2\theta^3}$	$+$

$\theta =$ customer acceptance of direct channel

$\hat{\theta} =$ cannibalistic threshold, $\hat{\theta} = \frac{(1+c_r)^2+(1-c_r)\sqrt{1+6c_r+c_r^2}}{4}$

$c_r =$ marginal cost incurred by the manufacturer for the product sold through retailer

Table 5.3. Price-Setting Game When $\theta < 1$.

considered in CCH. We summarize the results as well as the comparative statics in the following theorem and in Table 5.3.

THEOREM 5.2 (EQUILIBRIUM PRICES WHEN $\theta < 1$) *When consumers prefer the retail stores more than the direct channels ($\theta < 1$), the equilibrium prices of the stackelberg game are as follows:*

$$(p_d^*, w^*, p_r^*) = \begin{cases} (\frac{1+c_r}{2}, \frac{1+c_r}{2}, \frac{3+c_r}{4}) & \text{if } \theta \leq \hat{\theta} \\ (\frac{\theta+c_r}{2}, \frac{\theta+c_r}{2}, \frac{\theta+c_r}{2\theta}) & \text{if } \theta \geq \hat{\theta} \end{cases}, \tag{5.22}$$

where

$$\widehat{\theta} = \frac{(1+c_r)^2 + (1-c_r)\sqrt{1+6c_r+c_r^2}}{4}. \tag{5.23}$$

Proof See CCH.

Interestingly, as is evident from Table 5.3, when $\theta < 1$, no matter how well the direct channel is accepted by consumers, it is most profitable for the manufacturer to arrange prices so that nothing is ever sold in its own direct channel. When θ is not high enough ($\theta \leq \hat{\theta}$), adding a direct channel to the market will not affect the equilibrium prices[7]: the retailer can effectively ignore the potential cannibalization of customers by the direct channel. On the other hand, when θ is high enough ($\theta \geq \hat{\theta}$), to prevent its demand from being cannibalized by the direct channel, the retailer will lower the price to compete with the direct channel. Under the price competition, it is difficult for the manufacturer to drive traffic to the direct channel. Although no sales occur in the direct channel, the manufacturer's profit increases in θ because the direct channel helps to partially resolve *double marginalization*: the direct channel induces the retailer to lower its price, which in turn spurs demand in the retail channel. While operated by the manufacturer to constrain the retailer's pricing behavior, a direct channel may not always be detrimental to the retailer because it will be accompanied by a wholesale price reduction. In equilibrium, the combination of manufacturer pull and push can possibly increase the retailer's profit.

4.2.2 When Consumers Prefer the Direct Channel to the Retail Channel. When customers prefer the direct channel to the retail channel, solving the channel equilibrium problem given in (5.19)-(5.21) yields the following wholesale, retail, and direct channel prices.

THEOREM 5.3 (EQUILIBRIUM PRICES WHEN $\theta > 1$) *When consumers prefer the retail stores more than the direct channels ($\theta > 1$), the equilibrium prices of the stackelberg game are as follows:*

$$(p_d^*, w^*, p_r^*) = \begin{cases} \left(\frac{\theta + c_d}{2}, \frac{1 + c_r}{2}, \frac{2\theta + c_d + \theta c_r}{4\theta}\right) & \text{if } \theta \leq c_d/c_r \\ \left(\frac{\theta + c_d}{2}, \frac{\theta + c_d}{2}, \frac{c_d + \theta}{2\theta}\right) & \text{if } \theta \geq c_d/c_r \end{cases} \tag{5.24}$$

Proof See Appendix.

The related outcomes as well as the comparative statics when $\theta > 1$ are provided in Table 5.4. When $\theta > c_d/c_r$, the customers prefer the direct channel so much that it is simply not profitable for the retailer to be in business. All sales occur only in the direct channel and the solution is the same as the integrated firm operating only the direct channel.

[7]The equilibrium prices are exactly the same as those from the traditional *double marginalization* problem.

(a) Outcomes of Price-Setting Game When $\theta > 1$

	$1 < \theta < c_d/c_r$	$\theta > c_d/c_r$
Price		
Wholesale, w :	$\frac{1+c_r}{2}$	$\frac{\theta+c_d}{2}$
Direct Channel, p_d :	$\frac{\theta+c_d}{2}$	$\frac{\theta+c_d}{2}$
Retail Store, p_r :	$\frac{2\theta+c_d+\theta c_r}{4\theta}$	$\frac{c_d+\theta}{2\theta}$
Sales Volume		
Direct Channel, Q_d :	$\frac{2\theta^2-2\theta-2\theta c_d+c_d+\theta c_r}{4\theta(\theta-1)}$	$\frac{\theta-c_d}{2\theta}$
Retail Store, Q_r :	$\frac{c_d-\theta c_r}{4(\theta-1)}$	0
Total, $Q_d + Q_r$:	$\frac{2\theta-\theta c_r-c_d}{4\theta}$	$\frac{\theta-c_d}{2\theta}$
Profit		
Manufacturer, π_m :	$\frac{2\theta^3+\theta^2(c_r^2-4c_d-2)-2\theta c_d(c_r-c_d-2)-c_d^2}{8\theta(\theta-1)}$	$\frac{(\theta-c_d)^2}{4\theta}$
Retail Store, π_r :	$\frac{(c_d-\theta c_r)^2}{16\theta(\theta-1)}$	0
Total, $\pi_m + \pi_r$:	$\frac{3c_d^2+\theta^2c_r^2+4\theta^3-4\theta^2-2\theta c_d+2\theta^2c_r-4c_d^2\theta}{16\theta(\theta-1)}$	$\frac{(\theta-c_d)^2}{4\theta}$

(b) Comparative Statics

	$1 < \theta < c_d/c_r$		$\theta > c_d/c_r$	
	Derivative w.r.t. θ	Sign	Derivative w.r.t. θ	Sign
Price				
Wholesale, w :	0	0	$\frac{1}{2}$	+
Direct Channel, p_d :	$\frac{1}{2}$	+	$\frac{1}{2}$	+
Retail Store, p_r :	$\frac{-c_d}{4\theta^2}$	−	$-\frac{c_d}{2\theta^2}$	−
Sales Volume				
Direct Channel, Q_d :	$\frac{\theta^2(c_d-c_r)+(\theta-1)^2c_d}{4\theta(\theta-1)}$	+	$\frac{c_d}{2\theta^2}$	+
Retail Store, Q_r :	$-\frac{c_d-c_r}{4(\theta-1)^2}$	−	0	0
Total, $Q_d + Q_r$:	$\frac{c_d}{4\theta^2}$	+	$\frac{c_d}{2\theta^2}$	+
Profit				
Manufacturer, π_m :	$\frac{\theta^2(c_d+1)(c_d-c_r)+(\theta-1)^2\left(2\theta^2+c_d^2\right)}{8\theta(\theta-1)}$	+	$\frac{\theta^2-c_d^2}{4\theta^2}$	+
Retail Store, π_r :	$-\frac{(c_d-\theta c_r)(2c_d\theta-\theta c_r-c_d)}{16\theta^2(\theta-1)^2}$	−	0	0
Total, $\pi_m + \pi_r$:	$\frac{4\theta^2(\theta-1)^2+2\theta^2(c_d-c_r)}{16\theta^2(\theta-1)^2}+\frac{(c_d^2-\theta^2c_r^2)+2c_d^2(2\theta-1)(\theta-1)}{16\theta^2(\theta-1)^2}$	+	$\frac{\theta^2-c_d^2}{4\theta^2}$	+

$\theta =$ customer acceptance level of the direct channel

$c_d =$ marginal cost incurred by the manufacturer for the product sold through the direct channel

$c_r =$ marginal cost incurred by the manufacturer for the product sold through retailer

Table 5.4. Price-Setting Game When $\theta > 1$ and $c_d > c_r$.

When $1 < \theta < c_d/c_r$, both the direct channel and the retailer get non-zero demand. Compared to the integrated case, the retailer's price is higher, hurting its demand and profit. In fact, the retailer's demand is exactly half of what it would be if the channels were integrated. The manufacturer charges the same price in the direct channel as it does in the integrated case. Some of the customers switch to the manufacturer but others do not participate, resulting in a net drop in the total demand. The manufacturer makes higher profit from the direct channel, although the total profit of the two channels decreases due to double marginalization.

In the range $1 < \theta < c_d/c_r$, as θ increases, the retailer is forced to reduce its price while the manufacturer increases its price. But the retailer still cannot hold on to all its customers; the direct channel demand as well as the profit of the manufacturer increase. The customer's affinity for the direct channel, captured in higher θ, explains the curious phenomenon of both higher prices and higher demand. Unlike the case of $\theta < 1$, the retailer is never better off with the introduction of direct channel.

The value of c_d plays no role in the determination of prices, demand or profits when $\theta < 1$. Its role when $\theta > 1$ is quite prominent as Table 5.4 shows. As c_d increases, the manufacturer has to increase it direct channel price. This also prompts the retailer to increase its price but not as much ($\frac{\partial p_r}{\partial c_d} = \frac{1}{4\theta} < \frac{\partial p_d}{\partial c_d} = \frac{1}{2}$). As a result, direct channel demand drops but the retailer's demand increases. This increase does not offset the loss of profit from the direct channel for the manufacturer who suffers a net drop in profit with higher c_d. The retailer, on the other hand, benefits slightly from the manufacturer's cost inefficiency of selling direct.

4.3 Channel Design for the Decentralized Supply Chain

Recall that in the first stage of the price-setting game, the manufacturer has to decide whether to add a direct channel alongside an established retail channel. In other words, we implicitly assume that the manufacturer is contractually committed to retail distribution for the price-setting game. Without this assumption, there are three types of channel distribution strategies: retail-only distribution, direct-only distribution, and dual-channel distribution. What is the best channel strategy for the manufacturer when the supply chain is decentralized?

Let π_m^R, π_m^D, and π_m^B be the manufacturer's profit when retail-only, direct-only, and dual-channel distribution strategies, respectively, are applied. Recall that from 4.2.1, when $\theta \leq \hat{\theta}$, adding the direct channel

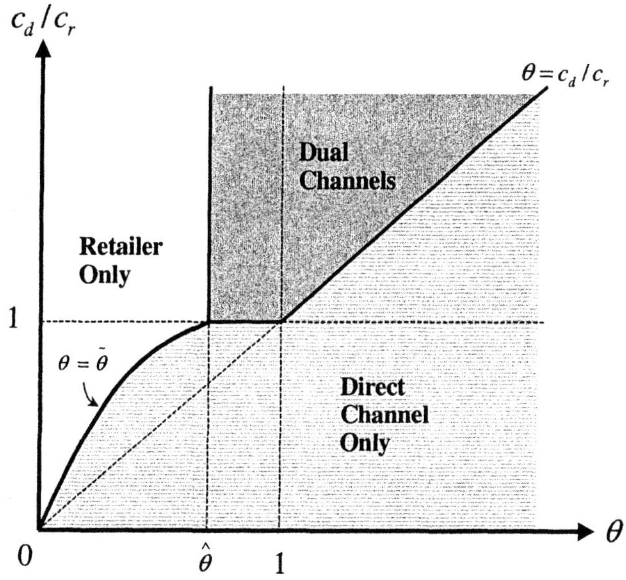

Figure 5.5. Channel Design for the Decentralized Supply Chain. $\widehat{\theta}=\frac{(1+c_r)^2+(1-c_r)\sqrt{1+6c_r+c_r^2}}{4}$ and $\widetilde{\theta}=\frac{(1+c_r)^2+(1-c_r)\sqrt{1+6c_r+c_r^2+8(c_d-c_r)}+4(c_d-c_r)}{4}$.

alongside the retail store will not affect the equilibrium prices; therefore, dual-channel distribution is not favorable for the manufacturer since it is a dominated strategy ($\pi_m^B \leq \pi_m^R$ and $\pi_m^B \leq \pi_m^D$). To find the best channel strategy for the manufacturer when $\theta \leq \hat{\theta}$, we only need to compare the following two profits: $\pi_m^R = \frac{(1-c_r)^2}{8}$ and $\pi_m^D = \frac{(\theta-c_d)^2}{4\theta}$. It is easy to show that $\pi_m^D \leq \pi_m^R$ if $\theta \leq \widetilde{\theta}$, where

$$\widetilde{\theta} = \frac{(1+c_r)^2 + (1-c_r)\sqrt{1+6c_r+c_r^2+8(c_d-c_r)} + 4(c_d-c_r)}{4}. \tag{5.25}$$

The result is plotted on the (θ, c_d/c_r) plane as illustrated in Figure 5.5. Not surprisingly, a direct-only distribution strategy will be preferred when the direct channel is logistically more efficient ($c_d \leq c_r$) and the customer acceptance level of the direct channel is high enough ($\theta \geq \widetilde{\theta}$).

When $\hat{\theta} \leq \theta < 1$, retail-only distribution strategy is dominated, and the manufacturer will use either direct-only or dual-channel distribution strategies. Clearly, when the retail channel is logistically more efficient ($c_r \leq c_d$), a dual-channel distribution strategy will be more profitable ($\pi_m^B = \frac{(\theta-c_r)^2}{4\theta} \leq \frac{(\theta-c_d)^2}{4\theta} = \pi_m^D$).

Channel Strategies for the Integrated and Non-Integrated Supply Chains

	$\theta < 1$ Direct channel has *lower* customer acceptance than the traditional retail store		$\theta > 1$ Direct channel has *higher* customer acceptance than the traditional retail store	
	Possible Channel Strategies		*Possible Channel Strategies*	
	Integrated	**Non-integrated**	**Integrated**	**Non-integrated**
$c_d/c_r > 1$ Direct channel is logistically *less efficient*	✓ Retailer Only	✓ Retailer Only ✓ Dual Channels	✓ Retailer Only ✓ Direct Only ✓ Dual Channels	✓ Direct Only ✓ Dual Channels
$c_d/c_r < 1$ Direct channel is logistically *more efficient*	✓ Retailer Only ✓ Direct Only ✓ Dual Channels	✓ Retailer Only ✓ Direct Only	✓ Direct Only	

Table 5.5. Channel Strategies for Integrated and Non-Integrated Supply Chains.

When $\theta > 1$, the best channel strategy conforms to the results discussed in Section 4.2.2. As shown in Figure 5.5, although it is disadvantageous to the retailer when $\theta > 1$, if the unit cost of distributing the product to customers through the direct channel is too high ($c_d/c_r \geq \theta \geq 1$), the manufacturer will still keep the retailer and use both channels. On the other hand, if $\theta \geq c_d/c_r$, the manufacturer will use its own direct channel to distribute the product and the retail channel will be disintermediated.

Direct channel is the dominant strategy whenever its cost is better than the preference adjusted retail cost (i.e. $\theta c_r > c_d$). Even without this cost advantage, a direct channel is more profitable for certain values of $\theta < 1$. On the other hand, the retailer-only strategy has advantages in more limited conditions, requiring low retail cost and low customer acceptance for the direct channel. Dual channels become more profitable when both the cost of operating the direct channel and customer acceptance of the direct channel are high.

How would the channel strategies of an integrated and a non-integrated supply chain differ under the same scenario (same values of θ, c_d, and c_r)? The answer follows from the comparison of figures 5.3 and 5.4, which is summarized in Table 5.5. If the direct channel is logistically more efficient, the non-integrated supply chain would never use a dual-channel distribution strategy but the integrated supply chain might for some values of $\theta < 1$.

5. Conclusion

Direct distribution facilitated by the Internet has led more manufacturers to adopt a multi-channel distribution strategy by using both integrated and non-integrated channels simultaneously. Is it always the best strategy to adopt a multi-channel distribution strategy? In this chapter, we investigate the impact of the interplay between customers' channel preference and distribution costs on the supply chain channel design for a manufacturer that can sell through a retailer and directly to consumers. Our analysis extends the problem considered by Chiang, Chhajed and Hess (2003), where they assume that customers prefer the traditional bricks-and-mortar retail store to the direct channel, which may not always be true. We relax this assumption and provide a more generic model that helps to generate comprehensive insights and implications for the supply channel design problem. By comparing the profitability of three types of channel distribution strategies (retail-only distribution, dual-channel distribution, and direct-only distribution) under different scenarios, we disclose the optimal supply-chain channel design from the manufacturer's perspective. Analytical results are presented for both centralized and decentralized supply chains.

It should be evident that a deeper understanding of the role of customer acceptance level of buying a product through different channels and the role of supply chain costs is critical for multi-channel management. Our purpose has been to highlight the main issues, but in the process we have omitted some important details. For example, the role of inventory could be understood by explicitly modeling demand uncertainty. Also, the single-period nature of the framework and the assumption of a bilateral monopoly environment have limitations. Clearly, studies seeking to tackle these issues would be valuable.

Appendix: Proof of Theorem 5.3

When $\theta > 1$, the subgame perfect equilibrium of the Stackelberg game corresponds to the solution of the manufacturer's profit optimization problem that involves two cases be considered.

Case 1: Optimal Solution in Region $R_1^{\theta>1}$

In Case 1, the manufacturer sets (p_d, w) in region $R_1^{\theta>1}$ so that the retailer chooses the retail price $p_r = (p_d/\theta + w)/2$. The quadratic programming problem in this region can be specified as:

$$\begin{aligned} \underset{p_d,\, w\geq 0}{\text{maximize}} \quad & \pi_m^{R_1^{\theta>1}} = (w - c_r)\frac{p_d - \theta p_r}{\theta - 1} + (p_d - c_d)(1 - \frac{p_d - p_r}{\theta - 1}), \\ \text{subject to} \quad & p_r = (p_d/\theta + w)/2, \text{ and} \\ & w \leq p_d \end{aligned} \qquad \text{(A1)}$$

If $\theta < c_d/c_r$, then the optimal solution is in the interior of region $R_1^{\theta>1}$

$$(p_d^*, w^*) = (\frac{c_d + \theta}{2}, \frac{c_r + 1}{2}).$$

The corresponding retail price is

$$p_r^* = \frac{p_d^*/\theta + w^*}{2} = \frac{2\theta + c_d + \theta c_r}{4\theta}.$$

If $\theta > c_d/c_r$, then (A1) binds and the solution is dominated by the optimal solution in region $R_2^{\theta>1}$.

Case 2: Optimal Solution in Region $R_2^{\theta>1}$

The quadratic programming problem in Case 2 can is given by:

$$\begin{aligned} \underset{p_d,\, w\geq 0}{\text{maximize}} \quad & \pi_m^{R_2^{\theta>1}} = (p_d - c_d)(1 - \frac{p_d}{\theta}), \\ \text{subject to} \quad & w \geq p_d/\theta \text{ and } w \leq p_d \end{aligned}$$

The optimal solution is:

$$(p_d^*, w^*) = (\frac{c_d + \theta}{2}, \frac{c_d + \theta}{2}).$$

In this case, it is not profitable to have any demand in the retail store. Thus, the corresponding retail price is

$$p_r^* = \frac{p_d^*}{\theta} = \frac{c_d + \theta}{2\theta}.$$

References

Bell, D.R., Y. Wang, V. Padmanabhan. 2002. An Explanation for Partial Forward Integration: Why Manufacturers Become Marketers. Working Paper, The Wharton School, University of Pennsylvania.

Carrie J., J. Walker. 2003. Q3 2003 Online Sales: Surprisingly Strong Growth. *Forrester Research*, October 22, 2003.

Chiang, W.K., D. Chhajed, J.D. Hess. 2003. Direct Marketing, Indirect Profits: A Strategic Analysis of Dual-Channel Supply Chain Design. *Management Science*, **49**(1) 1-20.

Chiang, W.K. 2004. Competitive and Cooperative Multi-Channel Inventory Policies in a Two-Echelon Supply Chain. Working Paper, Department of Information Systems, University of Maryland, Baltimore County. Baltimore, MD.

Chiang, W.K., G.E. Monahan. 2005. "Managing Inventories in a Two-Echelon Dual-Channel Supply Chain", *European Journal of Operational Research*, **162**(2) 325-341.

Hendershott, T., J. Zhang. 2001. A Model of Direct and Intermediated Sales. Working Paper, University of California at Berkeley and University of Rochester.

Kacen, J., J. Hess, W.K. Chiang. 2003. Shoppers' Attitudes Toward Online and Traditional Grocery Stores. Working Paper, University of Houston and University of Maryland, Baltimore County.

Kotler, P. 1997. Marketing Management: Analysis, Planning, Implementation, and Control. Englewood Cliffs, N.J.: Prentice-Hall.

Kumar, N., R. Ruan. 2002. On Strategic Pricing and Complementing the Retail Channel with a Direct Internet Channel, Working Paper, University of Texas at Dallas.

Liang, T., J. Huang. 1998. An Empirical Study on Consumer Acceptance of Products in Electronic Markets: A Transaction Cost Model. *Decision Support Systems.* **24** 29-43.

Peleg, B., H.L. Lee. 2002. Secondary Markets for Product Diversion with Potential Manufacturer's Intervention, Working Paper, Department of Management Science and Engineering, Stanford University.

Preston, L.E., A.E. Schramm. 1965. Dual Distribution and Its Impact on Marketing Organization. *California Management Review.* **8**(2) 59-69.

Spengler, J. 1950. Vertical Restraints and Antitrust Policy. *Journal of Political Economy.* **58**(4) 347-352.

Rhee, B. 2001. A Hybrid Channel System in Competition with Net-Only Direct Marketers. Working Paper, The Hong Kong University of Science & Technology.

Rhee, B., S. Park. 2000. Online Store as a New Direct Channel and Emerging Hybrid Channel System. Working Paper, The Hong Kong University of Science & Technology.

Tsay, A., N. Agrawal. 2003. Channel Conflict and Coordination in the eBusiness Era. Forthcoming, Production & Operations Management.

Tsay, A., N. Agrawal, 2004. Modeling Conflict and Coordination in Multi-Channel Distribution Systems: A Review. Forthcoming, Supply Chain Analysis in the eBusiness Era (International Series in Operations Research and Management Science), D. Simchi-Levi, D. Wu, and Z.-J. Shen (Eds.), Kluwer Academic Publishers.

Yao D.Q., Liu J.J. 2002. Channel Redistribution with Direct-Selling. *European Journal of Operational Research.* **144** 646-658 .

Chapter 6

USING SHAPLEY VALUE TO ALLOCATE SAVINGS IN A SUPPLY CHAIN

John J. Bartholdi, III
School of Industrial and Systems Engineering
Georgia Institute of Technology
765 Ferst Drive, Atlanta, GA 30332-0205

Eda Kemahlıoğlu-Ziya
Kenan-Flagler Business School
University of North Carolina at Chapel Hill
CB# 3490, Chapel Hill, NC

Abstract Consider two retailers, whose inventory is provided by a common supplier who bears all the inventory risk. We model the relationship among the retailers and supplier as a single-period cooperative game in which the players can form inventory-pooling coalitions. Using the Shapley value to allocate the profit, we analyze various schemes by which the supplier might pool inventory she holds for the retailers. We find, among other things, that the Shapley value allocations are individually rational and are guaranteed to coordinate the supply chain; but they may be perceived as unfair in that the retailers' allocations can, in some situations, exceed their contribution to supply chain profit. Finally we analyze the effects of demand variance and asymmetric service level requirements on the allocations.

1. An Inventory Centralization Model

Consider an electronics manufacturing services provider (EMS), who keeps inventory of cpu chips for two or more competing original equipment manufacturers (OEM). The current inventory policy dictated by the OEMs is to keep each company's inventory physically separated. Is this the most profitable inventory policy for the EMS? Furthermore, is the most profitable inventory policy for the EMS also the most profitable for her customers?

In general we are interested in knowing whether a supplier should pool inventory held for her customers (the retailers). If so, what will be the benefits and how should they be shared over the supply chain? Will a customer (retailer) who requires a higher level of service be indirectly subsidizing a competitor who would accept a lower level of service? We explore such questions in the following 2-echelon supply chain using a single-period model.

Consider two retailers selling a single product procured form a single, common supplier. Even though there may be more suppliers providing the same product in the larger supply chain, we consider a situation where the retailers already chose to work with a particular supplier. For example companies in the electronics industry prefer to have a sole supplier for each product whenever possible (Barnes et al. (2000)). The retailers face uncertain demand and do not carry inventory. When they observe demand, they place an order at the supplier and receive shipments without significant delay. Ownership passes from the supplier to a retailer after the retailer places the order and pays for the product and so the supplier bears all the inventory risk. Sales are lost to the retailers in case of a stock-out at the supplier. (There is no backlogging.) To service the retailers, the supplier either keeps inventory reserved for each of her customers or else pools inventory to share among all of her customers.

Inventory-pooling is known to reduce costs and so increases profits for the supply chain party that owns the inventory, in this case, the supplier (Eppen (1979)). However, the retailers may object to inventory-pooling because of two concerns. First is the concern of how inventory will be allocated among the retailers when there are shortages. With reserved inventory, the retailer can control his risk of stock-out by specifying minimum-inventory levels to be held by the supplier. But if the retailers draw on a common, pooled inventory, which of the competing retailers has priority when requesting the last of the inventory? Any inventory-pooling contract will need to address this issue either directly (by specifying a stock-rationing mechanism) or indirectly (by specifying reservation profits to the parties such that their profits are at least as much as their before-pooling profits).

The second concern is how much information should be shared in the supply chain to facilitate inventory-pooling. In the case of reserved inventories, each company shares demand information only with the supplier. However, in the case of inventory pooling, a company can, by observing his own service level, infer something about the demand faced by the competitor with whom he is sharing inventory.

In this paper, we first consider supply chain members with varying degrees of power, where we take power to be the ability to dictate a strategy of pooling or no pooling. We show that the supply-chain-optimal inventory level cannot be attained under powerful retailers who preclude pooling or a powerful supplier who pools inventory to maximize her profits. Furthermore, retailers may lose profits (compared to the case without inventory pooling) when the supplier pools inventory subject to the retailers' service constraints. We conclude that the frequently used service measure, probability of no stock-out, does not induce supply-chain-optimal inventory levels in the system.

Instead we propose a *value-sharing method* based on Shapley value from cooperative game theory and derive closed-form expressions of the Shapley values. We find that the Shapley value induces coordination and the allocations under this mechanism satisfy individual rationality conditions for all players and belong to the core of the game. Though stable, an allocation based on Shapley value may induce envy among some players. In particular, we find that the allocation mechanism may be interpreted as "unfair" by some players. We show that the mechanism favors retailers in the sense that retailer allocations may exceed their contribution to total supply chain profit at the expense of the supplier.

Under the proposed contract, the retailers prefer to form pooling coalitions with retailers with either very high or very low service requirements. Up to a threshold service level a retailer prefers to be the one requesting the higher service level because it ensures him the greater share of total profits. Beyond the threshold level a coalition partner with very high service requirements forces the supplier to overstock, increasing sales for both of the retailers. We also show that when the supplier has the power to maximize her profits by manipulating the service levels she provides for the retailers, the retailer with lower demand variance has a better chance of increasing his profits. The Shapley value scheme rewards the retailer introducing less risk into the supply chain and one can reasonably argue that this is "fair".

In the next section, we survey related literature and position our model. In Section 3 we analyze the supply chain profit and its distribution among parties of varying degrees of power. We then introduce the Shapley value profit allocation mechanism in Section 4 and explore the Shapley value allocations and their properties in Section 5. In Section 6, we discuss the possible instabilities that may be caused by the Shapley value allocation scheme. Finally, in Section 7, we analyze the question "With whom to form a coalition" from the (different) perspectives of a retailer and the supplier given the service level constraints of each of the retailers. We conclude with a discussion of our findings and future research directions.

2. Literature Review

Most of the cost models analyzed up to now are extensions of the classical news vendor problem, for which Porteus (1990) provides a review. The literature on inventory pooling (also known as risk pooling) can be classified under three headings.

- Component commonality
- Inventory rationing/transshipment in single echelon supply chains
- Inventory and risk pooling in multi-echelon supply chains

Component Commonality

If end products share common components, safety stock can be reduced and service levels maintained by pooling inventory of common parts. The work-to-date on component commonality concentrates merely on changes in safety stock levels and does not consider the benefits of pooling to different members of the supply chain nor how they should be shared. Baker, Magazine, and Nuttle (1986) consider a two product system with service level constraints and where the objective is to minimize total safety stock. They show that total safety stock (common and specialized) drops after pooling; however total stock of specialized parts increases. Gerchak, Magazine, and Gamble (1988) extend these results to a profit maximization setting. Finally, Gerchak and Henig (1986) extend these models to a multi-period setting and show that myopic policies are optimal for the infinite horizon models.

Inventory Rationing in Single Echelon Supply Chains

Inventory rationing defines the rules of how to allocate total inventory to n different members of the same echelon of a supply chain in case of a shortage (shortage for all members or shortage for some and overage for others). This can either be done through transshipments among supply chain members carrying decentralized inventory or by defining rules to allocate inventory when it is centralized at a single location. This approach is different from our work in that it concentrates on one of the echelons only.

One question regarding centralized inventories that has received attention in the literature is whether total inventory level in the supply chain decreases after pooling. Gerchak and Mossman (1992), Pasternack and Drezner (1991), and Yang and Schrage (2002) show that, contrary to intuition, this is not always the case. These papers present inventory increase as an undesired outcome of pooling. We show that increasing inventory may be beneficial for the supply chain as a whole because it also increases sales. In addition, we show that if the service level constraints are binding, inventory will not increase due to pooling. Conversely, Tagaras (1989) looks at a two retailer model and shows that if the total reserved safety stock for the two retailers is pooled and used to replenish both of the retailers from a central location, service levels at both of the retailers will increase.

One stream of papers analyzes the inventory-sharing problem as a transshipment problem among different players at the same echelon, possibly with positive transshipment costs. These papers are more closely related to our work in that they consider decentralized systems, but they differ from our work in that they concentrate on different players within the same echelon. Anupindi, Bassok, and Zemel (2001) analyze the problem in a cooperative game theoretic framework. They propose a modified duality-based allocation mechanism that achieves the profit level of the centralized system. Granot and Sosic (2002) extend their work by relaxing an assumption on the amount of residual inventory available for transshipments among the retailers. Rudi, Kapur, and Pyke (2001) analyze a similar problem with only two retailers. Instead of fixing the transshipment prices like Anupindi et al. do, they let the transshipment prices be variable and try to come up with prices that would coordinate the supply chain.

In addition to allocation of parts in case of shortages, allocation of costs to supply chain members is an important issue in centralized inventory systems. Gerchak and Gupta (1991) analyze this question for a system with an EOQ-based inventory policy and argue that allocating costs with respect to volume of demand or contribution to total cost may result in unacceptable cost allocations for some parties. They propose an allocation mechanism that allocates costs based on stand-alone costs. In his note on Gerchak and Gupta's paper, Robinson (1993) proposes the concept of core as a possible fair cost allocation scheme and provides a numerical example. Hartman and Dror (1986) and Hartman and Dror (2003a) also discuss core allocations and, in the former paper, compare several cost allocation methods (one of which is Shapley value) on a numerical example. However, Robinson (1993), Hartman and Dror (1986), and Hartman and Dror (2003a) do not analyze the operational properties of the proposed allocation mechanisms.

Inventory Pooling in Multi-Echelon Supply Chains

Of the existing literature, the work that is closest to our work is that of Anupindi and Bassok (1999). They consider a two level supply chain with a single manufacturer and two retailers. Unlike our model, the inventory decision is made by the retailers without constraining service levels and the retailers bear all the inventory risk. They model a system where only a fraction of the customers are willing to wait for a delivery from another retailer. They show that under this setting, the manufacturer may not always benefit from inventory pooling because total sales may drop. They discuss the possibility of optimizing wholesale prices or introducing holding cost subsidies as methods for coordinating the supply chain. Dong and Rudi (2002) extend the model of Anupindi, Bassok, and Zemel (2001) to a two echelon supply chain. Similar to our objective, they explore whether transshipments, which are beneficial for the retailers, are also beneficial for the upstream manufacturer. However, in their model the manufacturer does not hold inventory and the retailers make the transshipment decisions.

As in our work, Netessine and Rudi (2001) consider a model where the supplier bears all the inventory risk. Although they also consider a two-echelon system, the second echelon consists of a single retailer. In their model, the retailer is merely an intermediary between the end customer and the supplier and functions only to expand the customer base through marketing effort. The authors conjecture that the risk-pooling effect that will be observed in the case of multiple retailers will make this kind of business model even more profitable. However, we will show that a supplier who carries out inventory pooling in order to maximize her own profit may actually reduce the total supply chain profit.

Finally, Plambeck and Taylor (2003) consider capacity rather than inventory pooling. They consider a two-stage model where the first stage is a competitive game on capacity investment and the second stage is the cooperative stage where the firms pool inventory and determine the division of profit. The second stage of their model is similar to ours in that a cooperative game ensues from the capacity pooling interactions but different from ours in the profit-allocation rule used.

This paper may also be considered to lie within the literature on supply chain coordinating contracts, of which the chapter by Cachon (2002) provides an excellent review (see especially the second section). A recent paper by Raghunathan (2003) is relevant to this paper in terms of the methodology employed. Raghunathan also utilizes Shapley value as an allocation mechanism, but the subject of his paper is information sharing rather than inventory pooling.

3. Inventory Pooling: Definitions and Preliminary Results

Consider a supply chain with a single supplier and two retailers as in Figure 6.1. The retailers require a minimum service level from the supplier and the service level is defined as the probability of no stock-out. How the retailers' minimum service level requirements are set is exogenous to our model. For example the electronics-industry standard is that the supplier carries a minimum of two weeks' inventory for each customer (Barnes et al. (2000)). In industries where such standards exist, the minimum service level can be defined as one corresponding to this standard. Even when the supplier and each retailer rather negotiate on the service level, we only model the interactions that take place after the service levels are decided on. The minimum service level information is shared only with the supplier and since the service levels are set exogenous to our model, we assume the retailers cannot provide false information to gain advantages.

Each retailer observes local demand, places an order with the supplier, pays a per-unit-price, and receives the inventory immediately (zero lead-time). The supplier manufactures or buys the product and holds it in inventory at her expense until an order is placed from the retailer(s). The objective of each is to maximize her single period profits. Retailer profit only depends on expected sales since the retailers do not hold inventory.

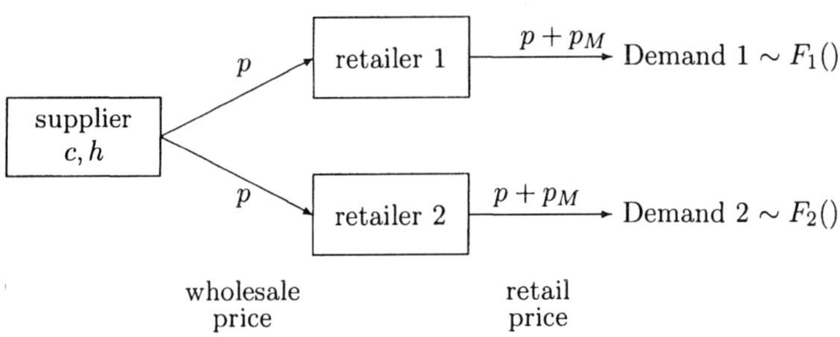

Figure 6.1. Sample 2-echelon supply chain and relevant cost and revenue parameters

Let p be the wholesale price the supplier charges to the retailers, c be the procurement/ manufacturing cost per unit, h be the holding cost per unit (or we can think of h as the disposal cost), and p_M be the markup on wholesale price the retailers charge. We assume that the cost and revenue parameters are common knowledge to all of the supply chain players. End customer demand is independent at the retailers and we assume the probability distributions of the demand functions are known. Let $F_i(\cdot)$ denote the cumulative demand distribution for retailer i ($i = 1, 2$). We assume that $F_i(\cdot)$ is strictly increasing and differentiable (with pdf $f_i(\cdot)$ over the interval $[0, \beta)$ where $\beta = \inf\{y : F(y) = 1\}$ (β can be ∞)) and has a finite mean.

We look at the inventory holding problem among the supplier and the two retailers in two different perspectives: the supplier holds *reserved* inventory separately for both of the players or inventory at the supplier is pooled and is *shareable* by the retailers. The total supply chain profit and its allocation among supply chain partners depend on who owns the supplier and the retailers and who makes the pooling decision. We consider the following scenarios:

- When powerful retailers forbid pooling
- When a powerful supplier pools inventory
- When a centralized supply chain makes globally optimal pooling decisions
- When a weak supplier pools inventory subject to a service contract

3.1 Powerful Retailers: No Inventory Pooling

In this scenario the retailers are powerful enough to prevent inventory pooling at the supplier. Retailers may insist on a reserved-inventory policy if the product in question is scarce (like Intel chips) and there is ambiguity about how the scarce product would be allocated or if they fear they may be underwriting the service level of a competitor. The objective of the supplier is the maximization of expected profit, which is defined as expected revenue less the expected holding (or disposal) cost and the procurement (or manufacturing) cost subject to the service level constraints. Let x_i be the stock level kept for retailer i, S_i be the expected sales at retailer i, and H_i be excess stock in retailer i's stock. For each retailer, the supplier sets inventory levels to maximize profit by solving the problem as stated in Expression 6.1.

$$\begin{aligned} &\max\ p\ S_i - h\ H_i - c\ x_i \\ &\text{s.t. } F_i(x_i) \geq \underline{\rho}_i \end{aligned} \tag{6.1}$$

where $\underline{\rho}_i$ = minimum acceptable probability of no-stockout for retailer i or "service level".

Since the retailers do not hold inventory, their expected profit is equivalent to markup times expected sales. Each retailer's expected profit is as given in Expression 6.2.

$$p_M S_i \tag{6.2}$$

Expression 6.1, without the service level constraint, is the news vendor problem (Silver, Pyke, and Peterson (1998)). It is well-known that the profit-maximizing stocking level for the supplier facing demand with distribution $F(\cdot)$ is $F^{-1}(\frac{p-c}{p+h})$. The optimal stocking level corresponds to a service level of $(\frac{p-c}{p+h})$, which we call the critical ratio. The critical ratio corresponds to the probability of no stock-out, also known as Type-1 service measure. In this paper, unless otherwise specified, service level always denotes Type-1 service level.

The optimal stocking level is $F^{-1}\left(\max\left(\underline{\rho}, \frac{p-c}{p+h}\right)\right)$ when service level constraints are present and the total stock supplier must hold is given by $\sum_i F_i^{-1}\left(\max\left(\underline{\rho}_i, \frac{p-c}{p+h}\right)\right)$. This means that if the required service level is higher than the critical ratio then the inventory level is found such that the service constraint is binding. Service level is an increasing function of inventory and expected profit is a concave function of inventory. Therefore, whenever the required service level is higher than the critical ratio, the supplier ends up with less than optimum profit. If the service level requirements of the retailers are in the range $\left(0, \frac{p-c}{p+h}\right)$ then it is optimal for the supplier to provide higher than required service. However, beyond $\frac{p-c}{p+h}$, the supplier loses money if she provides higher service to the retailers.

Examining the structure of the optimal decision, one may observe the following:

- When the profit margin of the supplier $(p-c)$ is small or when the holding cost h is large relative to the price p, it is costlier for the supplier to provide higher-than-required service to the retailers. Therefore, utilizing the "optimum" method of pooling becomes more important.

- Like service level, expected sales is an increasing function of total stock level. Therefore, in the region, $\rho \in (0, \frac{p-c}{p+h})$, the retailers' expected sales are greater than or equal to what their service level guarantees them. Beyond $\frac{p-c}{p+h}$, however, they get exactly what they ask for because higher stock levels are not optimal for the supplier.

3.2 Inventory Pooling by a Powerful Supplier

When the supplier pools inventory to be shared by the two retailers, she effectively makes the inventory decision based on the cumulative demand $F_c(\cdot) = F_1(\cdot) * F_2(\cdot)$. Let S_c be expected cumulative sales, H_c be the expected cumulative excess stock, and x_c be the stock level. The supplier's problem is

$$\max pS_c - hH_c - cx_c \tag{6.3}$$

which has the same news vendor structure as the no-pooling case. The optimum stock level the supplier will carry is $F_c^{-1}(\frac{p-c}{p+h})$. Under this scenario, the supplier sets the optimum stock level disregarding any service level requirements the retailers may have.

3.3 Centralized Supply Chain Makes Pooling Decision

If both the retailers and the supplier were owned by the same company, the resulting centralized problem would be

$$\max (p_M + p)S_c - hH_c - cx_c \tag{6.4}$$

The centralized system revenue on each unit sold is $p + p_M$. Expression 6.4 has the form of a news vendor problem and so the optimal stock level is $F_c^{-1}(\frac{p+p_M-c}{p+p_M+h})$. The following observation relates the total stock in the centralized system to the total stock in the decentralized system where the supplier decides on the size of pooled inventory.

Observation 3.1 *In a decentralized system, the supplier always stocks less than the system-optimum stock level.*

Comparison of the critical ratio for the centralized system, $\frac{p+p_M-c}{p+p_M+h}$, with the critical ratio for the supplier, $\frac{p-c}{p+h}$, yields that $\frac{p+p_M-c}{p+p_M+h} > \frac{p-c}{p+h}$, which is equivalent to Observation 3.1. This is not surprising since it is the supplier who incurs the procurement and holding costs and thus has incentive to understock. This observation also indicates that the decentralized system will not reach its total sales capacity. On the other hand if the stock level is set to that of the centralized system under coordination, the supplier will profit less than she would in a decentralized system, where she can set inventory levels optimally.

Another important point is that $F_c^{-1}\left(\frac{p+p_M-c}{p+p_M+h}\right)$ maximizes total supply chain profit profit but may not satisfy the service level requirements for the retailers. This means that enforcement of service level requirements may decrease total system profit. We explore this observation in the next section.

3.4 Weak Supplier, Weak Retailers: Inventory Pooling Subject to Service Constraints

Consider a supply chain where the supplier is too weak to make the pooling decision by herself and the retailers are too weak to preclude pooling. Instead, the retailers allow the supplier to pool inventory subject to the service level constraints they set.

Because of competition, retailers may be willing to share some but not all inventory. Thus we may consider the total stock to be broken up into four partitions. The supplier holds two types of inventory for each retailer: shareable and reserved. Shareable inventory may be used to satisfy the other retailer's demand once the demand of the primary inventory owner is satisfied; whereas reserved inventory cannot. For example, if the stock kept for retailer 1 runs out and there is stock available only in the reserved section of the inventory for retailer 2 then this cannot be used to satisfy the unsatisfied demand of retailer 1. Let us define the notation:

$$x_i^d \;=\; \text{amount of reserved stock for retailer } i$$
$$x_i^s \;=\; \text{amount of shareable stock for retailer } i$$

Total expected sales after pooling and total expected left-over inventory are simply the sum of the individual expected sales and expected left-over inventory figures. The problem of maximizing total profit may be formalized as

$$\begin{aligned}
&\max\, pS_c - hH_c - c(x_1^d + x_2^d + x_1^s + x_2^s)\\
&\text{subject to}\\
&\rho_i \geq \underline{\rho}_i,\; i \in \{1,2\}
\end{aligned}$$

When cost structures are symmetric and no extra incentives/costs exist regarding inventory sharing, we make the following observation.

Observation 3.2 *To maximize total expected profit, one need never hold reserved inventory.*

This result is easy to see since the supplier's profit when $x_1^d = x_2^d = 0$ is at least as much as her profit when $x_1^d > 0$ and $x_2^d > 0$. A model similar to our 4-partition model allows only a fraction, $f \leq 1$, of a retailer's demand to be met at another retailer (or in our case using his stocks). This restriction may be due to transshipment delays or a fraction of customers not willing to wait. This differs from our model in that if the extra demand at retailer i is large enough, regardless of how small f is, the spill-over demand can deplete all extra inventory at retailer j with positive probability. In our model, if $x_i^d > 0$ then whether it would be depleted or not depends only on the magnitude of demand at retailer i.

We drop the superscript notation differentiating between reserved and shareable inventory because by Observation 3.2 reserved inventory is zero in an optimal solution. Let D_i and D_j be the random variables representing the demand at retailers i and j respectively. Under this complete pooling scheme, the probability of no stock-out at retailer i is

$$\rho_i = P(D_i \leq x_i) + P(x_i \leq D_i \leq x_i + x_j - D_j) \tag{6.5}$$

In the remainder of this section, we concentrate on calculating stocking levels after pooling. We first analyze the supplier's problem and ignore the effects on the retailers. It is known that expected profit increases due to pooling. We would also expect total stock level to decrease. However, Gerchak and Mossman (1992) give a simple example in which total inventory level after pooling is *higher* than the total inventory level before pooling.

When stocking levels increase, the expected service level provided to the retailers and their expected sales also increase. If the required service level exceeds the critical ratio, the supplier loses money by providing a higher service level. Therefore it is important to calculate the stock levels so that the service level constraints are binding whenever the service level requirements exceed the critical ratio. When we calculate stock levels in this way, we can show that stock levels after pooling do not exceed those before pooling as formalized in Lemma 3.1.

	$\rho_2 \leq \rho_2^l$	$\rho_2^l < \rho_2 \leq \rho_2^u$	$\rho_1 > \rho_2^u$
$\rho_1 \leq \rho_1^l$	x_c^*	x_c^*	$F_2(x_2^*) = \rho_2, x_1^* = 0$
$\rho_1^l < \rho_1 \leq \rho_1^u$	x_c^*	Requires analysis (1)	Requires analysis (2)
$\rho_1 > \rho_1^u$	$F_1(x_1^*) = \rho_1, x_2^* = 0$	Requires analysis (2)	Solve service level equations

Table 6.1. Optimum pooled inventory level depending on service levels ρ_1 and ρ_2

Lemma 3.1 *The after-pooling stock level does not exceed the total before-pooling stock level if the probability of no-stockout after pooling is equal to the probability of no-stockout before pooling for each retailer.*

Proof See Appendix B for all proofs.

3.4.1 Supplier-Optimal Pooled Stock Size.

To avoid excessive inventory costs, the supplier should provide no more than the contracted service level when service level requirements are higher than the critical ratio. We make use of this fact to characterize the optimal solution for the supplier in case of pooling subject to service constraints. The characterization also determines the sizes of x_1 and x_2, shareable stock over which retailers 1 and 2 have priority respectively after pooling.

Define $x_c^* = F_c^{-1}(\frac{p-c}{p+h})$, the optimum pooled inventory in the absence of service level constraints. Even though we assume complete sharing of available stock by the two retailers, we still distinguish the levels, x_1 and x_2, over which retailers 1 and 2 have priority in case of a stockout, because these levels determine the respective service levels observed at the retailers. By letting $x_1 = x_c^*$ or $x_2 = x_c^*$, we can obtain the boundary values on service level at the two retailers. Further define for $i, j \in \{1, 2\}$

$$\rho_i^l = \text{service level at retailer } i \text{ when } x_i = 0 \text{ and } x_j = x_c^*$$
$$\rho_i^u = \text{service level at retailer } i \text{ when } x_i = x_c^* \text{ and } x_j = 0$$

With respect to these boundary values, the required service level pair (ρ_1, ρ_2) will fall in one of the nine regions depicted in Table 6.1. For three of the nine combinations, x_c^* is also a feasible total stocking level given the service level requirements. For the two cases, in which one requirement is below its corresponding lower bound and the other is above its corresponding upper bound, the optimal stocking level is found by solving the service level constraint for the higher service level and setting the other stocking level to zero. In this case, the retailer with the lower service level has no stock over which he has priority. The stock kept for the retailer with the higher service level is used to cover the other retailer. This situation, although optimal for the supplier, may create a conflict of interest between the retailers and therefore may be unacceptable because the retailer with the higher service level requirement is underwriting the service level of the other retailer. For the case where both service level requirements exceed their corresponding upper bounds, the stocking level is found by solving both of the service level constraints as equalities. The solution is optimal because it provides the least stock to satisfy both of the equations.

If the service level pair falls in the region marked by (1), the situation is more complicated: If $F_c^{-1}(\frac{p-c}{p+h})$ can be partitioned such that both of the constraints are satisfied then it is obviously the optimal stock level. This can be checked simply by finding the partition that would still satisfy the service level constraint at the retailer with the higher requirement and then verifying whether the same partition satisfies the service level constraint of the other retailer. If so, $F_c^{-1}(\frac{p-c}{p+h})$ is the optimal stocking level. If not, the second step is to set $x_i^* = 0$ where i is the retailer with the lower service level requirement and find x_j^* that satisfies retailer j's service level constraint. If x_j^* also satisfies retailer i's constraint then it is optimal, as established in Lemma 3.2. Otherwise, one needs to solve for x_i^* and x_j^* by setting the two service level constraints as equalities. Clearly, providing more service (higher stock levels) is suboptimal.

Lemma 3.2 *When $F_c^{-1}(\frac{p-c}{p+h})$ is not feasible, $x_c^* = x_j^* = F_j^{-1}(\rho_j)$, where j is the retailer with the higher service level, is optimal when it is feasible.*

If the the service level pair falls in the region marked by (2) then first find $x_j^* = F_j^{-1}(\rho_j)$ where j again is the retailer requiring the higher service level. If x_j^* is also feasible for retailer i then $x_j^* = x_c^*$ is the optimal stock level. If not, one needs to solve for x_i^* and x_j^* by setting the two service level constraints as equalities as in the case of (1).

3.4.2 Retailer Profits under Pooling. Retailer profits may decrease due to pooling because the total inventory in the supply chain decreases. This phenomenon was first observed by Anupindi and Bassok (1999) in a different setting, where the retailers pay the holding cost and it is their decision whether to pool inventory or not. We show by example that this loss cannot be prevented even with the introduction of Type-1 service measure constraints.

Example 3.1 *Consider a system with two retailers. Let both demand distributions be $U(0,1)$ and the critical ratio be $\frac{p-c}{p+h} = 0.9$. Then before pooling, the optimal stocking levels are $x_1 = x_2 = 0.9$ with total expected sales at 0.99. The before-pooling service levels at the retailers are each 0.9. The stock level corresponding to $F_c^{-1}(\frac{p-c}{p+h})$ is 1.55279. Using equation 6.5 and letting $x_1 = x_2 = 1.55279/2$, this stock level corresponds to a service level of approximately 0.92 at each of the retailers, which means service level constraints are more than satisfied. However, the total expected sales is 0.985. Therefore, the expected profits of the retailers drops even though the service level constraints are satisfied.*

Thus a simple contract between the retailers and the supplier, where the retailers only enforce their expected service levels, is not adequate to protect the retailers from losing sales when the supplier has the power to pool inventory.

In Appendix A we briefly discuss another service measure that can guarantee profits for the retailers; but one not frequently used because it is hard to measure. Like most other researchers we use the easier measure of service, probability of no-stockout; but we compensate to some extent for its deficiencies by proposing a profit allocation mechanism that ensures expected profits of all parties involved in the contract remain at before-pooling levels.

4. Coalitions in Cooperative Games and Shapley Value

We analyze the inventory pooling problem among the retailers and the supplier as a cooperative game, which allows for the possibility of coalitions among players. Coalitions are possible because players are assumed to negotiate effectively with each other (Myerson (1991)). Let $N = 1, 2, \ldots, n$ be the set of players. For each coalition, $J \subseteq N$, of supply chain partners let the value of the coalition $v(J)$ be the total expected profit of coalition J. For each coalition J, $v(J)$ consists of two parts: the total expected profit of the retailers and the supplier in the coalition and the total profit the supplier earns due to the retailers who are not in the coalition. By definition, $v(\emptyset) = 0$. We use the subscript notation to represent the elements of set J; that is if $J = \{1, 2, S\}$, $v(J) = v_{12S}$ denotes the expected profit of a coalition consisting of retailers 1 and 2 and the supplier, denoted by S.

An allocation ϕ is a vector, where each ϕ_i is the payoff to player i. Given that N represents the grand coalition, an allocation ϕ is said to be in the core of v if and only if

$$\begin{aligned} \textstyle\sum_{i \in N} \phi_i &= v(N) \\ \textstyle\sum_{i \in J} \phi_i &\geq v(J), \forall J \subseteq N \end{aligned}$$

If an allocation is not in the core there is incentive for some players to leave the coalition. A core solution is desirable because it is stable; but the core of a cooperative game may be empty. In addition, even when the core exists, an allocation in the core may have other undesirable characteristics. For example, it may be extreme and/or sensitive to system parameters (Myerson (1991), page 429) or may fail to satisfy coalitional monotonicity (Granot and Sosic (2002)). In general, it is hard to determine whether the core of a coalitional game exists or not. Even when it does, the more important question is whether the suggested value allocation scheme is actually in the core. While such issues can be important, we avoid them as unpromising in this context. Instead, we follow Shapley (1953) in representing the expected payoff to player i, $\phi_i(v)$, as the unique solution to the following axioms. For the second axiom, a *carrier* of v is any set $U \subseteq N$ with $v(S) = V(U \cap S)$, $\forall S \subseteq N$.

- **Symmetry** For all permutations $\Pi(N)$ of N, $\phi_{\pi i}(\pi v) = \phi_i(v)$ for each permutation π in $\Pi(N)$.
- **Efficiency** For each carrier U of v $\sum_U \phi_i(v) = v(U)$.
- **Law of aggregation** $\phi_i(v + w) = \phi_i(v) + \phi_i(w)$.

These axioms are meaningful and practical in terms of our problem. We would expect players of equal power to receive the same allocation and the first axiom ensures that the Shapley value allocation only depends on the contribution of the player to the coalitions. The second axiom makes sure that the Shapley value allocation mechanism allots the total worth of the coalition to the players and a player who is not in the carrier receives zero allocation. Again, in our context we would expect any reasonable allocation mechanism to exhaustively distribute the total profit of the system to the players and to assign zero value to a player who does not increase the value of a coalition. Finally, if the players play two different games with value functions v and w, then the total Shapley value allocation to player i is the same as if the players were to play a game with value function $v + w$. This axiom shows that Shapley value allocations are not dependent on the time of bargaining between the players.

The Shapley value as stated in Expression 6.6 may be interpreted as the expected marginal contribution of player i to a coalition. In Expression 6.6, the term $(v(J \cup \{i\}) - v(J))$ is the marginal contribution of player i to coalition J. We can interpret the fractional term as follows. There are $|N|!$ different ways all the players are ordered to enter the grand coalition and $| J |!(| N | - | J | -1)!$ different ways all the players in J enter the grand coalition before player i does. Assuming all orderings are equally likely, $\frac{|J|!(|N|-|J|-1)!}{|N|!}$ is the probability a coalition J is already formed before i enters the coalition (for a more detailed interpretation see Myerson (1991)).

$$\phi_i(v) = \sum_{J \subseteq N \setminus \{i\}} \frac{| J |!(| N | - | J | -1)!}{| N |!}(v(J \cup \{i\}) - v(J)) \tag{6.6}$$

In the inventory centralization context, coalitions are formed when a subset of players agree to pool inventory. We propose a value-sharing mechanism where each player's after-pooling profit allocation is equal to his Shapley value.

5. Shapley Value Allocations for Two-Retailer Games

For two retailers and one supplier, the value of the coalition increases only when all three players agree to inventory pooling. Therefore, the value of a 2-player coalition is the sum of the individual expected profits of the players before pooling. This simplifies the calculation of the Shapley value for player i ($i \in \{1, 2, S\}$) to

$$\phi_i(v) = \frac{2}{3}\, v_i + \frac{1}{3}\left(v_{12S} - \sum_{j \in \{1,2,S\}, j \neq i} v_j\right) \tag{6.7}$$

where v_{12S} is the value of the coalition when all three players agree to pooling and v_1,v_2, and v_S are the individual expected profits of the players before pooling. Equation 6.7 tells us that in the Shapley value allocation, for each player i, the weight of his contribution to the coalition is half the weight of his before-coalition payoff. The Shapley value formalizes the rule for the allocation of total profit to the three players. However to fully characterize the value-sharing mechanism we also need to define a rule for calculating the individual expected profits of the players without pooling. Without pooling, the supply chain has the structure described in Section 3.1. If the retailers do not agree to pooling under the Shapley value allocation rule, they will be reserved a stock level of $F_i^{-1}\left(\max\left(\underline{\rho}_i, \frac{p-c}{p+h}\right)\right)$. Therefore v_1, v_2, v_S are calculated with respect to the stock levels set at $F_i^{-1}\left(\max\left(\underline{\rho}_i, \frac{p-c}{p+h}\right)\right)$ for each retailer.

Writing Expression 6.7 in a different way, we obtain the equivalent expression

$$\phi_i(v) = v_i + \frac{1}{3}\left(v_{12S} - v_1 - v_2 - v_S\right) \tag{6.8}$$

which shows that for two retailers, the three players share the extra revenue due to pooling equally. Each player's expected payoff is his expected payoff before pooling plus one third of the increase in total expected system profit due to pooling.

We next establish some stability properties of the Shapley value allocations.

Theorem 5.1 *The Shapley value allocation scheme induces coordination of the supply chain.*

An allocation for player i is *individually rational* if it is at least as much as what the player would get if he had not participated in the coalition, that is $\phi_i(v) \geq v(\{i\})$.

Proposition 5.1 *The Shapley value allocations for the inventory holding game are individually rational for all of the players.*

The next proposition shows that the Shapley value allocations are in the core of the game and thus establishes that the core of the game is non-empty.

Proposition 5.2 *The Shapley value allocations are in the core of the inventory holding game.*

Thus when the Shapley value is used as the profit allocation scheme in a 2-retailer supply chain, the retailers and the supplier have incentive to form pooling coalitions. In addition, the resulting coalition is stable (in the core) and the total joint profit is the maximum the supply chain can attain.

6. Second-Order Instabilities

That the profit allocations under Shapley value allocation scheme are individually rational and in the core may not be adequate to prevent what we call *second-order instabilities.* These kinds of instabilities may arise if one or more of the players believe there is asymmetric, unfair profit allocation to some other player(s). In cooperative game theory, it is assumed that players would not be willing to deviate from coalitions if individual rationality constraints are satisfied and the allocations are in the core. However, players may hesitate to form coalitions if they believe their competitor benefits more than he should from the coalition. They may require further adjustments to the coalition contract, for example in the form of side payments.

In the remainder of this paper we use the BP and AP notation in the superscript to differentiate the values each variable (such as inventory level, expected sales) takes before pooling and after pooling respectively.

6.1 Shapley Value Allocations Favor Retailers

Retailer profit is the product of sales by the mark-up per item and so we define *effective sales* at retailer i as the Shapley value allocation to retailer i divided by the unit mark-up, and

$$\text{E[effective sales at retailer } i] = \frac{\phi_i}{p_M}$$

Comparing total expected effective sales by total expected actual sales after pooling, we can determine whether the retailers get more than their contribution to total after-pooling profit, in which case the supplier gets less than her contribution. More specifically, we are interested in knowing when the following inequality occurs:

$$\text{E[total effective sales]} = \frac{\phi_1 + \phi_2}{p_M} > \text{E[total sales after pooling]} \quad (6.9)$$

Theorem 6.1 *Total retailer allocations are greater than actual retailer contribution to after-pooling profit if and only if the expected change in supplier profit exceeds the expected change in average retailer profit.*

In other words, when the change in expected profit for the supplier after pooling is greater than the average change for the retailers, the supplier is forced to give up a portion of her extra profits to the retailers, the size of which is determined by the Shapley value calculations.

Even when Expression 6.9 holds, it is possible that only one of the retailers benefits from the extra allocation:

Example 6.1 *Consider two retailers with iid* $U(0,1)$ *demand. Service level is set at* 0.9 *by retailer 1 and at 0.65 by retailer 2. Let* $p = 4$, $p_M = 4$, $c = 2$, *and* $h = 0.1$. *The ex-post profit allocations are:* $\phi_1 = 2.367776$ *and* $\phi_2 = 1.911776$. *E[total sales after pooling]* $= 0.980813$ *and E[total effective sales] is* $(2.367776 + 1.911776)/4 = 1.069888$. *Comparing the two,* $1.069888 > 0.980813$ *implies that the retailers' total allocation is greater than their total expected profit. In addition, the effective sales for retailer 2 is* $1.911776/4 = 0.477944$. *However,* $0.980813 - 0.477944 > 0.5$, *which implies his effective sales is less than his expected sales (because expected sales at retailer 1 cannot exceed 0.5). Therefore retailer 2's allocation under Shapley value scheme is less than his expected sales revenue after pooling.*

In this example both retailer 2 and the supplier get allocations less than their individual contributions to total after pooling profit, while retailer 1 gets a higher allocation. In this example, this is a fair allocation because retailer 1 requests a higher service level before pooling. Retailer 2, by forming a pooling coalition with retailer 1, gains access to a larger stock but has to to give up some of his profits to retailer 1.

Proposition 6.1 *Given E[total effective sales]* $\geq$ *E[total sales after pooling], if the change in expected sales at retailer* i *is greater than or* *equal to the change in expected sales at retailer j then E[effective sales at retailer j] $\geq$ E[sales at retailer j after pooling].*

Proposition 6.1 says that the expected change in retailer i's sales after pooling is greater than the change in retailer j's sales ensures that retailer j's final profit allocation will correspond to an effective sales level higher than his expected sales. However the same condition is not adequate to ensure the same for retailer i. This result is counterintuitive because we would normally expect retailer i would be ensured a greater portion of the extra profit due to pooling since he is making the more positive impact on expected sales.

The Shapley value allocation rule, since it is in the core, guarantees that none of the supply chain players can be better off by breaking away from the coalition. However, while one player may be only infinitesimally better off when compared to the no-pooling scenario, another player may receive a significantly high allocation, an allocation that is more than that player's contribution to total supply chain profit. This inequitable distribution of savings is in the core and so is stable in a technical sense. But many people would find it well within the range of human behavior for the player receiving the lower allocation to refrain from pooling and forgo his minuscule extra profits. This illustrates a weakness of the concept of "core".

7. With Whom to Form a Coalition?

In the previous section, we have shown that even though the Shapley value allocation scheme ensures profit allocations higher than before-pooling profit levels for all players, some players may get more favorable allocations. Therefore it is important for all players to know with whom it is most advantageous to form pooling coalitions. In this section, we analyze this question from the points of view of the retailers and the supplier separately. We take required service level and the demand distribution as the defining characteristics of the retailers. Cost and revenue parameters are still assumed to be identical for both of the retailers.

7.1 The Retailer's Perspective

The question we seek to answer is: "Given a fixed service level for retailer i, at what service level for retailer j would retailer i form a coalition with retailer j?" Throughout this section we make use of the following rule in the contract: before-pooling profit levels, v_i, v_j, and v_S are calculated with respect to the stock levels set at $F_i^{-1}\left(\max\left(\underline{\rho}_i, \frac{p-c}{p+h}\right)\right)$ for each retailer. Therefore, our region of interest is $\rho_j \in \left(\frac{p-c}{p+h}, 1\right)$ because in the region $\left(0, \frac{p-c}{p+h}\right]$ the stock level is set at $F_j^{-1}(\frac{p-c}{p+h})$ regardless of the service level requirement. When the stock level for retailer j is fixed at $F_j^{-1}(\frac{p-c}{p+h})$, the service level requirement of retailer j does not have an impact on the ex-post profit allocation to retailer i. The following theorem establishes that the profit allocation to one retailer is unimodal in the service level requirement of the other retailer.

Theorem 7.1 *The Shapley value profit allocation to retailer i is a unimodal function of service level ρ_j of retailer j. In addition, $\rho_j^* = \frac{p+p_M-c}{p+p_M+h}$ is the global minimizer of the payoff to retailer i.*

In all examples we studied, $\phi_i(\rho_j)$ has always been a convex function. However, we could not prove this in general because $\frac{\partial^2 \phi_i(\rho_j)}{\partial \rho_j^2}$ is a function of $\frac{\partial^2 F_j^{-1}(\rho_j)}{\partial \rho_j^2}$, which is difficult to sign. However, proving unimodality is sufficient for our purposes because the interesting point in this theorem is that the ex-post profit allocation to a retailer decreases if he forms a coalition with a retailer with service level in the range $(\frac{p-c}{p+h}, \frac{p+p_M-c}{p+p_M+h})$.

The next natural question is whether there is a threshold service level ρ_j in the region $(\frac{p+p_M-c}{p+p_M+h}, 1)$ beyond which $\phi_i(\rho_j)$ is greater than $\phi_i(\frac{p-c}{p+h})$. The answer is "not necessarily".

Proposition 7.1 *When the demand distribution for retailer j has infinite support, then the ex-post profit allocation for retailer i goes to infinity as ρ_j goes to 1.*

Thus when $F_j(\cdot)$ has infinite support there is a range of ρ_j beyond $\frac{p+p_M-c}{p+p_M+h}$, where $\phi_i(\rho_j)$ is greater than $\phi_i(\frac{p-c}{p+h})$, and retailer i always prefers to form a pooling coalition with a retailer requiring a high service level. However, when $F_j(\cdot)$ has finite support, whether such a region exits or not depends on the system parameters as we demonstrate with the following example.

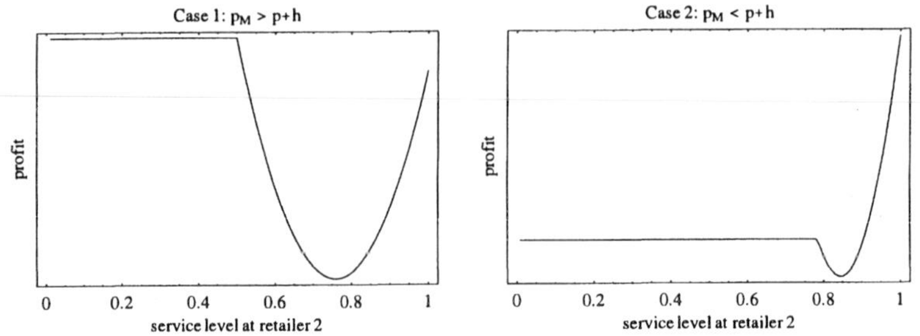

Figure 6.2. Profit allocations to retailer 1 as retailer 2's service changes (graphs not to scale)

Example 7.1 *Let the demand function for retailer 2 be U(0,1). The demand function for retailer 1 is arbitrary but independent from that of retailer 2. Let* $\rho_1 = 0.96, p = 5, c = 2, h = 0.1, p_M = 5.5$.

In Figure 6.2: Case 1, the highest value $\phi_1(\rho_2)$ *attains beyond* $\frac{p+p_M-c}{p+p_M+h}$ *is still lower than* $\phi_1(\frac{p-c}{p+h})$. *However, if we change* p_M *to* 2, *Figure 6.2: Case 2 shows that higher profit allocations are possible for retailer 1 beyond* $\frac{p+p_M-c}{p+p_M+h}$.

If the demand distribution of retailer 2 is U(0,1), $\lim_{\rho_2 \to 1} \phi_1(\rho_2) > \phi_1(\frac{p-c}{p+h})$ when $p_M < p + h$. This condition does not depend on the value of c. We can interpret this result if we consider p_M to be the potential profit to the whole supply chain from the sale of a single item and $p + h$ to be the potential loss to the supplier when an item does not sell. When the potential loss to the supplier is large, she will tend to under-stock and this hurts the retailers. However, when the service level requirement of one or both of the retailers is very high, the supplier will have to stock enough to cover the requirement even if it is suboptimal for herself. Therefore, when the overage cost is very high, it is better for a retailer to form a coalition with a retailer with a high service level requirement since this would force the supplier to stock more.

The retailers share their minimum service level requirements with the supplier and not necessarily with each other. However a retailer may still infer information regarding the service levels at other retailers by observing other properties such as small versus large retailer, small versus large market share. The retailers can differentiate more favorable pooling partners based on this type of prediction of service level requirements.

7.2 The Supplier's Perspective

In section 3.1 we set the contract such that the before-pooling profits are calculated to maximize the before-pooling supplier profit as long as the service level constraints set by the retailers are satisfied. This means that the before-pooling inventory levels are calculated using the equation $F_i^{-1}(\max(\underline{\rho_i}, \frac{p-c}{p+h}))$ for each retailer i. Although this maximizes the supplier profit before pooling and guarantees at least ρ_i level of service for each retailer, this calculation may not maximize the supplier's after-pooling profit according to the Shapley value allocation scheme. The next theorem shows that the Shapley value allocation to the supplier is a unimodal function of the service level requirements of the retailers. Figure 6.3 is an example of how supplier profit changes as the service level requirement of one of the retailers changes.

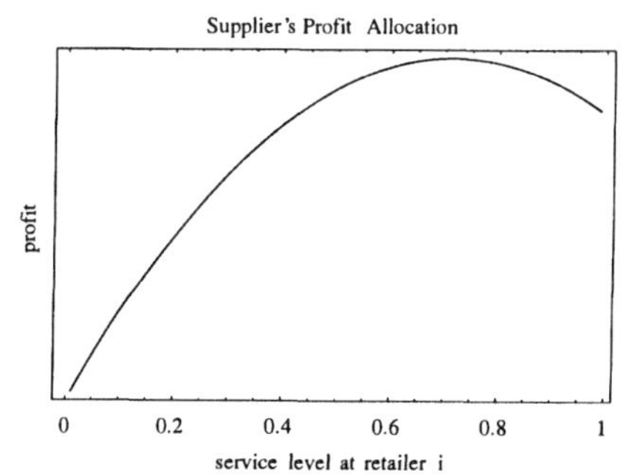

Figure 6.3. Supplier profit allocation as a function of service level

Theorem 7.2 *The Shapley value allocation to the supplier is unimodal in the service level requirements of the retailers and the global maximum occurs at*

i. $(\rho_1, \rho_2) = \left(\max\left(0, \frac{2(p-c)-p_M}{2(p+h)-p_M}\right), \max\left(0, \frac{2(p-c)-p_M}{2(p+h)-p_M}\right)\right)$ *if* $\frac{2(p-c)-p_M}{2(p+h)-p_M} < 1$,

ii. $(\rho_1, \rho_2) = (0, 0)$ *otherwise.*

Theorem 7.2 states that the supplier has incentive to relax the terms of the contract. The current contract calculates before-pooling profits using $x_i = F_i^{-1}(\max(\underline{\rho_i}, \frac{p-c}{p+h}))$ for each retailer. Hence the supplier guarantees each retailer a service level of at least $\frac{p-c}{p+h}$, which is higher than the service level that maximizes her Shapley value allocation. Therefore the supplier prefers a contract that calculates before-pooling profits based on $x_i = F_i^{-1}(\underline{\rho_i})$ — a contract that does not place a lower bound on the service she provides. Then she is allowed to maximize her after-pooling profits by setting the service level at $\left(\frac{2(p-c)-p_M}{2(p+h)-p_M}\right)$ for the retailer(s) requiring a service level that is less than or equal to $\left(\frac{2(p-c)-p_M}{2(p+h)-p_M}\right)$. Unless at least one of the retailers requires a service level smaller than $\left(\frac{2(p-c)-p_M}{2(p+h)-p_M}\right)$, the supplier does not have room for manipulation since the contract still guarantees that the after-pooling profit allocations are at least as much as the before-pooling profits (as set through the service level constraint $\underline{\rho_i}$).

7.3 Conflict Between Retailers and Supplier

In the previous two sections, we looked at how service level requirements can be used to optimize profits by both the retailers and the supplier. However, we did not analyze the effects of these decisions on the other parties in the coalition. The total supply chain profit does not increase when the supplier maximizes her profits by varying the terms of the contract and relaxing the lower bound on service. Therefore the Shapley value allocation to one or both of the retailers must be reduced. We would like to know "what happens to the profits of the retailers when the supplier maximizes her profit?". Note that if both of the retailers are identical then the profit allocations to both will decrease when the supplier maximizes her profit allocation.

We define the *base case* as the case where the stocking levels are determined by $F^{-1}(\frac{p-c}{p+h})$. From Theorem 7.2 we know that supplier profit is maximized at either $(\rho_1, \rho_2) = (0,0)$ or $(\rho_1, \rho_2) = \left(\max\left(0, \frac{2(p-c)-p_M}{2(p+h)-p_M}\right), \max\left(0, \frac{2(p-c)-p_M}{2(p+h)-p_M}\right)\right)$. Clearly $(\rho_1, \rho_2) = (0,0)$ is not implementable. Therefore the supplier wants to set $(\rho_1, \rho_2) = \left(\frac{2(p-c)-p_M}{2(p+h)-p_M}, \frac{2(p-c)-p_M}{2(p+h)-p_M}\right)$ and this requires $p - c \geq p_M/2$. The supplier's per-unit profit $(p-c)$ needs to be at least as much as half the retailers' total per-unit profit

(p_M) for the supplier to be able to maximize her after-pooling profits. We can interpret this condition as a measure of the relative power of the supplier. If the supplier is making a high per-unit margin on each item she sells, she has the ability to manipulate the contracted service levels whenever the retailer requirements allow it.

For two random variables X and Y with distribution functions $F(\cdot)$ and $G(\cdot)$, X is said to be larger than Y in *dispersive order* if $F^{-1}(\beta) - F^{-1}(\alpha) \geq G^{-1}(\beta) - G^{-1}(\alpha)$ whenever $0 < \alpha \leq \beta < 1$ (denoted as $X \geq_{disp} Y$) (Shaked and Shanthikumar (1994)). Dispersive order requires the difference between two quantiles of X_i to be smaller than the difference between the corresponding quantiles of X_j; therefore dispersive order compares the variability of the two distributions. Assuming there is dispersive order between the demand distributions, the following theorem identifies which one of the retailers (if either) will be better off when compared to the base case.

Theorem 7.3 *Assume $D_i \geq_{disp} D_j$. When the supplier maximizes her own after-pooling profit allocation by changing (ρ_i, ρ_j), either the after-pooling profit allocations to both of the retailers are reduced or the profit allocation to the one with smaller demand in dispersive order is increased while the profit allocation to the other is reduced when compared to the allocations under the base case.*

The next result directly follows from Theorem 7.3 since for two random variables Y and Z, $Y \leq_{disp} Z$ implies $\text{Var}(Y) \leq \text{Var}(Z)$.

Corollary 7.1 *If the demand of one retailer is greater than the demand of the other retailer in dispersive order and the supplier maximizes her own after-pooling profit allocation, either the profit allocation to the retailer with the smaller demand variance will increase or the profit allocations to both of the retailers will decrease when compared to the allocations under the base case.*

This result is intuitive in terms of the supply chain because when the supplier maximizes her profits, if the Shapley value allocation to one of the retailers will increase then it will be the one with smaller demand variance. This result is not surprising because the retailer with the smaller demand variance brings less risk into the pooling coalition and we would expect that retailer to receive a higher allocation.

The next theorem states that convolutions of random variables with logconcave densities can be ordered in the dispersive sense. This result implies that assuming dispersive order between the demand variables is not very restrictive.

Theorem 7.4 *[Shaked and Shanthikumar (1994), Thm. 2.B.3] The random variable X satisfies $X \leq_{disp} X + Y$ for any random variable Y independent of X if and only if X has a logconcave density.*

Normal and gamma (with $p \geq 1$) distributions are frequently invoked models of demand distributions and they have logconcave densities Bagnoli and Bergstrom (1989). Therefore, by Theorem 7.4 normal and gamma demands with different shape parameters can be ordered in the dispersive sense and thus satisfy the condition on Theorem 7.3.

Another interesting property of the dispersive order is $X \leq_{disp} Y$ if and only if $X + c \leq_{disp} Y$ for any real number c. This means that the dispersive order between two random variables is preserved even if there is a shift in the mean(s). This property has an interesting implication on our results: the retailer whose profits decrease due to the supplier maximizing her profits cannot reverse the situation (become the retailer whose profits increase) even if his mean demand increases and thus creates more sales. However he can reverse the situation by changing the shape of his demand distribution by reducing the demand variance, because the allocation mechanism favors the retailer with lower risk. Note that the coefficient of variation of $X + c$ is less than that of X when c is positive (that is when mean demand increases). The surprising result is that the ordering of the variances rather than that of the coefficients of variation determines which retailer is more likely to lose profits.

8. Conclusions

In an interesting recent survey on game theory as a tool in supply chain analysis, Cachon and Netessine (2003) emphasize that cooperative game theory has not received much attention in the supply chain literature in spite of its potential usefulness. In the same chapter, Cachon and Netessine also indicate that the Shapley value has not yet been employed in supply chain research in spite of its desirable characteristics such as uniqueness. Robinson (1993) and Hartman and Dror (1986) consider Shapley value as a cost-allocation scheme but do not analyze the operational implications of using it. Granot and Sosic (2002) appear to have been the first to mention Shapley value as a profit-allocation mechanism that may induce supply-chain-optimal inventory decisions but, as far as we know, this idea has not been followed up. We offer the present paper as an initial step in understanding the uses of Shapley value as a value-sharing mechanism to affect the operational decisions of supply chain partners.

Our model shares some limitations with most work in this area. For example, like others (Anupindi and Bassok (1999); Rudi, Kapur, and Pyke (2001); Tagaras (1989)), we are limited by analytic tractability mostly to 2-retailers. In other work we have been able to extend some analysis to arbitrary numbers of retailers (Bartholdi and Kemahlıoğlu-Ziya (2003)). Similarly, to derive more particular results we have to make some simplifying assumptions about the demand distributions experienced by the retailers. We also assume that the service levels are determined exogenously to the cooperative game. They are either industry-driven or set through negotiations that are beyond the scope of our model. This assumption allows us to ignore incentives to set service levels strategically.

We have analyzed the behaviors of the supply chain members under the proposed value-sharing mechanism. It is important to compare various mechanisms for coordinating the supply chain by studying the strategic behavior that they might induce. For example, how will supply chain players answer such questions as with whom to form a coalition or whether one can game the system?

We are assuming a long-term relationship among the supply chain partners because we model the pooling problem as an *allocation game in expectation* (AGE) (Anupindi, Bassok, and Zemel (1999)). Another approach is a *snapshot allocation game* (SAG), which is used by Anupindi, Bassok, and Zemel (2001). In SAG, the value of the game is calculated based on each realization of random demand. While allocations in the core of SAG are renegotiation proof, allocations for AGE implicitly assume the players will not break from the contract based on individual realizations of demand (Anupindi, Bassok, and Zemel (1999)) (See Hartman and Dror (2003b) for a discussion on allocations based on actual demand realizations).

The Shapley value allocations for the 2-retailer supply chain correspond to equal sharing of extra revenue due to pooling. Cachon and Lariviere (2000) analyze revenue-sharing contracts and identify their limitations. They conclude that revenue sharing is not prevalent in practice partly because of high administrative costs and difficulties in monitoring revenues of retailers. Similar shortcomings apply to our value-sharing mechanism as well. We are proposing a contract where the three players first pool their profits and then the total is redistributed to them according to the Shapley values. We can think of this as a taxing mechanism where some players pay their taxes (return some of their profit) and some players get refunds (receive payments). This framework would work best if the supply chain members are in a long-term relationship, which is also the implicit assumption underlying AGE. All members are better off pooling inventory and sharing it based on Shapley value; however the mechanism will not work if there is doubt some player will break away from the coalition after getting a refund and will not be there to pay his tax when it is his turn.

As Cachon and Lariviere (2000) emphasize, to share value, it must be possible to monitor revenues of the retailers. The Shapley-value mechanism, in addition, requires visibility of both the stocking level of the supplier and her costs. Our proposed value-sharing mechanism also raises the issue of information guessing at the retailers: Can players infer information about their coalition partners that might allow them to gain advantages? To answer this and similar questions we plan further research on the truth-inducing properties of our model.

Appendix A: Service Contracts and Fill Rate as an Alternative Service Measure

A service contract based on probability of no stock-out does not always guarantee profits for the retailers. Although a more sophisticated service contract based on *fill rate* can achieve this, fill rate has weaknesses that render it less attractive as a basis for sharing than Shapley value. Fill rate β is defined as the fraction of demand routinely satisfied from shelf:

$$\beta = 1 - \frac{\mathrm{E[shortage]}}{\mathrm{E[demand]}}$$

We also differentiate between fill rate observed at retailer i before and after pooling, β_i^b and β_i^a respectively. Define $E[x_{ji}]$, the expected size of retailer j's shareable stock used by retailer i. Expected before and after-pooling fill rates are

$$
\begin{aligned}
\beta_i^b &= 1 - \frac{\int_{x_i}^{\infty}(y_i - x_i) f_i(y_i) dy_i}{\mu_i} \\
&= \frac{S_i^b}{\mu_i} \\
\beta_i^a &= 1 - \Big[\int_{x_i}^{\infty}\int_{x_j}^{\infty}(y_i - x_i) f_i(y_i) dy_i f_j(y_j) dy_j + \\
&\qquad \int_0^{x_j}\int_{x_i+x_j-y_i}^{\infty}(y_j - (x_i + x_j - y_i)) f_i(y_i) dy_i f_j(y_j) dy_j\Big] / \mu_i \\
&= \frac{S_i^b + \int_0^{x_j}(1 - F_i(x_i + x_j - y_i)) F_j(y_i) dy_i}{\mu_i} \\
&= \frac{S_i^b + E[x_{ji}]}{\mu_i} \\
&= \frac{S_i^a}{\mu_i}
\end{aligned}
$$

Expected fill rate is a function of expected sales when unsatisfied demand is lost. Therefore contracting to assure a minimum expected fill rate guarantees a minimum expected sales level, and thus a minimum profit level, for the retailers. In addition, after pooling, fill rate at retailer i increases by the expected size of retailer j's shareable stock used by retailer i scaled by expected demand.

Fill rate, but not probability of no stock-out, ensures minimum expected sales because fill rate takes into account the size of a shortage when it happens, whereas probability of no stock-out does not. The size of shortages becomes important in evaluating expected sales when sales are lost in case of a stock-out. In addition, the magnitude of probability of stock-out is not a good estimate of the ratio of unsatisfied demand to expected demand (Porteus (1990)).

Even though a service contract based on fill rate guarantees a minimum profit level for the retailers in case the supplier pools inventory, it does not necessarily induce the supplier to hold the supply-chain-optimal level of inventory. In addition, due to the dependencies in service levels after pooling, calculations become complicated especially as the number of retailers in the supply chain increases. Therefore we find the Shapley value allocation mechanism to be more useful than a service contract based on fill rate.

Appendix B: Proofs

Proof of Lemma 3.1 We find the before-pooling inventory levels x_1 and x_2 as solutions to

$$\rho_i = F_i(x_i) \qquad i = 1, 2 \tag{6.10}$$

Then defining x_1' and x_2' as the after pooling inventory levels and using Equation 6.5, we obtain the following two equations.

$$\begin{aligned} \rho_1 &= F_1(x_1') + P(x_1' \leq D_1 \leq x_1' + x_2' - D_2) \\ &= F_1(x_1') + \int_0^{x_2'} \int_{x_1'}^{x_1'+x_2'-y_2} f_1(y_1)\, f_2(y_2)\, dy_2\, dy_1 \end{aligned}$$

$$\begin{aligned} \rho_2 &= F_2(x_2') + P(x_1' \leq D_2 \leq x_1' + x_2' - D_1) \\ &= F_2(x_2') + \int_0^{x_1'} \int_{x_2'}^{x_2'+x_1'-y_1} f_2(y_2)\, f_1(y_1)\, dy_1\, dy_2 \end{aligned}$$

For each of these equations, the second term is greater than or equal to zero. By Expression 6.10 and the fact that $F_1(\cdot)$ and $F_2(\cdot)$ are non-decreasing functions of inventory level $x_1' \leq x_1$ and $x_2' \leq x_2$, which proves the claim. □

Proof of Lemma 3.2 Since the supplier profit is maximized beyond $\frac{p-c}{p+h}$ for the smallest stock level that satisfies the service level constraints, all we need to show is that $x_j^* = F_j^{-1}(\rho_j), x_i^* = 0$ gives a smaller inventory level than having both $x_j^* > 0, x_i^* > 0$. Consider two cases. In the following proof, we use the additional 1 or 2 in the subscript to denote the inventory levels under cases 1 and 2 respectively.
Case 1: Let $x_{i1} = 0$. The inventory level pair (x_{i1}, x_{j1}) are set so as to satisfy the service level constraints. The service level expressions are:

$$\begin{aligned} F_j(x_{j1}) &= \rho_j \\ \int_0^{x_{j1}} F_i(x_{j1} - y_j)\, f_j(y_j)\, dy_j &\geq \rho_i \end{aligned} \tag{6.11}$$

Case 2: Let $x_{i2} > 0$. The corresponding service level expressions are:

$$\begin{aligned} F_j(x_{j2}) + \int_0^{x_{i2}} F_j(x_{i2} + x_{j2} - y_i)\, f_i(y_i)\, dy_i - F_i(x_{i2})F_j(x_{j2}) &= \rho_j \\ F_i(x_{i2}) + \int_0^{x_{j2}} F_i(x_{i2} + x_{j2} - y_j)\, f_j(y_j)\, dy_j - F_i(x_{i2})F_j(x_{j2}) &\geq \rho_i \end{aligned} \tag{6.12}$$

The assumption $x_{i2} \geq 0$ implies $\int_0^{x_{i2}} F_j(x_{i2} + x_{j2} - y_i)\, f_i(y_i)\, dy_i - F_i(x_{i2})F_j(x_{j2}) \geq 0$. Therefore $x_{j1} \geq x_{j2}$. Now let $x_{i2} = x_{j1} - x_{j2}$ and compare the left hand sides of Equations 6.11 and 6.12.

$$\begin{aligned}
F_j(x_{j2}) + \int_0^{x_{j1}-x_{j2}} & F_j(x_{j1} - y_i)\, f_i(y_i)\, dy_i - F_i(x_{j1} - x_{j2})F_j(x_{j2}) \\
\leq \; & F_j(x_{j2}) + (F_j(x_{j1}) - F_j(x_{j2}))\, F_i(x_{j1} - x_{j2}) \\
= \; & F_j(x_{j1})\, F_i(x_{j1} - x_{j2}) + F_j(x_{j2})\, (1 - F_i(x_{j1} - x_{j2})) \\
\leq \; & F_j(x_{j1})
\end{aligned}$$

which implies that $x_{i2} \geq x_{j1} - x_{j2}$ and thus proves our claim. □

Proof of Theorem 5.1 Using Expression 6.8, one can see that ϕ_i for $i = 1, 2, S$ is maximized when $v(N) = v_{12S}$ is maximized, which happens when the pooled-inventory level for the 2-retailer coalition is set at the supply chain optimum level. □

Proof of Proposition 5.1 Employing the no-pooling strategy is one possible inventory management policy available to the coalition of two retailers and the supplier and therefore $v_{12S} \geq v_1 + v_2 + v_S$. The proposition follows from Expression 6.8. □

Proof of Proposition 5.2 By using Expression 6.8 we can easily verify that the allocations add up to v_{12S}, the value of the grand coalition. In addition, by Theorem 5.1 and Proposition 5.1, the second condition on the definition of core is satisfied. □

Proof of Theorem 6.1 In terms of S_1^{AP} and S_2^{AP} Expression 6.9 is:

$$\frac{\phi_1 + \phi_2}{p_M} \;\geq\; S_1^{AP} + S_2^{AP} \tag{6.13}$$

An equivalent expression to (6.13) is:

$$\begin{aligned}
\frac{S_1^{BP} + S_2^{BP}}{3}\left(1 - \frac{2p}{p_M}\right) + \frac{2c}{3p_M}\left[x_1^{BP} + x_2^{BP} - x_1^{AP} - x_2^{AP}\right] \qquad (6.14)\\
+ \tfrac{2h}{3p_M}\left(H_1^{BP} + H_2^{BP} - H_1^{AP} - H_2^{AP}\right) \;\geq\; \frac{S_1^{AP} + S_2^{AP}}{3}\left[1 - \frac{2p}{p_M}\right]
\end{aligned}$$

Change in expected supplier profit exceeding average change in total expected retailer profit is represented as

$$\Delta\text{E[supplier profit]} \geq \frac{\Delta\text{E[total retailer profit]}}{2} \tag{6.15}$$

Using the definition of E[profit], we can rewrite inequality 6.15 as follows

$$2p(S_1^{AP} + S_2^{AP} - S_1^{BP} + S_2^{BP}) - 2c(x_1^{AP} + x_2^{AP} - x_1^{BP} - x_2^{BP}) \\ -2h(H_1^{AP} + H_2^{AP} - H_1^{BP} - H_2^{BP}) \\ \geq p_M(S_1^{AP} + S_2^{AP} - S_1^{BP} + S_2^{BP}) \quad (6.16)$$

Algebraic manipulation reveals that inequality 6.15 is equivalent to Expression 6.14, which proves the claim. □

Proof of Proposition 6.1 Let Λ be the change in the supplier's expected cost and Δ_i be the change in expected sales at retailer i due to pooling.

$$\begin{aligned} \Lambda &= c\left(x_1^{BP} + x_2^{BP} - x_1^{AP} - x_2^{AP}\right) + h\left(H_1^{BP} + H_2^{BP} - H_1^{AP} - H_2^{AP}\right) \\ \Delta_i &= S_i^{AP} - S_i^{BP}, \quad i \in \{1,2\} \end{aligned}$$

In the proof of Theorem 6.1 we have established the equivalency of $\frac{\phi_1+\phi_2}{p_M} \geq S_1^{AP} + S_2^{AP}$ to Expression 6.15. Now rewriting Expression 6.15 using the new notation, we obtain

$$\frac{\phi_1+\phi_2}{p_M} \geq S_1^{AP} + S_2^{AP} \Leftrightarrow 2\Lambda \geq (p_M - 2p)(\Delta_1 + \Delta_2)$$

Similarly, we can write the following equivalent conditions.

$$\begin{aligned} \frac{\phi_1}{p_M} &\geq S_1^{AP} \Leftrightarrow \Lambda \geq (2p_M - p)\Delta_1 - (p_M + p)\Delta_2 \\ \frac{\phi_2}{p_M} &\geq S_2^{AP} \Leftrightarrow \Lambda \geq (2p_M - p)\Delta_2 - (p_M + p)\Delta_1 \end{aligned}$$

Without loss of generality, assume $\Delta_1 \geq \Delta_2$. The proposition states $2\Lambda \geq (p_M - 2p)(\Delta_1 + \Delta_2)$. This inequality along with $\Delta_1 \geq \Delta_2$ implies $\Lambda \geq (2p_M - p)\Delta_2 - (p_M + p)\Delta_1$ which proves the result. □

Proof of Theorem 7.1 Let π_c^{AP} be the expected supply chain profit after pooling and π_S^{BP} be expected supplier profit before pooling. Rewriting Expression 6.7, the Shapley value allocation to retailer i is

$$\phi_i = \tfrac{2}{3}S_i^{BP} + \tfrac{1}{3}(\pi_c^{AP} - \pi_S^{BP} - p_M S_j^{BP})$$

By definition, only the last two terms of the above equation depend on ρ_j. Let $x_j(\rho_j)$ be the before-pooling stocking level for retailer j as a function of the service level. Then, $x_j(\rho_j) = F_j^{-1}(\rho_j)$. Let $\Omega = \frac{\partial F_j^{-1}(\rho_j)}{\partial \rho_j}$. Then,

$$\frac{\partial \phi_i}{\partial \rho_j} = -\frac{\Omega}{3}(-c - h\rho_j + (p + p_M)(1 - \rho_j))$$

Due to the assumptions we made on $F(\cdot)$, Ω is always positive. When $\rho_j < \frac{p+p_M-c}{p+p_M+h}$, then $\frac{\partial\phi_i(\rho_j)}{\partial\rho_j}$ is negative which means the function is decreasing and when $\rho_j > \frac{p+p_M-c}{p+p_M+h}$, the derivative is positive, which means the function is increasing. Therefore, the function is unimodal and $\rho_j^* = \frac{p+p_M-c}{p+p_M+h}$, is the global minimizer. □

Proof of Proposition 7.1 The Shapley value allocation to retailer i as a function of the service level of retailer j is

$$\begin{aligned}
\phi_i(\rho_j) &= \frac{2}{3}S_i^{BP} + \frac{1}{3}(\pi_c^{AP} - \pi_S^{BP} - p_M S_j^{BP}) \\
&= \frac{2}{3}S_i^{BP} + \frac{1}{3}\left(\pi_c^{AP} - (p(S_i^{BP} + S_j^{BP}) - h(H_i^{BP} + H_j^{BP})\right. \\
&\quad \left. -c(x_i(\rho_i) + x_j(\rho_j))) - p_M S_j^{BP}\right) \\
&= K - \frac{1}{3}\left[(p + p_M - c)F_j^{-1}(\rho_j) - \right. \\
&\quad \left. (p + p_M + h)\int_0^{F_j^{-1}(\rho_j)} F_j(x)dx\right]
\end{aligned}$$

where the term K represents the part of the $\phi_i(\rho_j)$ expression that does not depend on ρ_j and K is a function of ρ_i, p, p_M, h, and c. We can find the limit of the term in the parenthesis when $F_j(\cdot)$ has infinite support as follows:

$$\begin{aligned}
&\lim_{\rho_j\to 1}\left[(p + p_M - c)F_j^{-1}(\rho_j) - (p + p_M + h)\int_0^{F_j^{-1}(\rho_j)} F_j(x)dx\right] \\
&= \lim_{\rho_j\to 1}\left[(p + p_M + h)\int_0^{F_j^{-1}(\rho_j)} (1 - F_j(x))\,dx - (h + c)F_j^{-1}(\rho_j)\right] \\
&= (p + p_M + h)E[D] - (h + c)\lim_{\rho_j\to 1} F_j^{-1}(\rho_j) \\
&= -\infty
\end{aligned}$$

This implies $\lim_{\rho_j\to 1}\phi_i(\rho_j) = \infty$. □

Proof of Theorem 7.2 Let $\Omega_i = \frac{\partial F_i^{-1}(\rho_i)}{\partial\rho_i}$. Then

$$\frac{\partial\phi_S(\rho_i,\rho_j)}{\partial\rho_i} = \frac{\Omega_i}{3}(2(p - c) - p_M - (2(p + h) - p_M)\rho_i)$$

Since $F_i(\cdot)$ is a cumulative distribution function $\Omega_i > 0$ for $i \in \{1, 2\}$. It is sufficient to consider the following three cases.

- Case 1: $2(p-c) - p_M \geq 0$ and $2(p+h) - p_M \geq 0$
  In this case both $\frac{\partial \phi_S(\rho_1,\rho_2)}{\partial \rho_1}$ and $\frac{\partial \phi_S(\rho_1,\rho_2)}{\partial \rho_2}$ are positive over the interval $\left(0, \frac{2(p-c)-p_M}{2(p+h)-p_M}\right)$ and negative over the interval $\left(\frac{2(p-c)-p_M}{2(p+h)-p_M}, 1\right)$. Therefore both $\phi_S(\rho_1)$ and $\phi_S(\rho_2)$ are increasing over the interval $\left(0, \frac{2(p-c)-p_M}{2(p+h)-p_M}\right)$ and decreasing over the interval $\left(\frac{2(p-c)-p_M}{2(p+h)-p_M}, 1\right)$, which shows $\phi_S()$ is unimodal in both ρ_1 and ρ_2. For this region, the global maximum is at $(\rho_1, \rho_2) = \left(\frac{2(p-c)-p_M}{2(p+h)-p_M}, \frac{2(p-c)-p_M}{2(p+h)-p_M}\right)$.

- Case 2: $2(p-c) - p_M < 0$ and $2(p+h) - p_M \geq 0$
  In this region, for $\rho_1 \geq 0$ $\frac{\partial \phi_S(\rho_1,\rho_2)}{\partial \rho_1}$ is negative meaning $\phi_S(\rho_1)$ is decreasing. The same argument is true for $\frac{\partial \phi_S(\rho_1,\rho_2)}{\partial \rho_2}$ and $\phi_S(\rho_2)$. Therefore in this region $(\rho_1, \rho_2) = (0,0)$ is the global maximum.

- Case 3: $2(p-c) - p_M < 0$ and $2(p+h) - c_M < 0$
  In this region $\frac{2(p-c)-p_M}{2(p+h)-p_M} \geq 1$ and beyond the meaningful service level region $[0,1)$. For $\rho_1 \in [0,1)$ $\frac{\partial \phi_S(\rho_1,\rho_2)}{\partial \rho_1}$ is negative so $\phi_S(\rho_1)$ is decreasing. The same argument is true for $\frac{\partial \phi_S(\rho_1,\rho_2)}{\partial \rho_2}$ and $\phi_S(\rho_2)$. Therefore in this region $(\rho_1, \rho_2) = (0,0)$ is the global maximum.

□

Proof of Theorem 7.3 Let ϕ_i' and ϕ_j' denote the profit allocations to retailer i and j after the supplier maximizes her profit allocation. Define the following notation:

$$\begin{aligned} \alpha &= \frac{2(p-c) - p_M}{2(p+h) - p_M} \\ \beta &= \frac{p-c}{p+h} \\ \nu &= F_i^{-1}(\alpha) \\ \eta &= F_i^{-1}(\beta) \\ \varepsilon &= F_j^{-1}(\alpha) \\ \gamma &= F_j^{-1}(\beta) \end{aligned}$$

That the profit allocation to retailer i after the supplier maximizes her profit allocation is greater than or equal to retailer i's allocation under the base case, that is $\phi_i' \geq \phi_i$, is equivalent to

$$3p_M\left(\nu-\eta+\int_\nu^\eta F_i(x)\,dx\right)$$
$$\geq(-p-p_M+c)(\gamma-\varepsilon+\eta-\nu)$$
$$+(p_M+p+h)\left[\int_\varepsilon^\gamma F_j(x)dx+\int_\nu^\eta F_i(x)\,dx\right]$$

and similarly $\phi_j' \geq \phi_j$ is equivalent to

$$3p_M\left(\varepsilon-\gamma+\int_\varepsilon^\gamma F_j(x)\,dx\right)$$
$$\geq(-p-p_M+c)(\gamma-\varepsilon+\eta-\nu)$$
$$+(p_M+p+h)\left[\int_\varepsilon^\gamma F_j(x)\,dx+\int_\nu^\eta F_i(x)\,dx\right]$$

The total after-pooling profit of the supply chain does not increase when the supplier maximizes her own after-pooling profit allocation. Then both of the inequalities cannot hold at the same time. Either neither of the equalities will hold or only one of them will hold. Therefore we need to compare $\nu-\eta+\int_\nu^\eta F_i(x)dx = \int_\nu^\eta(-1+F_i(x))\,dx$ and $\varepsilon-\gamma+\int_\varepsilon^\gamma F_j(x)dx = \int_\varepsilon^\gamma(-1+F_j(x))\,dx$ to find which retailer's profit allocation increases, if any. Since $D_i \geq_{disp} D_j$, we have $\eta-\nu \geq \gamma-\varepsilon$.

Since $D_i \geq_{disp} D_j$, we have

$$F_i^{-1}(1-y)-F_j^{-1}(1-y) \geq F_i^{-1}(1-x)-F_j^{-1}(1-x) \tag{6.17}$$

for $y \leq x$ and $y, x \in [1-\beta,\ 1-\alpha]$. Expression 6.17 implies that $1-F_i(\nu+\delta) \geq 1-F_j(\varepsilon+\delta)$ for $\delta \in [0,\gamma-\varepsilon]$ and that $\eta-\nu \geq \gamma-\varepsilon$. Then $\int_\nu^\eta(-1+F_i(x))\ dx \leq \int_\varepsilon^\gamma(-1+F_j(x))\ dx$, which concludes the proof. □

References

R. Anupindi and Y. Bassok. 1999. Centralization of Stocks: Retailers vs. Manufacturer. *Management Science.* **45**(2) 178–191.

R. Anupindi, Y. Bassok, E. Zemel. 2001. A General Framework for the Study of Decentralized Distribution Systems. *Manufacturing and Service Operations Management.* **3**(4) 349–368.

R. Anupindi, Y. Bassok, E. Zemel. 1999. Study of Decentralized Distribution Systems: Part I – A General Framework. Working Paper.

Kellogg Graduate School of Management. Northwestern University, Evanston, IL.

M. Bagnoli and T. Bergstrom. 1989. Log-Concave Probability and Its Applications. Working Paper. University of Michigan, Ann Arbor, MI.

K. R. Baker, M.J. Magazine, H.L.W. Nuttle. 1986. The Effect of Commonality on Safety Stock in a Simple Inventory Model. *Management Science.* **32**(8) 982–988.

E. Barnes, J. Dai, S. Deng, D. Down, M. Goh, H.C. Lau, M. Sharafali. 2000. Electronics Manufacturing Service Industry. Research Report. The Logistics Institute–Asia Pacific, Georgia Tech and The National University of Singapore.

J.J. Bartholdi, III and E. Kemahlioglu Ziya. 2003. Inventory Pooling and Profit Allocation in Multi-Retailer - Single Supplier Supply Chains. Working Paper. School of Industrial and Systems Engineering, Georgia Institute of Technology, Atlanta, GA.

G. Cachon. 2002. Supply Chain Coordination with Contracts. To appear in *Handbooks in Operations Research and Management Science: Supply Chain Management.* eds. S. Graves and T. de Kok, North-Holland.

G. Cachon and M. Lariviere. 2000. Supply Chain Coordination with Revenue-Sharing Contracts: Strengths and Limitations. Working Paper. The Wharton School. University of Pennsylvania, Philadelphia, PA.

G. Cachon and S. Netessine. 2003. Game Theory in Supply Chain Analysis. To appear in *Supply Chain Analysis in the eBusiness Era.* eds. D. Simchi-Levi, S.D. Wu, and Z.-J. Shen, Kluwer Academic Press.

L. Dong and N. Rudi. 2002. Supply Chain Interaction under Transshipments: Exogenous vs. Endogenous Wholesale Prices. Working Paper. Olin School of Business, Washington University, St. Louis, MO.

G. Eppen. 1979. Effects of Centralization on Expected Costs in Multi-location Newsboy Problem. *Management Science.* **25**(5) 498–501.

Y. Gerchak and D. Gupta. 1991. On Apportioning Costs to Customers in Centralized Continuous Review Systems. *Journal of Operations Management.* **10**(4) 546–551.

Y. Gerchak and M. Henig. 1986. An Inventory Model with Component Commonality. *Operations Research Letters.* **5**(3) 157–160.

Y. Gerchak, M.J. Magazine, A.B. Gamble. 1988. Component Commonality with Service Requirements. *Management Science.* **34**(6) 753–760.

Y. Gerchak and D. Mossman. 1992. On the Effect of Demand Randomness on Inventories and Costs. *Operations Research.* **40**(4) 804–807.

D. Granot and G. Sosic. 2002. A Three-Stage Model for a Decentralized Distribution System of Retailers. *Operations Research* **51**(5) 771–784.

B. Hartman and M. Dror. 1996. Cost Allocation in Continuous-Review Inventory Models. *Naval Research Logistics* **43** 549–561.

B. Hartman and M. Dror. 2003a. Optimizing Centralized Inventory Operations in a Cooperative Game Theory Setting. *IIE Transactions* **35** 243–257.

B. Hartman and M. Dror. 2003b. Allocation of Gains from Inventory Centralization in Newsvendor Environments. To appear in *IIE Transactions.*

R.B. Myerson. 1991. *Game Theory Analysis of Conflict*, Harvard University Press, Cambridge, Massachusetts.

S. Netessine and N. Rudi. 2001. Supply Chain Structures on the Internet: Marketing-Operations Coordination under Drop-shipping. Working Paper. Simon Graduate School of Business, University of Rochester.

B.A. Pasternack and Z. Drezner. 1991. Optimal Inventory Policies for Substitutable Commodities with Stochastic Demand. *Naval Research Logistics.* **38** 221–240.

E. Plambeck and T. Taylor. 2003. Sell the Plant? The Impact of Contract Manufacturing on Innovation, Capacity and Profitability. Working Paper. Graduate School of Business, Stanford University.

E.L. Porteus. 1990. Stochastic Inventory Theory. *Handbooks in OR&MS.* eds: D.P. Heyman and M.J. Sobel. **2** 605–652.

S. Raghunathan. 2003. Impact of Demand Correlation in the Value of and Incentives for Information Sharing in a Supply Chain. *European Journal of Operational Research.* **146** 634–649.

L. Robinson. 1993. A Comment on Gerchak and Gupta's "On Apportioning Costs to Customers in Centralized Continuous Review Systems". *Journal of Operations Management.* **11** 99–102.

N. Rudi, S. Kapur, D. Pyke. 2001. A two-location inventory model with transhipment and local decision making. To appear in *Management Science.*

M. Shaked and J.G. Shanthikumar. 1994. *Stochastic Orders and Their Applications.* Academic Press, San Diego, CA.

L.S. Shapley. 1953. A Value for N-Person Games. *Contribution to the Theory of Games*, Princeton University Press, Princeton, NJ. **2** 307-317.

E. Silver, D. Pyke, and R. Peterson. 1998. *Inventory Management and Production Planning and Scheduling.* John Wiley & Sons, New York.

G. Tagaras. 1989. Effects of Pooling on the Optimization and Service Levels of Two-Location Inventory Systems. *IIE Transactions.* **21**(3) 250–257.

H. Yang and L. Schrage. 2002. An Inventory Anomaly: Risk Pooling May Increase Inventory. Working Paper, Graduate School of Business, University of Chicago.

Chapter 7

SERVICE FACILITY LOCATION AND DESIGN WITH PRICING AND WAITING-TIME CONSIDERATIONS

Michael S. Pangburn
Lundquist College of Business
University of Oregon
Eugene, OR 97403

Euthemia Stavrulaki
McCallum School of Business
Bentley College
Waltham, MA 02452

1. Introduction

The strategic role of effective supply chain design has been well recognized in recent years by both academics and practitioners (see, for example, Tayur et al. 1998). Locating and sizing facilities to serve customers is one aspect of supply chain design that presents a number of challenges, due to the recent emphasis on time-based competition. Customers are sensitive to the *total cost* of interacting with a firm's service, including queuing time and access costs, in addition to price. Therefore, when setting up new service facilities, managers must carefully weigh capacity and location decisions, and choose an appropriate corresponding price. In this chapter, we formally address the interrelated location, capacity, and pricing decisions for a firm's service facilities, via analytical (nonlinear) optimization methodologies.

Several research streams have addressed a subset of these interrelated decisions by employing network optimization models. Network models incorporating *congestion effects* address the impact of queuing delays on customers' waiting costs but ignore pricing concerns, focusing instead on minimizing customer travel times (e.g., see Bolch et al. 1998, and Daskin 1995 for comprehensive reviews of linear and non-linear facility location models). We will explore an alternative methodology for understanding the interdependent decisions related to service-facility design, in the presence of time sensitive customers and congestion delays. To gain managerial insights regarding the interactions among these important supply chain decisions and enhance our ability to generate solutions, we assume that consumers are continuously dispersed over a single location dimension (as in Hotelling's 1929 "linear city" model), rather than employ a network representation of consumer locations. Relative to more detailed location models (e.g., a network topology), this approach permits us to analytically assess the structure of the firm's optimal pricing and capacity strategy. A related benefit is that we can perform comparative statics analysis to infer the direction of change in the optimal decision variables as problem parameters change. In addition, by simplifying the location model, we are able to extend the analysis to address both consumer segmentation and competition.

The location and capacity issues we address are especially relevant in settings for which service capacity is relatively expensive, implying that the firm cannot practically afford to install enough capacity to eliminate customer waiting. Businesses such as car wash/oil-change services, or tax preparation services, are representative contexts. In these examples, a consumer's total transactions cost is influenced by both the inconvenience of traveling to the service facility, and the waiting time at the facility. The significance of the access and waiting costs, relative to the firm's cost of capacity, will determine whether the firm's strategy should employ many small facilities, rather than fewer, larger facilities. For example, if consumers' access costs are high and the firm's capacity costs are low, then the firm can maximize profits by creating relatively small facilities in close proximity to customers. Maximizing profits thus requires that the firm consider pricing in conjunction with the interrelated issues of facility-location and capacity.

In this chapter, we develop a modeling framework that provides insights regarding the firm's location, capacity, and pricing decisions, for consumers who are sensitive to both waiting-time and transportation (i.e., location related delays). We begin, in the next section, by formulating the firm's decision problem for a setting with a single customer segment. Later, we extend the basic decision context to consider heterogeneous consumers (implying segmentation opportunities), non-uniform dispersion of consumers (e.g., a metropolitan area), and competition. Throughout, we maintain a common model structure, which captures the complex interactions between two important factors: *negative congestion externalities* and *queuing economies of scale.* When a consumer must share service capacity with other customers, the resulting waiting (queuing) times imply negative congestion externalities—reflecting the interdependency between customers' waiting times. To mitigate the negative effects of congestion, the firm can appropriately plan its service capacity, and take advantage of scale economies. The interplay between these two factors (consumers' waiting costs and the firm's capacity costs) underlies the firm's optimal decisions in each of the ensuing model variations.

We next describe the core set of assumptions that define the modeling approach we employ throughout this chapter. For simplicity, we begin by assuming that consumers are dispersed along a single location dimension, with a density of l consumers per unit distance. (Later, we will consider non-uniform consumer densities, to better reflect the dispersion of consumers in metropolitan areas.) For a unit interval of *active* customers (i.e., those who choose to use the firm's facility), we assume that the corresponding demand process (with an average of one order per customer, per unit time) is Poisson with rate of l. Formally, if we let I denote the interval of locations of all customers who decide to purchase from the firm's facility, then customer orders arrive according to a Poisson process with a *mean rate* of $\lambda = \int_I l dx$. Intuitively, the interval I will be *centered* around the firm's facility, with the customers defining the edges of that interval receiving zero net surplus (therefore, more distant consumers will opt *not* to purchase, due to their higher access costs).

Each consumer values the firm's service/product according to a known utility value. This *reservation price*, which we denote as $\hat{p}$, reflects the inherent value of the product to the consumer, exclusive of any purchase transactions costs. The *net* utility a consumer realizes is therefore the product utility minus the associated transactions costs (e.g., any waiting time or other access costs, and the purchase price). Because we do not address bundling issues, we assume consumers visit the firm's facility to purchase a single service or product. The distance separating the firm's facility and a consumer implies a facility access cost. We assume the access time is linear in the distance between the consumer and the firm's facility; similarly, we assume the consumer's monetary access costs (if any) are also a linear function of distance. Given this assumption of linearity, we can define a single affine function that subsumes both of these potential access cost components. We denote this affine access cost as $g(s) = G_o + Gs$, where s represents distance. With respect to the scaling of $g(s)$, we choose units of *time*, and separately apply a scaling factor α representing the cost per unit time—i.e., a time delay of $g(s)$ implies a cost of $\alpha \cdot g(s)$ to the consumer. Therefore, $\alpha \cdot g(s)$ captures both the financial and time related aspects of accessing the firm's facility (e.g., travel costs in service contexts that require direct customer access, or shipping costs/delays for facilities which ship packages to customers).

In addition to the facility access cost, consumers incur processing delays at the facility. The total processing delay within a facility should incorporate all the elements of the order fulfillment process. For example, in a distribution context, in which the facility represents an order-fulfillment center, processing may involve a customization operation as well as steps for preparing an order for shipment (e.g., credit checking, packaging, etc.). For simplicity, we do not model the internal workings of the facility, but rather employ a standard *M/M/1 queuing model* to determine the expected sum of the queuing delay and processing time. We assume that the facility operating cost, per unit time, is an affine function of capacity, equal to $c\mu + B$, where μ is the processing capacity, c is the variable capacity cost, and B subsumes all scale-independent (i.e., "overhead" type) costs.

The remainder of this chapter is organized as follows. In Section 2, we present the fundamental problem of optimizing price and capacity for a single facility with time-sensitive consumers. In Section 3, we expand our discussion to address heterogeneous consumers. Section 4 generalizes the decision problem to permit multiple facilities, thus requiring that the firm determine both the optimal number of facilities and the appropriate inter-facility spacing, to maximize profits per unit distance. Section 5 relaxes the assumption of uniformly located customers and explores an alternative consumer density function resembling a metropolitan area of consumers. In Section 6, we analyze a competitive scenario between two firms, and discuss the existence of equilibrium strategies under which the firms choose location and capacity strategies that partition the interval of consumers. Section 7 provides a brief summary and conclusion.

2. Serving homogeneous customers with one facility

We initially consider the problem of designing a single service facility for consumers that are evenly dispersed along a single location dimension. We could equivalently interpret consumer locations as being in $\Re^2$ space, assuming that the consumer density at a radius r around the facility is proportional to $1/r$. Although our analysis can apply to either of these two consumer-location models (i.e., a constant density over a linear region, or a density proportional to $1/r$ in $\Re^2$ space), for simplicity we emphasize the "linear city" perspective.

In the current context, consumers are homogeneous (with respect to their product valuations and time sensitivity), differing only in their locations. Although we relax the assumption later, we begin by assuming that the region of consumers is sufficiently large that the firm will not completely exhaust (i.e., cover) the entire interval of consumers—this must hold true, for example, if there are consumers located at a distance s such that their access costs exceed their product valuation, i.e., $\alpha \cdot g(s) \geq \hat{p}$.

The firm must address the following decisions: (i) what should be the facility's capacity, and (2) what is the optimal price to charge customers? Given the firm's price p (we assume throughout that $p > c > 0$) and capacity μ, and a corresponding (aggregate) Poisson expected arrival rate of λ, the firm's net profit (rate) will equal:

$$\pi(\lambda, \mu) = p\lambda - c\mu - B.$$

Notice that since the facility cost term B is independent from μ (and p), its magnitude will not influence the optimal decision variables. Later, however, when we extend this formulation to permit multiple facilities, the scalar term B will not only influence the objective value, but the optimal solution as well. Observe that if the cost B is sufficiently large and the proposed facility cannot achieve positive profits, then the optimal solution is to not operate the facility. Since this case is uninteresting, we assume henceforth that the fixed cost of opening a facility is not so large that positive profits are infeasible.

To maximize profits, the firm must analyze how the mean demand λ relates to both price and capacity. Defining this relationship requires that we model consumers' purchase decisions. Each consumer will choose to purchase from the firm only if the firm's product offers positive net value, considering not only price, but also expected waiting and access costs. We denote the expected processing delay by $W(\lambda, \mu)$; for the M/M/1 queueing system, the expected wait is $W(\lambda, \mu) = 1/(\mu - \lambda)$. Given the facility access cost function $g(s)$, we can thus express *the expected net surplus* for a consumer at a distance s from the facility as $\hat{p} - \alpha W(\lambda, \mu) - \alpha g(s) - p$. Notice that although we have implicitly assumed that the same *cost-of-time* rate (α) pertains to both the time-related constructs $W(\lambda, \mu)$ and $g(s)$, if a consumer values time differently for either of these constructs, then we can suitably adjust α and $g(s)$.

Each consumer's decision regarding whether to purchase from the firm depends not only on their *individual* distance-and-time related costs, but also on the waiting time $W(\lambda, \mu)$ which is a function of the aggregate number of customers using the facility. From the expected net surplus expression, $\hat{p} - \alpha W(\lambda, \mu) - \alpha g(s) - p$, we can see that for a sufficiently large distance away from the facility, access costs will be too large for consumers to justify incurring the price p and the expected wait $W(\lambda, \mu)$. For fixed values of the price p and capacity μ, the *threshold distance* S defines the precise distance at which the consumer surplus is zero, i.e., $\hat{p} - \alpha W(\lambda, \mu) - \alpha g(s) - p = 0$. Since consumers are uniformly located along the line, the facility will attract an equal number of consumers in both directions. For example, if the facility is located at the point zero, then the interval of served customers is $I = [-S, S]$. Therefore, the mean arrival rate of customers to the facility is $\lambda = 2lS$. For the firm to induce demand over the interval $I = [-S, S]$, price and capacity must be set to satisfy the constraint $\hat{p} - \alpha W(\lambda, \mu) - \alpha g(s) \geq p$. Solving the firm's decision problem to determine the optimal capacity and price requires that the firm "internalize" this consumer surplus constraint. Formally, the firm's *Single Facility Problem* (SFP) is:

(SFP)

$$\begin{aligned} & \max\{\pi(\lambda,\mu) = p\lambda - c\mu - B\} \\ \text{subject to:} \quad & \hat{p} - \alpha W(\lambda,\mu) - \alpha g(s) \geq p, \\ & \lambda = 2lS, \\ & \text{where } \mu > \lambda \geq 0. \end{aligned}$$

It is instructive to examine in more detail the influences that determine the choice of the optimal decision variables for the SFP. Each consumer's purchase decision depends on the queuing delay, which is a function of the decisions of other consumers. This dependence implies a direct relationship (through the surplus constraint) between the firm's capacity, price, and the interval of served customers. Moreover, this model captures the notion of negative congestion externalities, since individual consumers' purchase decisions impact the expected waiting of all consumers. To illustrate the complex interdependencies between these factors, consider the example of a firm wishing to lower its price as a means to attract more customers. As a result of a price drop, S increases, and thus λ increases, causing the wait $W(\lambda, \mu)$ to increase. To compensate for this waiting time increase, the firm might accompany the price drop with a simultaneous capacity increase, and attempt to leverage scale economies—i.e, operating at higher utilizations as capacity increases, without larger (expected) waits.

Figure 7.1 provides an intuitive visualization of the mathematical structure of the SFP model. The triangle supported by the shaded base defines the region of *purchasing* consumers (i.e., those for whom the surplus constraint is satisfied, and thus lie within the threshold distance S from the facility). As we move from the center of the triangle towards its leftmost or rightmost corners, consumer utility falls to zero. Therefore, the outer corners of the triangle define the threshold distance S, beyond which the firm does not attract customers. As discussed above, if price were to drop below the level shown in the figure, two effects would result: (1) the sloping sides of the triangle representing total consumer surplus would extend further (until reaching the new lower price level), and (2) the *apex* of the triangle would fall, due to an increase in the wait $W(\lambda, \mu)$ resulting from a larger number of customers—served with the same capacity.

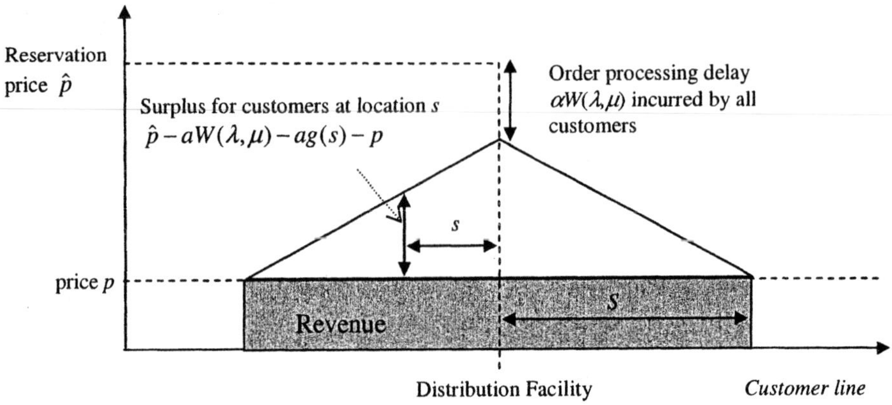

Figure 7.1. Consumers served by a single facility.

For this single-facility problem with homogeneous consumers, specified by the above SFP formulation, we next address the optimal policy, defined by the optimal price $p^*$and capacity μ^*.

Result 1. The SFP has a unique optimal solution (p^*, μ^*) such that $\mu^* = 2lS^* + \sqrt{\alpha 2lS^*/c}$, and $p^* = \hat{p} - \alpha W(2lS^*, \mu^*) - \alpha g(S^*)$, where S^* is the optimal threshold distance.

The proof of this result (a detailed version of which appears in Dobson and Stavrulaki 2004) follows from observing that the surplus constraint of the SFP is binding and so price can be cast as a function of S and μ. Then, applying first order conditions with respect to μ permits us to express μ as a function of S (yielding the optimal capacity expression in Result 1). Thus, the SFP problem reduces to a single-variable, unconstrained problem with respect to S, which has a unique feasible solution (the optimal threshold distance S^*). Interestingly, the relationship between the average arrival rate and the optimal capacity shown in Result 1 is broadly consistent with the "square-root" functional form encountered in related contexts. For instance, Halfin and Whitt (1981) have shown that economies of scale in large processing systems are of the order of the square root of capacity. Similarly, we find that as the total arrival rate λ increases, the optimal excess capacity (i.e., the capacity in excess of the arrival rate) increases proportionally with the square root of the arrival rate.

As we discussed in the beginning of this section, the optimal policy defined by Result 1 applies when consumers are dispersed over an area larger than can be served effectively by a single facility. We now allow for the possibility that the range of consumers is restrictive. Let the range of consumers be an interval of length M. If the range of consumers is longer than $2S^*$ (i.e., if $M \geq 2S^*$), then the finite range of consumers does not represent a binding limit, and the optimal policy as defined in Result 1 continues to apply. In contrast, if the limited range of consumers covers a distance less than $2S^*$ (i.e., if $M < 2S^*$), then the firm will optimally plan to serve only that limited range, implying $S^* = M/2$.[1]

The SFP formulation, which addresses the simplest version of the single-facility service design problem, relates to several models in the literature. Congested network location models, for instance, focus on minimizing travel costs rather than maximizing a firm's profit and do not consider pricing decisions (e.g., Ghosh and Harche 1993, Brandeau 1992). In contrast, Mendelson (1985), Dewan and Mendelson (1990), Stidham (1992), and Ha (1998) address pricing and capacity decisions, but focus on distributed-computing environments, for which location is not a significant factor (since travel times across an electronic network are typically negligible). In contrast, since we consider the processing of physical customers and their orders, the customer-to-facility distance cannot be ignored. The SFP provides the basis for the various modeling extensions we discuss in subsequent sections. We next extend our scope to address contexts with multiple customer segments. For the time being, we will retain our restriction of a single facility, although we later relax that assumption as well.

[1]In this case, $S^* = M/2$ is optimal because the SFP profit function, when expressed in terms of the single variable S, is unimodal when profits are positive—implying that the binding constraint (created by introducing the limited customer interval M) will define the optimal solution.

3. Serving heterogeneous customers with one facility

In the prior section, we considered the firm's optimal strategy (encompassing decisions regarding both price and facility size) under the assumption that the only differentiating characteristic for consumers is *location*. In that context, after gaining access to the facility, consumers are non-differentiated, and therefore the firm sets a single price, applicable to all consumers. We now assume that the firm serves *two* distinct consumer segments, which may differ both in terms of their product and time valuations. With heterogeneous consumers, effective segmentation may be feasible. As above, we consider consumers (now, of both segments) to be evenly dispersed, and we continue to employ the one-dimensional "linear city" location model. To distinguish between the two segments, we employ the subscript i, where $i = 1, 2$; thus, orders are generated by segment i according to a Poisson distribution with a mean rate of l_i per unit distance.

The specific nature of the consumer heterogeneity will determine what form of segmentation, if any, is feasible for the firm to implement. *Enforcing* segmentation requires an appropriate distinguishing consumer characteristic (e.g., academic versus non-academic customer status). If the distinguishing characteristic does not support enforced segmentation (e.g., income level, or other such hidden information), then the firm can still offer multiple services, and allow consumers to self-select their preferred option. These considerations imply three distinct service offering scenarios. The first scenario corresponds to settings in which enforced segmentation is feasible. The second scenario applies when enforced segmentation is not feasible, and so the firm simply offers a *single* service process and price for all customers. The third case applies when a firm offers *two* distinct service options (and prices), and consumers self-select their preference. Pangburn and Stavrulaki (2004) refer to these scenarios as the segment-restricted, segment-pooled, and segment-selected designs, and we highlight the differences between the three cases in this section.

3.1 Segment-restricted service

We assume, for the segment-restricted scenario, that the firm designs a dedicated service offering for each of the two segments, with corresponding prices and service capacities—which we denote as p_i and μ_i, for segment i. We consider two different customer segments with reservation prices $\hat{p}_i$ and time sensitivities α_i, respectively. The corresponding arrival rate for segment i is λ_i, and the expected order-fulfillment processing time is thus $W(\lambda_i, \mu_i)$. Since, in this scenario, consumers are split into two distinct segments and are served by dedicated service capacity, the firm's decision problem decouples into two single-segment formulations. Therefore, the structure of the earlier SFP solution applies to determine the optimal policy for serving each consumer segment in the segment-restricted service scenario.[2]

Why might the firm forgo queuing scale economies and offer a distinct service for each segment? The incentive is the potential to price-discriminate against the consumer segment with the higher reservation price, and/or waiting sensitivity. But, in some contexts, the option of serving the distinct customer segments with dedicated capacity may not be desirable (or even possible), and therefore we next address the alternative of pooling both segments and serving all customers with a single price and service process.

3.2 Segment-pooled service

In this scenario, the firm offers consumers a single process with price p and service-rate μ. Let S_i denote the distance from the firm's facility to the furthest participating customer from segment i. As we explained (in Section 2), due to symmetry the threshold distance S_i applies to consumers on both sides of the facility, and therefore the total mean arrival rate is equal to $\lambda = 2l_1S_1 + 2l_2S_2$. We can now formulate the profit maximization problem for the *Segment-Pooled Design* (SPD) case.

[2]Because the problem decouples into independent single-segment problems, the same approach would also hold for any number (i.e., beyond two) consumer segments.

(SPD)

$$\begin{aligned} & \max\{p\lambda - c\mu - B\} \\ \text{subject to}: \quad & \hat{p}_1 - \alpha_1 W(\lambda, \mu) - \alpha_1 g(S_1) \geq p, \\ & \hat{p}_2 - \alpha_2 W(\lambda, \mu) - \alpha_2 g(S_2) \geq p, \\ & \mu > \lambda = 2l_1 S_2 + 2l_2 S_2 \geq 0. \end{aligned}$$

The SPD formulation closely parallels the SFP problem of the prior section, except that there are two consumer surplus constraints—one for each segment. These two constraints implicitly define the threshold distances S_1 and S_2 for the two consumer segments. We employ the same access cost $g(\cdot)$ for both segments, although more generally a distinct function might apply to each segment. Pangburn and Stavrulaki (2004) verify that the SPD has a unique (globally optimal) solution. The following result compares the optimal pooled price with the optimal segment-specific prices for the SRD problem (i.e., with segmentation).

Result 2. Let p^*_{pooled} denote the optimal price for the SPD, and let p^*_i denote the optimal price for segment i when offering segment-restricted services. Then, $p^*_{pooled} > \min\{\mathrm{p}^*_1, \mathrm{p}^*_2\}$, but it is *not* necessarily the case that $p^*_{pooled} < \max\{\mathrm{p}^*_1, \mathrm{p}^*_2\}$.

When contrasting the pooled and segment-restricted service designs, we might intuitively expect the pooled price to represent an average of the segment-specific prices p^*_i. Such intuition would suggest merely that $p^*_{pooled} > \min\{\mathrm{p}^*_1, \mathrm{p}^*_2\}$; however, it is possible for p^*_{pooled} to be larger than *both* p^*_1 and p^*_2. Thus, when pooling the customer segments, the resulting optimal price does not necessarily "mediate" between the segment specific prices. To understand this possibility, we must recall that consumers are time sensitive, and therefore the waiting-time reductions resulting from scale economies can enable the firm to increase price. Notice that this finding qualifies the conventional wisdom that segmentation should permit the firm to charge higher prices for at least some consumers.

3.3 Segment-selected service

We now address a scenario in which the firm offers two distinct service options, (p_1, μ_1) and (p_2, μ_2), and allows consumers to *self select* their preferred option. Because consumers' access costs are sunk upon reaching the facility, all customers within a segment will choose identically. That choice is dictated by a comparison of the expected surpluses $\hat{p}_i - \alpha_i W(\lambda_1, \mu_1) - p_1$ and $\hat{p}_i - \alpha_i W(\lambda_2, \mu_2) - p_2$. Note that if either of these two surplus expressions dominates the other for *both* consumer segments, then the firm's service offering cannot be optimal (since, in that case, no consumers will use the less-preferred service). Therefore, with self-selection, the following two *self-selection constraints* must hold:

$$\hat{p}_1 - \alpha_1 W(\lambda_1, \mu_1) - p_1 \geq \hat{p}_1 - \alpha_1 W(\lambda_2, \mu_2) - p_2,$$
$$\hat{p}_2 - \alpha_2 W(\lambda_2, \mu_2) - p_2 \geq \hat{p}_2 - \alpha_2 W(\lambda_1, \mu_1) - p_1.$$

In these two constraints, without loss of generality, we denote segment 1 as the consumer segment that prefers the option (p_1, μ_1), whereas segment 2 prefers the option (p_2, μ_2). Thus, these constraints also reflect the need for the firm to avoid product-cannibalization losses, which occur when a consumer—given the choice—switches to a less profitable option than they would have purchased otherwise.

With self-selection in effect, we have the following formulation for the *Segment-Selected Design* (SSD) problem:

(SSD)

$$\begin{aligned}
& \max\{p_1\lambda_1 + p_2\lambda_2 - c(\mu_1 + \mu_2) - B\} \\
\text{subject to :}\quad & \hat{p}_1 - \alpha_1 W(\lambda_1, \mu_1) - \alpha_1 g(S_1) \geq p_1, \\
& \hat{p}_2 - \alpha_2 W(\lambda_2, \mu_2) - \alpha_2 g(S_2) \geq p_2, \\
& \hat{p}_1 - \alpha_1 W(\lambda_1, \mu_1) - p_1 \geq \hat{p}_1 - \alpha_1 W(\lambda_2, \mu_2) - p_2, \\
& \hat{p}_2 - \alpha_2 W(\lambda_2, \mu_2) - p_2 \geq \hat{p}_2 - \alpha_2 W(\lambda_1, \mu_1) - p_1, \\
& \mu_1 > \lambda_1 = 2l_1 S_1 \geq 0, \\
& \mu_2 > \lambda_2 = 2l_2 S_2 \geq 0.
\end{aligned}$$

The next result describes the structure of the corresponding optimal policy.

Result 3. If consumers can self-select, then the lower-price service must apply to the less time-sensitive segment, irrespective of the two segments' relative reservation prices.

This result implies that when consumers can self-select, the firm cannot successfully segment consumers based on their willingness-to-pay, as in the SRD scenario. With self-selection, because the firm can only leverage the segments' distinct time-sensitivities to partition consumers, it follows that a higher-price service (with shorter wait) can only appeal to the more time-sensitive segment. In contrast, if *enforced* segmentation is feasible, then the firm has the option of segmenting consumers directly based upon their willingness-to-pay, in which case the *less* time-sensitive segment might be targeted with the higher-price (and shorter wait) service.

Providing consumers with the freedom to self-select reduces the firm's segmentation "leverage", and thus profits decrease. In the formulation, this decrease is caused by the presence of the added self-selection constraints, which are not present in the SRD case. Even though the optimal profit with self-selection will always be lower than the optimal profit with enforced segmentation, it can be either higher or lower than the optimal profits with pooling, depending on the significance of the queuing scale-economies (e.g., the variable capacity cost c). Profit comparisons over a range of problem instances indicate that the revenue benefits from self-selected segmentation will outweigh the loss of queuing economies if the time-sensitivities of the segments differ substantially (Pangburn and Stavrulaki 2004). Interestingly, such comparisons also show that properly designing distinct service offerings can increase profits even in settings where consumers would prefer (i.e., gain higher net surplus) that the firm choose the segment-pooled design.

In this section, we have addressed several issues relating to segmentation and price-discrimination, when a monopolistic firm serves distinct consumer segments. We have, until this point, maintained the assumption that the firm locates its service offerings at a single location. Next, we relax that assumption and permit the firm to operate multiple facilities.

4. Serving homogeneous customers with multiple facilities

Let us now consider a firm serving an extended area of customers, implying the need for multiple facilities. In addition to the previous capacity and pricing decisions, the firm must now analyze how widely dispersed its facilities should be. Is it optimal to have many small facilities, or a relatively small number of large facilities? Because we wish to emphasize the location dimension, rather than the segmentation issue of the prior section, we assume consumers are homogeneous. To begin with, we will assume these homogeneous consumers are dispersed uniformly over an unbounded (linear) region; subsequently, we relax that assumption and consider a finite interval of customer locations.

When a firm serves a large number of widely dispersed consumers via many facilities, the firm must attempt to maximize the *aggregate* profit from all its areas of operation, rather than maximize the profit *per location* (the latter objective corresponds to maximizing profit for a single facility, which we discussed in Section 2). Consider, for example, a retailer planning multiple stores within a single city. Adding an nth location might actually decrease the per-store profits of the existing $(n-1)$ locations, while simultaneously increasing overall profits. Therefore, when relaxing the assumption of a single facility, the firm's appropriate objective is to maximize total profit over *all* locations, rather than simply the per-location profits.

We assume that consumers are uniformly dispersed along a line representing customer locations. We also assume that capacity costs are not facility-dependent, and therefore, given the symmetrical cost and demand economics, all facilities should be identical—Dobson and Stavrulaki (2004) formally prove this property. To optimize total profits across all locations, the firm equivalently maximizes *per-unit-distance* profits, implying the following *Unbounded Multi-Facility Problem* (UMFP):

(UMFP)

$$\max \left\{ \frac{(p\lambda - c\mu) - B}{S} \right\}$$

$$\text{subject to:} \quad \hat{p} - \alpha W(\lambda, \mu) - \alpha g(S) \geq p,$$

$$\mu > \lambda = 2lS \geq 0.$$

The UMFP's objective function is (strictly) concave, ensuring a global maximum. We denote the optimal UMFP solution as (μ_A, p_A), with corresponding coverage region $2S_A$; we use the subscript A to emphasize that this solution maximizes the *average* (per unit distance) profit. For example, if $S_A = 15$, then the firm will optimally replicate that facility design with an inter-facility spacing of exactly 30 miles. Of course, the firm *could* choose to set the distance between facilities larger than each facility's coverage breadth of 30 miles, but doing so would cause unnecessary "coverage gaps" which decrease profits.

In reality, firms do not serve unbounded (i.e., infinite) regions of consumers, and thus a more realistic model would consider a finite region of consumers, which we represent as the interval $[0, M]$. For example, consider a firm that wants to design a set of facilities to maximize profits within a 100-mile region of consumers (i.e., $M = 100$ miles). Can we use the solution of either the single facility problem (SFP), or the unbounded multi-facility problem (UMFP), to define the optimal number of facilities? Let us assume, for instance, that maximizing per-facility profits (i.e., employing the SFP solution) yields a coverage region of $2S^*$ $= 50$ miles, whereas maximizing per-unit-distance profit implies $2S_A =$ 30 miles. Since in our example $3(2S_A) = 90 < M = 100$, we can conclude that the profit-maximizing strategy suggests *at least* three facilities. In general, since the UMFP's objective function is strictly concave, the optimal number of facilities must be equal to either $\lfloor M/2S_A \rfloor$ or $\lceil M/2S_A \rceil$, assuming that the facilities are identical.

It is not yet clear, however, whether the firm *should* (optimally) use identical facilities in the bounded setting. In our current example, maximizing the profit per unit distance for each of these three facilities would suggest an inter-facility spacing of 30 miles, with an optimal capacity and price defined by (μ_A, p_A). However, since those three facilities would generate demand from only a $3(2S_A) = 90$ mile interval of consumers, this strategy forfeits 10% of the potential customer base. The profit maximizing strategy might entail lowering the price or raising capacity, so as to capture the remaining 10% of consumers. A related issue concerns how the customer base should be shared between the facilities. For example, should the three facilities' coverage regions be adjusted to [30, 40, and 30] miles respectively, or perhaps [35, 30, and 35] miles, or should the 100 mile interval be evenly split between the three facilities? Alternatively, the firm might consider using four slightly smaller service facilities, each with a coverage region of 25 miles. We next address these questions, by more generally formulating the decision problem with multiple service facilities. For details regarding the following results and their proofs, refer to Dobson and Stavrulaki (2004).

Let the subscript i denote individual service facilities, each having the same per-unit capacity cost. For any facility i, let (μ_i, p) denote the price and capacity decisions, with a corresponding coverage region of $2S_i$ for the facility; observe that the firm sets a common price p across its facilities.

Result 4: Consider an arbitrary interval of customers [0, M]. The optimal strategy for serving this interval from multiple facilities is such that each customer gets a strictly positive amount of surplus from *at most one* distribution facility (i.e., the coverage regions for the distribution centers are non-overlapping).

Knowing that the coverage regions, equal to $2S_i$ for all i, should not overlap, we can now formulate the *Generalized Multi-Facility Problem* (GMFP), which we can use to determine the optimal number of facilities, denoted by n, for serving the customers located over [0, M].

(GMFP)

$$\begin{aligned} & \max \left\{ \sum_{i=1}^{n} (p\lambda_i - c\mu_i - B) \right\} \\ \text{subject to :} \quad & \hat{p} - \alpha W(\lambda_i, \mu_i) - \alpha g(S_i) \geq p, \ \forall i \\ & \sum_{i=1}^{n} 2S_i \leq M, \\ & \mu_i > \lambda_i = 2lS_i \geq 0, \ \forall i \\ & n \in Z^+. \end{aligned}$$

The decision variables in this problem are the price and capacity variables (i.e., p and μ_i, for i = 1, 2), and also the integral number of facilities, n.

Result 5: The optimal solution to the GMFP is such that all facilities have identical capacities.

This result is consistent with the strategy that many large firms appear to practice, replicating "cookie cutter" facilities throughout their regions of operation. Result 5 allows us to further simplify the multi-facility profit maximization problem as:

(BMFP)

$$\max\{n[p\lambda - c\mu - B]\}$$

subject to:

$$\begin{aligned} &n2S \leq M, \\ &\hat{p} - \alpha W(\lambda, \mu) - \alpha g(S) \geq p, \\ &\mu > \lambda = 2lS \geq 0, \;\; n \in Z^+. \end{aligned}$$

We refer to this decision problem as the *Bounded* (region) *Multi-Facility Problem*, (BMFP). Since Result 5 establishes that the optimal approach is to employ identical facilities, we can use the unbounded problem (UMFP) solution to prescribe the BMFP solution, with either $\lfloor M/2S_A \rfloor$ or $\lceil M/2S_A \rceil$ facilities (as mentioned above). Using $\lfloor M/2S_A \rfloor$ facilities will not cover the full region of length M, unless the span of each facility is "stretched" beyond $2S_A$—by lowering price or increasing capacity. Recall, however, that the profit *per facility* will decrease if the coverage region is expanded beyond its optimal region $2S^*$, because the SFP profit function (when expressed in terms of S) decreases for $S > S^*$. Therefore, with $\lfloor M/2S_A \rfloor$ facilities, the optimal "stretching" of *each* facility is bounded from below by $2S_A$, and from above by $2S^*$, and so we can employ both bounds to prescribe the optimal coverage region for each facility. Moreover, if $\lfloor M/2S_A \rfloor \cdot 2S^* < M$, then the upper bound implies that the optimal strategy is to serve less than the full consumer base.

If the firm uses $\lceil M/2S_A \rceil$ facilities, a per-facility span of $2S_A$ would imply a consumer region of *greater* than length M, and therefore the BMFP solution optimally "shrinks" the identical facilities. In this case, the optimal strategy for the firm will, necessarily, span the entire interval [0,M], since the firm should not shrink the facility size more than is necessary to span the region size M.

In summary, when consumers are evenly dispersed over the bounded region, the optimal number of facilities must equal either $\lfloor M/2S_A \rfloor$ or $\lceil M/2S_A \rceil$. An important insight from the analysis is that serving the entire customer base can be suboptimal, even with multiple facilities. We also find that when serving uniformly distributed customers, all the facilities should be identical. However, when consumers are not evenly dispersed, but instead reflect a metropolitan area with high population densities at a central location, then we will show next that identical facilities are not necessarily optimal.

5. Locating facilities in areas with non-uniform customer densities

Our discussion in the prior section suggests that a firm serving an evenly dispersed population should employ identical facilities. In this section, we consider a consumer dispersion pattern that reflects a metropolitan area. Specifically, we use a triangular model of population density, with the apex of the triangle representing the center of the metropolitan area; we refer to this form as the *metropolitan density function.* We denote the height of the triangle as H, and the base as M, so the density of consumers at any location x, for $x \in [0, M]$, is equal to $l(x) = (H/M)(M - x)$. We are again interested in understanding the firm's decision problem, which specifies the number of facilities, and the associated pricing and capacity strategy. Throughout this section, we assume that the values of H and M (defining the metropolitan area) are sufficiently large to support positive profit—otherwise, the firm should cease operations. We also assume, as in the prior section, that consumers are homogenous, and the firm's price is consistent across locations.

We begin by assuming that the firm chooses to operate a single facility. Optimally, the firm should place the facility at the point which minimizes consumers' average access costs, and therefore the facility will be located at the heart of the populated area (a common solution in practice).[3] Subsequently, we will consider the strategy of developing multiple service facilities to serve the metropolitan region.

---

[3]Our discussion assumes that a facility incurs the same (scale independent) capacity cost component B irrespective of the particular facility location. Permitting location-specific capacity costs would extend the modeling approach in this section.

5.1 Single facility problem

The optimal location for a single facility is the central location coinciding with the apex of the consumer density function $l(x)$. By again denoting consumers' threshold distance as S, we can express the total arrival rate as $\lambda = 2 \cdot \int_0^S l(x)dx$, and applying the metropolitan density function $l(x) = (H/M)(M - x)$ yields $\lambda = HS(2 - S/M)$. We can now formulate the *Metropolitan 1-Facility Problem* (M1FP) as:

(M1FP)

$$\max\{p\lambda - c\mu - B\}$$

subject to:

$$\hat{p} - \alpha W(\lambda, \mu) - \alpha g(S) \geq p,$$
$$\mu > \lambda = 2\int_0^S l(x)dx \geq 0.$$

Using a methodology similar to the one underlined in Result 1 of Section 2, we conclude that there exists a unique solution to the M1FP, which we denote as (p_{M1}, μ_{M1}).

Result 6. The M1FP has a unique optimal solution (p_{M1}, μ_{M1}) such that $\mu_{M1} = (H/M)(2M - S_{M1})S_{M1} + \sqrt{(\alpha HM^2(2M - S_{M1})S_{M1})/c}$, and $p_{M1} = \hat{p} - \alpha W(2lS_{M1}, \mu_{M1}) - \alpha g(S_{M1})$, where S_{M1} is the corresponding threshold distance.

The derivation of this result follows from first eliminating price as a decision variable (by recognizing the participation constraint is binding at optimality), and then applying first order conditions with respect to μ, thus yielding the optimal capacity expression. Second order conditions hold, ensuring the solution is optimal.

Result 6 defines the profit-maximizing facility design, given that the firm will build a single facility. However, the single-facility solution might not be the best strategy, particularly when consumers' access costs (e.g., driving times) are high and new-facility costs are low. Therefore, we next address the two-facility problem, and we subsequently discuss the three-facility problem.

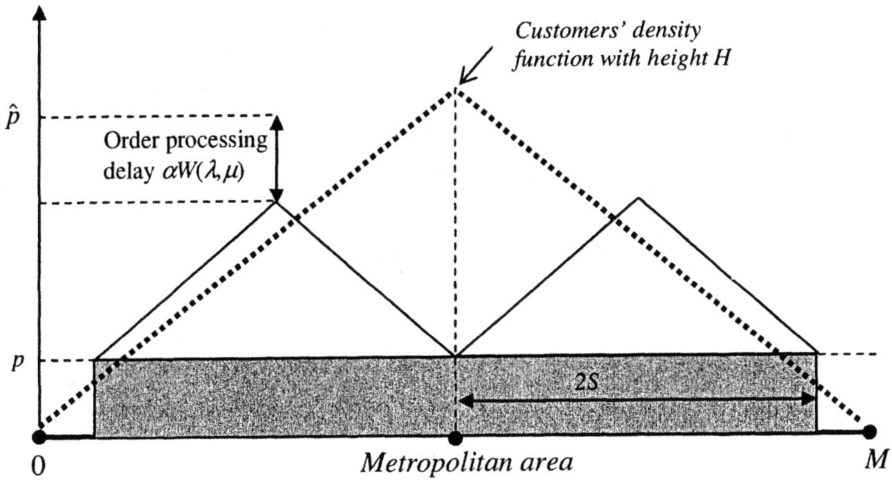

Figure 7.2. Two-facility problem with a metropolitan density function.

5.2 Two-facility problem

We now consider serving the metropolitan area with two facilities, which must optimally be equidistant from the metropolitan center.[4] Since the two facilities have symmetric locations, with each addressing mirror-image halves of the triangular distribution, we need only analyze one of the two facilities. Consider, for example, the right facility in Figure 7.2 below, which will serve all consumers located on the interval $[0, 2S]$, with corresponding arrival rate $\lambda = \int_0^{2S} l(x)dx = 2HS(1-S/M)$.

For each of the two (identical) facilities, the formulation for the *Metropolitan 2-Facility Problem* (M2FP) is:

(M2FP)

$$\max\{p\lambda - c\mu - B\}$$

[4]A non-symmetrical solution cannot be optimal, since in that case relocating (while maintaining fixed price and capacity) the less-centrally located facility to a symmetric (i.e., same distance but to the opposite side of the center) and more central location will increase profits—because the customer base increases.

subject to:

$$\hat{p} - \alpha W(\lambda, \mu) - \alpha g(S) \geq p,$$
$$\mu > \lambda = \int_0^{2S} l(x)dx \geq 0.$$

Due to the problem symmetry with two facilities, as shown in Figure 7.2, the coverage area for each facility begins at the population center. Conceivably, the firm could choose to leave an un-served gap at the population center (implying consumers in those locations would associate negative surplus with the firm's service, due to access cost considerations), or, alternatively, provide strictly positive surplus at the population center. However, neither of these options is optimal, since in either case the firm can increase profits by perturbing the facilities' locations (in the former case, by shifting the facilities inwards, keeping price and capacity fixed; in the latter case, by shifting the facilities outwards). By applying first-order conditions, we can derive the optimal price and capacity, (p_{M2}, μ_{M2}), for the M2FP formulation.

***Result** 7*. The M2FP has a unique optimal solution (p_{M2}, μ_{M2}) such that $\mu_{M2} = 2(H/M)(M - S_{M2})S_{M2} + \sqrt{2(\alpha HM^2(M - S_{M2})S_{M2})/c}$, and $p_{M2} = \hat{p} - \alpha W(2lS_{M2}, \mu_{M2}) - \alpha g(S_{M2})$, where S_{M2} is the optimal threshold distance.

Interestingly, even in the absence of facility costs (i.e., $B = 0$), the two-facility solution is not always preferred to the single-facility solution. For example, consider the following baseline values: consumers have a reservation price of $\hat{p} = \$100$, a time sensitivity $\alpha = \$0.1$ per unit time, capacity cost $c = \$25$, and access cost $g(s) = s$ (i.e., simply proportional to distance). Assume the metropolitan density function parameters are $H = 1$ and $M = 10$. In this case, the one and two facility solutions yield equivalent profits (specifically, \$730 per unit time). Thus, relative to this case, if we either increase capacity costs (c), or decrease consumers' time-sensitivity (α), then the single-facility solution will begin to dominate. Conversely, if we decrease capacity costs, or increase consumers' time-sensitivity, then the two-facility solution dominates. These results are intuitive; as α increases, the two-facility solution becomes favorable because it enables the firm to locate nearer to consumers (on average). For example, as the magnitude of the time sensitivity α increases from

0.1 to 1.0, the two-facility profit per unit time decreases from \$730 to \$656, whereas the single-facility solution yields a profit per unit time of only \$622 (a 5.2% reduction).

Above, in Section 3, we investigated the optimal solution for multiple facilities with evenly dispersed consumers. With the metropolitan density function, although we again find that employing multiple facilities can be optimal, defining the solution for more than two facilities is difficult, because for $n > 2$ the facilities need not (optimally) be identical. For $n > 2$, although all facilities are no longer identical, the problem structure is symmetrical around the metropolitan center, thus simplifying the ensuing analysis and discussion of the three-facility problem.

5.3 Three-facility problem

We now consider the problem of determining the optimal design (capacity, price, and location) for three facilities with the metropolitan consumer density. Consistent with our above approach, we continue to assume that the firm charges a uniform price across its facilities. In Figure 7.3, the three solid triangles graphically depict the coverage regions for three representative service facilities. Because the three facilities charge the same price, the slopes of these three triangles are identical. In the figure, the triangle corresponding to the middle facility is depicted with larger height, implying that this facility provides the shortest expected waiting time (graphically, the expected waiting cost is the difference between the value $\hat{p}$ and the peak of the triangle for that facility). Moreover, since the middle facility spans a larger region than the outlying facilities (and in a more populated area), that facility must also have more capacity.

Since we must now permit the central and peripheral facilities to have distinct sizes, we use the subscript "mid"/"out" to denote the middle/outer facility. As Figure 7.3 illustrates, the arrival rate of the middle facility is $\lambda_{mid} = 2\int_0^{S_{mid}} l(x)dx$, whereas the arrival rate of each of the outer facilities is equal to $\lambda_{out} = \int_{S_{mid}}^{S_{mid}+2S_{out}} l(x)dx$. By leveraging the problem symmetry (i.e., the two "outer" facilities are identical and located at a point $S_{mid} + S_{out}$ on either side of the central facility), we can formulate the *Metropolitan 3-Facility Problem* (M3FP) as:

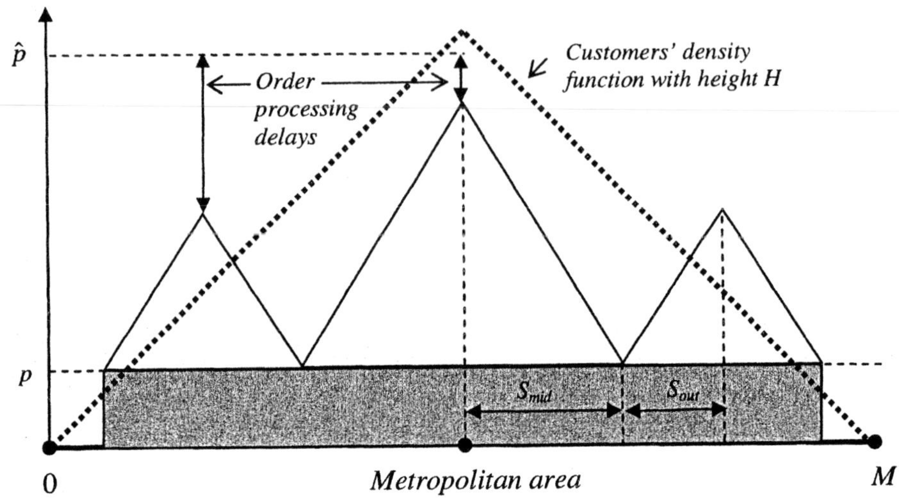

Figure 7.3. Three-facility problem with a metropolitan density function.

(M3FP)

$$\max\{(p\lambda_{mid} - c\mu_{mid} - B) + 2(p\lambda_{out} - c\mu_{out} - B)\}$$

subject to:

$$\begin{aligned}
&\hat{p} - \alpha W(\lambda_{mid}, \mu_{mid}) - \alpha g(S_{mid}) \geq p, \\
&\hat{p} - \alpha W(\lambda_{out}, \mu_{out}) - \alpha g(S_{out}) \geq p, \\
&\mu_{mid} > \lambda_{mid} = 2 \int_{0}^{S_{mid}} l(x)dx \geq 0, \\
&\mu_{out} > \lambda_{out} = \int_{S_{mid}}^{S_{mid}+2S_{out}} l(x)dx \geq 0.
\end{aligned}$$

As was the case when choosing between one or two facilities, we again expect that the tension between the firm's capacity costs and consumers' waiting costs will dictate whether the optimal 3-facility solution provides higher profits than the optimal 2-facility solution. Observe that the M3FP formulation does not decouple into independent single-facility subproblems, as was the case for the M2FP. Therefore, the M3FP retains more decision variables, and is harder to solve. Thus, we employ computational methods for the purpose of identifying profit-maximizing solutions for this problem.

As in the example of subsection 5.2, we expect that the optimal number of facilities will increase as we decrease capacity costs or increase the cost of consumers' time. We again consider the baseline values: reservation price $\hat{p}$ = \$100, travel time $g(s) = s$ (proportional to distance), height $H = 1$, and width $L = 10$ units. Recall that with $c =$ \$25 and $\alpha =$ \$1, the two-facility solution dominated the single-facility solution. For this setting, we find that the optimal three-facility solution (determined via numerical analysis) yields *even higher profits* (equal to \$662, versus \$656 for the two-facility problem). However, by further increasing the capacity cost, we find that the two-facility solution again becomes optimal (e.g., at $c =$ \$75); moreover, as we discussed earlier, the single-facility solution is optimal for even higher capacity costs.

In summary, our discussion of the M3FP formulation suggests that the optimal service-design problem becomes significantly more complex as the number of facilities increases, given a non-uniform dispersion of consumers. The complexity arises because the non-uniform consumer density implies that the resulting facilities need not (optimally) be identical—except for the $n = 2$ case, which we discussed in the prior subsection. Our investigations in this section yield three broad insights regarding the structure of the optimal solutions. Firstly, we expect "coverage gaps" of un-served customers only at the outskirts of the metropolitan area (as illustrated in Figure 7.3). Secondly, larger facilities should coincide with regions of high consumer density.[5] Finally, strategies involving fewer facilities are favored when the firm's service capacity is expensive, or consumers' are relatively less time-sensitive.

6. Extensions to consider competition

We now extend the scope of our analysis to consider two firms launching competing service facilities. To facilitate our analyses of the competitive scenario, we consider a single segment of consumers, dispersed uniformly over the $[0, M]$ consumer interval. We assume that the interval length M is sufficiently long to permit positive profits for two service facilities. Furthermore, we assume that the interval length will be fully

[5]Higher consumer densities imply higher capacity utilization due to larger queuing scale economies and lower *average* consumer access costs (lower access costs permit consumers to endure somewhat longer expected waits, with resulting higher utilization, *ceteris paribus*).

covered by the two (profit-maximizing) facilities. In other words, we assume that firms' capacity costs are low enough to ensure that each firm, acting as a monopolist, would want to cover more than half the interval; otherwise, firms would choose not to interact with each other and become "local monopolists". We will analyze both location and capacity decisions, for a fixed price; see Kwasnica and Stavrulaki (2004) for a more thorough development of results and extensions relating to this competitive context.

Several papers have analyzed capacity choices (with queuing effects) in the presence of competition, assuming price to be fixed but without considering consumers' locations (e.g., Kalai, Kamien, and Rubinovitch 1992, and Gilbert and Weng 1998). Sequential and simultaneous pricing and capacity decisions have also been addressed (e.g., Reitman 1991, and Cachon and Harker 2001). In a complementary stream of research, capacity is fixed and the price variable is endogenous (e.g., Lederer and Li 1997, and So 2002). Chayet and Hopp (2002) provide a comprehensive review of these research streams.

Several prior papers, reviewed by Eiselt, Laporte and Thisse (1993), have discussed location and pricing decisions within competitive contexts, but without addressing the impact of capacity and queuing effects. For example, in the classic competitive location framework proposed by Hotelling (1929), two competing firms set their respective facility locations, but there is no capacity decision. Hotelling's decision framework suggested the so-called *principle of minimum differentiation*, implying the two firms' equilibrium strategies would be to locate at the middle of the consumer interval. Later, D'Aspermont et al. (1979) showed that this principle does not hold when price becomes a decision variable, but rather that competitors would, in equilibrium, attempt to *maximize* their differentiation. We now present a competitive model that addresses both capacity and consumer waiting within the linear-city context.

Recall that the net surplus $\hat{p} - \alpha W(\lambda, \mu) - \alpha g(s) - p$ reflects a consumer's product valuation net of the expected waiting, access costs, and product purchase price. When there are two competing firms, consumers will opt to purchase from the facility offering the highest expected surplus (provided that the expected surplus is nonnegative). Hotelling assumed that each consumer would *necessarily* visit the most-preferred location. We extend that basic context to include capacity and queuing effects, and also permit consumers with negative expected surplus to withhold from purchasing. In this generalized setting, we can demonstrate that Hotelling's principle of minimum differentiation does not hold.

We present a two-stage game in which two firms *simultaneously* choose capacities (μ_1, μ_2) and then locations (x_1, x_2), where $x_i \in [0, M]$ for i=1,2; for simplicity, we assume for the remainder of this section that the consumer interval is [0, 1], i.e., we normalize $M = 1$. After the firms implement their strategies, consumers make their optimal purchase decision (i.e., they decide from which facility to purchase, if any). Since we do not address here the issue of price competition, we assume the same price p holds for both firms. Figure 7.4 illustrates a feasible solution for two competing service facilities. In the figure, the expected wait $W(\lambda_1, \mu_1)$ for the facility at the left is smaller than $W(\lambda_2, \mu_2)$ at the right facility, despite the larger threshold distance corresponding to the left facility, i.e., $S_1 > S_2$; therefore, we can conclude that the left facility has higher capacity. Notice in the figure that these facility locations and capacities (given their fixed price p) provide a strictly positive amount of utility to all consumers in an interval of width $2 \cdot z$ between the facilities. We refer to this interval as the "*region of overlap*" for the two facilities. Since we assume that the problem parameters (e.g., capacity and waiting costs) are such that the interval is fully served by the two (profit-maximizing) facilities, we have $2S_1 + 2S_2 = 1 + 2z$. Notice also that in Figure 7.4 the customers at the end points (i.e., at locations zero and one) receive zero surplus. Indeed, under the condition that a monopolist firm would not serve the entire [0, 1] interval, the second stage of the competitive game always satisfies this property.

Result 8: Given fixed capacities (μ_1, μ_2), the corresponding locations (x_1, x_2) represent a Nash equilibrium for the second-stage game if consumers at the endpoints $[0, 1]$ receive zero surplus.

We can explain Result 8 via an intuitive argument, by first recognizing that the $[0, 1]$ endpoints correspond to zero-surplus locations if the facilities are located at exactly the threshold distance (S_i) from the boundaries (as shown in Figure 7.4). When located a distance of S_i from the boundary, unilateral movement of the facility i location (in either direction) reduces the number of consumers visiting that facility, and

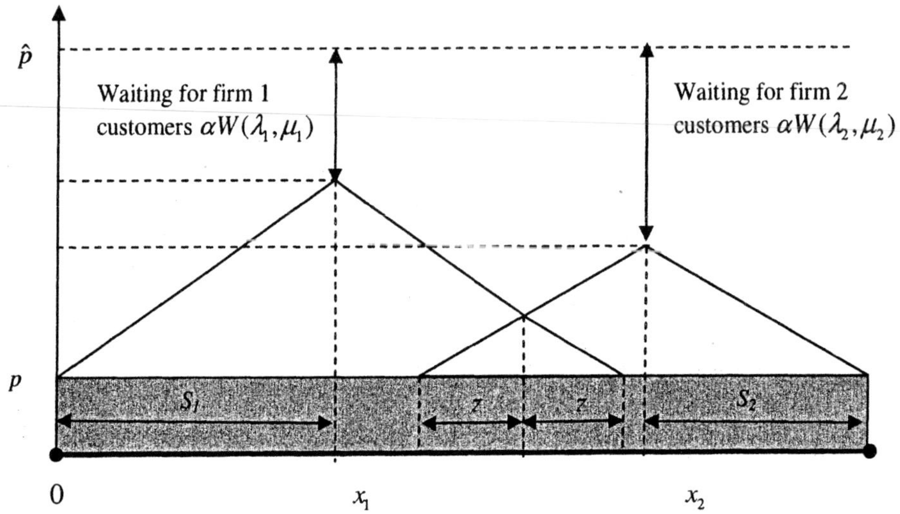

Figure 7.4. Two competing firms located at x_1 and x_2.

thus the firm has no incentive to perturb the location. Since neither firm has an incentive to deviate from its associated strategy, a *Nash equilibrium* exists. Based on Figure 7.4 and Result 8, we can use the threshold distances S_1 and S_2 to infer these equilibrium locations for the two firms (i.e., $x_1 = S_1$ and $x_2 = 1 - S_2$); therefore, we can frame the firms' location decisions using S_1 and S_2.

Having characterized the second stage of the game (i.e., the location decision), we next explore the existence of a Nash equilibrium in the first-stage decision problem (the capacity decision). This is the standard backward-induction approach for multi-stage games, yielding a *subgame perfect equilibrium.* For a given firm 2 capacity μ_2 and corresponding span S_2, we next formulate the *Constrained Competition Problem* (CCP), which determines the best-response for firm 1 (i.e., the capacity μ_1 with corresponding span S_1):

(CCP)

$$\max\{p\lambda_1 - c\mu_1 - B\}$$

subject to:

$$\begin{aligned}
&\hat{p} - \alpha W(\lambda_1, \mu_1) - \alpha g(S_1) \geq p, \\
&\hat{p} - \alpha W(\lambda_2, \mu_2) - \alpha g(S_2) \geq p, \\
&\mu_1 > \lambda_1 = l(S_1 - S_2 + \tfrac{1}{2}) \geq 0, \\
&\mu_2 > \lambda_2 = l(S_2 - S_1 + \tfrac{1}{2}) \geq 0.
\end{aligned}$$

Firm 2's maximization problem, given a choice of capacity by firm 1, is analogous. As mentioned above, we consider only problem contexts for which $2S_1 + 2S_2 > 1$ at optimality, since the two firms simply operate as local monopolists if $2S_1 + 2S_2 \leq 1$. In the formulation, the first constraint is the customer participation constraint for firm 1. The second participation constraint is also needed, because assessing the firm 1 load (λ_1) requires both S_1 and S_2, and this (second) constraint defines S_2. The last two constraints simply assess the demands on each facility. For example, to determine the demand for facility 1, we need only calculate $2S_1 - z$, where $z = S_1 + S_2 - \frac{1}{2}$ denotes the region of overlap (as illustrated in Figure 7.4). The following result addresses the optimal capacity decision of one firm, given the capacity choice of the other firm.

Result 9. Given a feasible capacity choice by one firm, the profit-maximizing solution to the other firm's problem, based upon that information, is unique (i.e., the CCP solution is unique).

This result implies that, for any choice of μ_2 (μ_1), firm 1 (2) has a unique best response. Therefore, a Nash equilibrium exists for the first (capacity) stage of the competitive game. Moreover, we know that for any feasible outcome of the first-stage game, the second-stage game has a unique Nash equilibrium (from Result 8). Therefore, we have established the existence of a subgame perfect equilibrium. Although asymmetric equilibria can potentially occur, we next focus on the existence of *symmetric* subgame perfect equilibria.

We noted above that the CCP is uninteresting if the problem parameters are such that the two firms' can partition consumers into two non-overlapping regions, operating as monopolists. This outcome will occur if, for a given set of problem parameters, the optimal single facility problem (SFP) solution yields $2S^* < 1/2$, in which case a monopolist firm spans less than half the consumer interval. Because we wish to consider settings for which competition is a factor, we restrict our attention to contexts yielding the SFP solution $2S^* > 1/2$.

Result 10. If the monopolistic single-facility problem yields $2S^* > 1/2$, implying that two profit-maximizing facilities would span more than the [0,1] interval, then in the competitive context, the two feasible forms of symmetric subgame perfect equilibria are:

i) $2S_1 + 2S_2 > 1$, implying that consumers at the midpoint of the consumer region derive strictly positive (and equal) surplus from both facilities; and,

ii) $2S_1 = 2S_2 = 1/2$, implying that consumers at the midpoint of the consumer region derive precisely zero surplus from both facilities.

We refer to the two cases in this result as the "overlap" and "standoff" strategies, respectively. Thus, when competition is a factor (i.e., when the SFP yields $2S^* > 1/2$), both the overlap and standoff cases are possible outcomes. The standoff case implies that both firms set their capacity so that their coverage spans exactly half the interval. Effectively, this result proves that competition might cause two firms to settle for half the market share, even though their ideal monopolistic strategy would suggest using a larger facility size to serve a larger number of consumers. The overlap case is less intuitive, since it may seem to imply that each firm "needlessly" offers positive surplus to some strictly positive interval of consumers (specifically, adjacent to the midpoint of the consumer region) who will only visit the competing facility—because that competing facility provides higher utility for those consumers. The underlying rationale for each firm's location and capacity choices in the overlap case is to ensure that the other firm does not have an incentive to increase its market-share by perturbing its corresponding (symmetric) strategy. Moreover, it is possible to predict when the overlap and standoff strategies will apply. For example, we can consider the impact of progressively reducing consumers' time-sensitivity parameter α, which causes corresponding increases to consumers' threshold-distance values.

Therefore, as α decreases, we can ensure that consumers' threshold distances S_1 and S_1 are sufficiently large so that $2S_1 + 2S_2 > 1$. This low time-sensitivity scenario provides one example for which the symmetric subgame perfect equilibria can correspond to the overlap case. Alternatively, for a fixed (positive) value of α, it is possible to define a threshold condition for the capacity cost c, below which the overlap case applies.

It is important to recognize that neither the overlap nor standoff strategies correspond to the principle of minimum differentiation, since the firms do not necessarily locate at the midpoint, but rather choose locations (as per Result 8) which yield zero net surplus at the boundary of the consumer interval. In Hotelling's classic model, there was a significant assumption that consumers *must* purchase from the most convenient firm (i.e., the firm offering the lowest total cost, including the distance inconvenience), in which case the firms have no incentive to move away from the midpoint—thus yielding the principle of minimum differentiation. In contrast, we have included customer participation constraints (capturing both access and waiting costs), and thus we find support for the existence of alternative equilibrium strategies.

7. Summary and conclusions

We have addressed various aspects of the service-facility location and design problem, taking into consideration pricing issues and customer access (i.e., waiting times). To simplify the analysis, we employed the linear city paradigm, following the approach of Hotelling (1929). Using this approach, we developed a model structure that permits analytic solution methodologies. Although we initially assumed a simple setting in which a firm utilized a single facility to serve homogeneous consumers, we generalized that initial model to consider multiple problem extensions and variations. We first relaxed the assumption that consumers are homogenous, and discussed segmentation issues. We subsequently relaxed the assumption that the firm would offer only a single facility, and addressed the optimal spacing between adjacent service facilities. We also relaxed the assumption that consumers be uniformly dispersed, to consider a triangle dispersion that more accurately reflects a metropolitan area, and analyze strategies for single and multiple facilities. Finally, we focused on a competitive scenario with two firms, and discussed conditions under which the classic principle of minimum differentiation does not hold.

These investigations have demonstrated the potential for addressing issues relating to service-facility location via nonlinear optimization techniques. Posing the service-design decision context as a problem in continuous variables (i.e., the capacity, location, and pricing decisions) enables us to derive insights that would otherwise be difficult to develop. There remain a number of interesting questions pertaining to the service-facility design problem that we have not considered. For example, alternative approaches could permit more sophisticated queuing disciplines (we assumed a basic M/M/1 system) or access cost functions (we assumed a simple linear function), treat location decisions within a two-dimensional space, or examine a competitive model with customer segmentation.

References

Brandeau, M.L. 1992. Characterization of the stochastic median queue trajectory in a plane with generalized distances. *Operations Research* **40**(2), 331-341.

Bolch, G., S. Greiner, H. De Meer and K. Trivedi. 1998. *Queueing networks and Markov chains*, John Wiley and Sons, New York.

Cachon, G.P. and P.T. Harker. 2001. Competition and outsourcing with scale economies. *Management Science* 48(10), 1314-1333.

Chayet, S. and W.J. Hopp. 2002. Sequential entry with capacity, price, and lead-time competition. Working paper, The University of Chicago, Chicago, IL.

D'Aspermont, C., J.J. Gabszewicz, and J.F. Thisse. 1979. On Hotelling's "stability in competition". *Econometrica* **47**(5), 1145-1150.

Daskin, M.S. 1995. Network and discrete location: models, algorithms, and applications. John Wiley & Sons.

Desai. P.S. 2001. Quality segmentation in spatial markets: when does cannibalization affect product lime design? *Marketing Science* **20**(3), 265-283.

Dewan, S. and H. Mendelson. 1990. User delay costs and international pricing for a service facility. *Management Science* **36**, 1502-1517.

Dobson, G. and E. Stavrulaki. 2004. Simultaneous price, location, and capacity decisions on a line of time sensitive customers. Working paper, Bentley College, Waltham, MA.

Eiselt, H.A., G. Laporte and J.F. Thisse. 1993. Competitive location model: A framework and bibliography. *Transportation Science* **27**, 44-54.

Gilbert, S.M. and Z.K. Weng. 1998. Incentive effects favor nonconsolidating queues in a service system: the principal-agent perspective. *Management Science* **44**(12) 1662-1669.

Ghosh, A.V. and F. Harche. 1993. Location-allocation models in the private sector: progress, problems, and prospects. *Location Science* **1**, 88-106.

Ha, A.Y. 1998. Incentive-compatible pricing for a service facility with joint production and congestion externalities. *Management Science* **44**(12), 1623-1636.

Halfin, S. and W. Whitt. 1981. Heavy-traffic limits for queues with many exponential servers. *Operations Research* **29**, 567-588.

Hotelling, H. 1929. Stability in competition. *Economic Journal* **39**, 41-57.

Kalai, E., M.I. Kamien and M. Rubinovitch. 1992. Optimal service speeds in a competitive environment. *Management Science* **38**(8), 1154-1163.

Kwasnica A., and E. Stavrulaki. 2004. Competitive location and capacity decisions for facilities serving time-sensitive customers. Working paper, Bentley College, Waltham, MA.

Lederer, P.J. and L. Li. 1997. Pricing, production, scheduling, and delivery time competition. *Operations Research* **45**(3), 407-420.

Mendelson, H. 1985. Pricing computer services: queueing effects. *Comm. ACM* **28**(3), 312-321.

Pangburn, M.S., and E. Stavrulaki. 2004. Capacity Setting with Pricing for Dispersed, Time-Sensitive Segments. Working paper, University of Oregon, Eugene, OR.

Reitman, D. 1991. Endogenous quality differentiation in congested markets. *Journal of Industrial Economics* **39,** 621-647.

Stidham, S. Jr. 1992. Pricing and capacity decisions for a service facility: Stability and multiple local optima. *Management Science* **38**(8), 1121-1139.

So, K.C. 2002. Price and time competition for service delivery. *Manufacturing Service Operations Management* **2**(4), 392-409.

Tayur, S., R. Ganeshan, and M.J. Magazine. 1998. *Quantitative models for supply chain management*, Kluwer Academic Publishers.

Chapter 8

A CONCEPTUAL FRAMEWORK FOR ROBUST SUPPLY CHAIN DESIGN UNDER DEMAND UNCERTAINTY

Yin Mo and Terry P. Harrison
Department of Supply Chain and Information Systems
Penn State University, University Park, PA 16802

1. Introduction

The concept of *robust design* was first introduced by Genuchi Taguchi in the 1960s, and was subsequently accepted in the field of experimental design and quality control. The basic idea of robust design is to make a manufacturing process insensitive to noise factors. Taguchi divided variables into two categories: design factors and noise factors. Design factors are controllable decisions affecting a process. Noise factors are those variables representing field sources of variation. The goal is to design a product or process to be *robust* to noise. One way to determine a robust design is to find a set of design variables that provides the minimum deviation from a target value of the response when noise variables are considered at different levels.

We propose that a similar idea can be applied to the design of supply chains, namely *robust supply chain design.* Supply chain design models determine strategic decisions, such as the most cost-effective location of facilities (including plants and distribution centers), flow of goods, services, information and funds throughout the supply chain, and assignment of customers to distribution centers. We next briefly and selectively review deterministic analytical models and stochastic analytical models for strategic supply chain design.

Williams (1983) creates a dynamic programming model for simultaneously determining the production and distribution batch sizes for all points within a supply chain. The objective of the model is to minimize average cost (including processing cost and inventory holding cost) per period over an infinite horizon. One of the important assumptions is that at each retail node the demand rate is known, constant and continuous. Breitman and Lucas (1987) develop a Production Location Analysis NETwork System (PLANETS, originally implemented in 1974) for General Motors to decide what products to produce, which market to pursue and which resources to use, etc. They use a mixed integer programming model and assume fixed demand. Cohen and Lee (1987) consider a global, deterministic, periodic mixed integer programming model with a nonlinear objective function (PILOT). Arntzen et al. (1995) develop the Global Supply Chain Model (GSCM) for Digital Equipment Corporation. It is a large scale, multi-product mixed integer programming model. The major contribution of this work is considering trade balance, local content, duty and duty drawback in an international supply chain design model. Camm et al. (1997) decompose Procter and Gamble's supply chain problem into two subproblems: a distribution-location problem and a production-sourcing problem. They develop an integer programming model for the first subproblem, and a linear programming model for the second. Again, demand for each product is assumed to be known. Their method allows quick evaluation of alternative solutions and is more interactive than a complex integrated mathematical programming model.

Many papers have concentrated on the stochastic aspects of a supply chain. For example, Cohen and Lee (1988) develop a stochastic optimization supply chain model, which incorporates a series of stochastic submodels (material control, production, stockpile inventory, and distribution). These submodels are optimized individually, and are linked together by a heuristic "optimization" routine. Cohen et al. (1990) build a multi-echelon inventory system to control service levels and spare parts inventory for IBM. They developed a system based on stochastic inventory control models. Lee and Billington (1993) introduce a heuristic model for managing material flows in decentralized supply chains on a site-by-site basis. They rely on a stochastic inventory model and assume that demand is normally distributed. In a later paper, Lee and

Billington (1995) report on the Worldwide Inventory Network Optimizer (WINO) developed for Hewlett-Packard. This model captures material flows and the associated uncertainties of the Vancouver supply chain. They develop single-site inventory models and then integrate all individual site models to cover the complete supply chain. Pyke and Cohen (1993) present a model of a simple integrated production-distribution system. They model a three-level production-distribution system using a Markov chain. Stochastic submodels are used to calculate the values of the included random variables.

Beamon (1998) provides a more comprehensive review of supply chain models and approaches. Typically, these models focus on either the strategic or the operational aspect of a supply chain. However, a paper by Sabri and Beamon (2000) is an exception. They developed a supply chain model that facilitates simultaneous strategic and operational planning. Their model incorporates production, delivery, and demand uncertainty, and provides a performance measure by using multi-objective analysis for the entire supply chain network. Their basic idea is to model the supply chain on two levels using two sub-models. The strategic sub-model optimizes the supply chain configuration and material flow. Uncertainty is incorporated in the operational level sub-model. Various sources of uncertainty are considered, including customer demand, production lead-time, and supply lead-time. An iterative procedure was developed to combine the strategic-level optimization sub-model with the operational-level optimization sub-model to determine the optimal supply chain configuration. At each iteration the unit variable cost (including unit production cost, unit cost of throughput and unit transportation cost) is determined by the operational-level sub-model. The costs are then passed to the strategic level sub-model and a new optimal supply chain configuration is determined. This iterative procedure continues until convergence is achieved for the binary decision variables in the strategic-level sub-model

Recently, Van Landeghem and Vanmaele (2002) describe an approach that they call *robust planning.* They develop a method which uses simulation to develop risk profiles based on uncertain values of various parameters. The outcome is a planning process for tactical demand chain planning. A method for general optimization problems which incorporates noisy, erroneous, or incomplete data is the concept of "robust optimization" by Mulvey et al. (1995).

Although uncertainty has not been often addressed in supply chain design models, it is an important issue in a different class of models-location-allocation models. These models locate plants or warehouses in such a way as to best balance fixed costs and variable transportation costs plus possibly variable operation costs. This class of models is in essence very similar to supply chain design models except that a supply chain design model usually includes more than one echelon of facilities (e.g. at least including plants and warehouses), so the network that is modeled is more complex and hence the problem size is much larger.

The stochastic aspect of the location-allocation model has been studied extensively and various methods have been developed to solve the problem. The uncertain variables may include production and distribution costs, and future demands for the product. Most of these studies modeled demand uncertainty only, and two-stage stochastic programming is the common solution method.

As Louveaux (1993) pointed out, most location-allocation models consider the location and the size of the facilities as the first-stage decisions. Various models include the decision of allocation of clients to facilities as the first-stage decision. For example, Laporte et al. (1994) assume that first-stage decisions consist of determining which facilities to open and the allocation of clients to facility while quantities are decided in second stage. Based on the assumption that a customer's demand can be satisfied from only one facility, they solved the problem using an exact procedure called the "integer-L-shaped method", a branch and cut based procedure. In Logendran and Terrell's (1988) model, demands are stochastic and price-sensitive, and plants are uncapacitated. They used a heuristic to select facilities to open and the allocation of clients to facilities. The quantities to be transported from plants to customers are optimized separately for each plant-customer combination in view of the random demand of each client to each plant. Louveaux and Peeters (1992) addressed the stochastic uncapacitated facility location problem in which demands, variable production and transportation costs as well as selling prices can be random. In addition, they modeled the uncertainty of the random variables via scenarios. They developed a dual-based procedure to solve the problem. Owen and Daskin (1998) and

Louveaux (1993) provide a comprehensive review of stochastic location problems.

A review of the supply chain literature reveals that almost all supply chain design models are deterministic, treating customer demand, production, and transportation processes as known. Although this simplification has dramatically reduced the complexity of modeling the supply chain, the usefulness of these models may also be dramatically reduced.

Uncertainty is one of the most challenging but important aspects of supply chain management. How to model uncertainty in the supply chain design context remains an important and yet unresolved problem. Since supply chain design involves decisions at the strategic level, it is desirable to keep the supply chain configuration unchanged over a relatively long period of time once it is determined. This is a key reason why robust supply chain design can be useful. A robust supply chain design finds a supply chain configuration (or perhaps a group of supply chain configurations) that provides robust and attractive performance while considering many sources of uncertainty. Since demand uncertainty is the major source of supply chain uncertainty, we focus on robust supply chain design methods under demand uncertainty. We also assume that the demand uncertainty can be modeled as a discrete probability distribution, with all possible demand scenarios and their corresponding probabilities known. In the next two sections, we propose a conceptual framework for robust supply chain design under demand uncertainty. In Section 2 we develop the corresponding performance measures and in Section 3, we discuss solution methods.

2. A Framework for Robust Supply Chain Design Under Demand Uncertainty: Performance Measures

There are a variety of performance measures of a supply chain at the strategic level, such as total cost (including fixed and variable cost) or total profit. Ideally, we would like to design a supply chain that has the lowest total cost (or highest total profit) under all possible demand scenarios, and therefore is "robust". However, this is usually unachievable. Therefore, we face tradeoffs when selecting the most "robust" supply chain. A supply chain that has the lowest total cost (or highest total profit) for some demand scenarios may not perform well for others. Hence, we must clearly define "robustness" before discussing any solution methods. We propose the following measures of the "robustness" of a supply chain design.

1 *Minimum total expected cost.* Using expected value as the performance measure when uncertainty appears is very common. This measure leads to a solution that guarantees optimal long-run performance when the potential demand scenarios are encountered repeatedly, with the frequency of appearance of each scenario according to the assumed probability distribution.

2 *Minimum variance of the total cost.* Variance is a standard measure of risk. For those firms that are risk averse, the optimal supply chain design which incorporates attitudes toward risk may be different than when expected cost is the decision criterion.

3 *Minimum total deviation from firm's target value.* This is a slight variant of the second measure. It can be used when a firm has a certain target value and the performance of a supply chain would be regarded as satisfactory as long as the target value can be achieved.

4 *Maximum* $z = E - \lambda V$. This is the mean-variance criterion, where E is the expected value of total profit, V is the variance of the profit, and λ is a non-negative parameter that represents the rate at which the firm is willing to substitute variance for expected value (Jucker and Carlson, 1976). This is a more sophisticated approach which combines the expected value and variance into a single measure. As Jucker and Carlson pointed out, this measure is approximately consistent with the principle of maximizing expected utility of a risk-averse firm if (a) the firm's utility function can be represented by a quadratic function of profit, or (b) the subjective probability distribution of profit is a two-parameter distribution, such as the normal distribution. The difficulty of using this measure lies in the determination of λ, which is subjective and not unique. Different techniques, some of which are borrowed from the field of multi-criteria optimization have been used to find the appropriate value of λ. Although there is no definitive way of choosing the value of λ, people favoring this measure believe that

even the crudest technique of finding the value of λ is likely better than forcing the problem into a single criterion formulation.

5 *Minimum of the maximum deviation* (Gutierrez and Kouvelis, 1995). Define the difference in total cost between the solution from one supply chain configuration and the solution from the optimal supply chain configuration for a given demand scenario $s \in S$ as: $D_s(Y) = Z_s(Y) - Z_s(Y_s^*)$, where Y is any one supply chain configuration and Y_s^* is the optimal supply chain configuration under scenario s. Then for this approach, the robust supply chain configuration is: $Y^{**} = \min_{Y}\{\max_{s \in S} D_s(Y)\}$, implying that the robust supply chain configuration gives the minimum of the maximum deviation over all demand scenarios. This criterion selects the supply chain configuration that performs best under the worst scenario. We may also have a variant of this criterion, which may not select the single best supply chain configuration under the worst scenario, but selects a group of supply chain configurations that guarantee reasonably good performance under all scenarios. It is defined as the following: $PD_s(Y) = \frac{Z_s(Y) - Z_s(Y_s^*)}{Z_s(Y_s^*)}$, $PD_s(Y) \leq p$, where p is a pre-specified number. This criterion selects the supply chain configuration(s) that guarantee(s) the difference in total cost from the optimal value for each demand scenario does not exceed $p\%$.

6 *Multiple criteria.* Each criterion listed from 1 to 5 emphasizes a different perspective of a robust supply chain (except 4, the mean-variance criterion, which combines two criteria), and they are not substitutes for one another. In reality, an ideal robust supply chain design may have to consider more than one criterion. For example, the firm may want to find a supply chain configuration that has good long-run performance (e.g., low total expected cost) and in the short term performs reasonably well under the worst scenario (e.g., low maximum deviation from the optimal). In such cases, multiple criteria methods may be used to select a robust supply chain configuration. More sophisticated optimization techniques developed in the area of multicriteria optimization may be used.

Once the meaning of "robustness" has been clearly defined in the problem context, the next step is to develop solution methods.

3. A Framework for Robust Supply Chain Design Under Demand Uncertainty: Solution Methods

We discuss three different approaches for robust supply chain design. They are *explicit enumeration*, *stochastic programming* and *an enumeration based stochastic programming method*.

3.1 Explicit Enumeration

Conceptually, one may find the best supply chain configuration (i.e. the collection of facilities) that satisfies the criteria of "robustness" by enumeration. This does assume that one can unambiguously order all possible configurations by some scoring mechanism. First, enumerate all possible supply chain configurations and all possible levels of demand, with corresponding probability of occurrence. Second, calculate the total cost for each supply chain configuration under each demand level. Finally, compare the performance of each supply chain configuration based on the criteria of "robustness" selected. For example, we may define the best robust supply chain as the one with the lowest total expected cost. This enumeration method directly follows the standard robust design procedure that is most often used in the area of experimental design or quality control, where demand is treated as a noise factor.

Although the above idea is straightforward conceptually, complete enumeration may only work when the number of potential supply chain configurations and the number of demand scenarios are small. The size of a real supply chain design problem is usually too large for the designer to evaluate the performance of all possible supply chain configurations under all possible levels of demand. Therefore, we must modify the basic idea to reduce the size of the problem. The reduction of the problem size contains two parts: a reduction in the total number of candidate supply chain configurations being considered, and a reduction in the total number of demand scenarios being considered.

(1) *Reduce the number of supply chain configurations being evaluated*

We consider an adjusted random selection procedure, which involves the decision maker's *a priori* beliefs about the importance of each facility. Assume that there are N candidate facilities in the supply chain, implying there are a total of 2^N possible facility configurations. Instead of exploring all possible configurations, we assume that we only select M out of 2^N possible configurations. The selection procedure is as follows:

Step One: Ask the decision maker to assess the probability of including a given facility within the supply chain. We assume that the decision maker has a good idea of which facilities are important based on past experience or managerial insight. To reduce the burden on the decision maker, we may list several choices of probabilities and ask them to select from those. For example, we may ask them: "What is the likelihood of using facility A? Please choose from the following: a) very likely (above 90%); b) moderate (around 50%); and c) very unlikely (below 10%)". Since there are N facilities, the decision maker is expected to answer N such questions.

Step Two: Make a decision on whether a facility should be open or not by a standard random number generating procedure. For example, to decide whether facility A should be included in a particular configuration, we randomly generate a number between 0 and 1. If the generated number is less than the subjective probability that the decision maker assigns to this facility, we assign the value '1' to the binary variable corresponding to facility A (i.e. open facility A). Following the same procedure, we decide whether to include the remaining $(N-1)$ facilities. Thus, we obtain one candidate supply chain configuration. Additional potential configurations are obtained in a similar manner.

Step Three: Adjust the candidate supply chain configurations we obtained in step 2. We eliminate those configurations that are easily identified as inferior or identical choices. We may also add other constraints on the candidate configurations; for example, we may want to open at least one warehouse near major customer zones. After we filter the candidate configurations obtained in step 2, we may generate additional random configurations until the total number of qualified configurations reaches the predetermined number M.

Clearly the proposed method does not guarantee that the "best" supply chain configuration is included in the pre-selected set of candidate configurations. Whether an acceptable configuration can be found may depend on the knowledge of the decision maker and the number of candidate configurations being generated. However, nothing short of complete enumeration will guarantee that the optimal configuration is included. The above method attempts to increase the probability that the optimal configuration is included presumably by use of the decision maker's insight. A major advantage of using explicit enumeration is that it is easy to incorporate different performance measures into the analysis.

(2) *Reduce the total number of demand scenarios being evaluated.*

In a typical supply chain, the number of customer regions may be large and each customer may demand various types of products, hence, the number of demand scenarios that the entire supply chain encounters may explode. Therefore, any solution procedure that is based on complete enumeration of all demand scenarios may be impractical due to enormous computation time. One way to reduce the number of demand scenarios being evaluated is to use sampling. That is, we can take a sample of size n of the demand scenarios and evaluate each supply chain configuration based on the sample we select. Different sampling techniques (random, stratified, etc.) may be used to select the representative sample.

3.2 Stochastic Programming

Another approach for incorporating uncertainty in an optimization problem is the use of stochastic programming. Stochastic programming uses expected value as the performance measure for finding the optimal solutions when uncertainty exists. In the current problem context, a robust supply chain is defined as the configuration that has the lowest expected cost under demand uncertainty. When this performance measure is justified, the problem of finding a robust supply chain configuration can be modeled as a classic two-stage stochastic program with fixed recourse, where decisions are made sequentially. Decisions that have to be made before demand is realized are the first-stage decisions, and decisions that are made after demand is realized are the second-stage decisions.

The standard formulation of a two-stage stochastic linear program with fixed recourse is:

$$\min \quad z = c^T x + E_\xi[\min q(w)^T y(w)] \tag{8.1}$$

$$s.t. \quad Ax = b, \tag{8.2}$$

$$T(w)x + Wy(w) = h(w), \tag{8.3}$$

$$x \geq 0, y(w) \geq 0 \tag{8.4}$$

The first-stage decisions are represented by the $n_1 \times 1$ vector x. Corresponding to x are the first-stage vectors and matrices c, b and A of sizes $n_1 \times 1$, $m_1 \times 1$, $m_1 \times n_1$, respectively. In the second stage, a number of random events $w \in \Omega$ may be realized. For a given realization of w, the second-stage problem data $q(w)$, $h(w)$ and $T(w)$ become known, where $q(w)$ is $n_2 \times 1$, $h(w)$ is $m_2 \times 1$, and $T(w)$ is $m_2 \times n_2$. ξ is the random vector. The above formulation is the simplest form of a two-stage stochastic program, since both the first stage decision variables x, and the second stage decision variables $y(w)$, are linear. However these variables need not be restricted to be linear. They can be integer variables or nonlinear variables, with a corresponding increase in the difficulty of solving the stochastic program. In our current problem context, x refers to the supply chain configuration variables, which are decided in the first stage and have to be integers, w refers to one demand scenario realization which belongs to the set Ω, the complete set of all possible demand scenarios.

The use of two-stage stochastic programming has been discussed extensively in the literature (for example, see Birge and Louveaux (1997); Kall and Wallace (1994); Van der Vlerk (2001)). Solving problem (8.1)-(8.4) directly usually is not very efficient. The more efficient solution methods of a two-stage linear stochastic program are most frequently based on a cutting plane technique called the L-shaped method. This method is based on building an outer linearization of the recourse cost function and a solution of the first-stage problem plus this linearization (Birge and Louveaux, 1997). Birge and Louveaux (1997) also discuss alternative algorithms, including one method based on Dantzig-Wolfe decomposition and another based on generalized programming.

When the decision variables at the first stage or/and at the second stage are integers, the two stage stochastic program is a stochastic integer program. An efficient solution method for this case is the integer L-shaped method, which is a combination of the regular L-shaped method and branch and bound.

We formulate the two-stage stochastic supply chain design problem based on the following setting. Consider a simple supply chain that comprises a number of plants, warehouses and customer zones. Each customer may order different types of products. Products are distributed to customers from open warehouses and warehouses receive products from open plants. The objective function is to minimize total expected cost, which includes the fixed cost of opening plants and warehouses, expected shipping cost from plants to warehouses and from warehouses to customers, and expected outsourcing cost when customer demands cannot be satisfied from warehouse shipments. The problem is formulated as follows:

Index Sets:

$\mathcal{I}$: customer zones $\mathcal{I} = \{i \mid 1, \ldots, I\}$

$\mathcal{J}$: warehouses $\mathcal{J} = \{j \mid 1, \ldots, J\}$

$\mathcal{L}$: products $\mathcal{L} = \{l \mid 1, \ldots, L\}$

$\mathcal{K}$: plants $\mathcal{K} = \{k \mid 1, \ldots, K\}$

$\mathcal{S}$: demand scenarios $\mathcal{S} = \{s \mid 1, \ldots, S\}$

Parameters:

f_k: fixed cost for the kth plant

g_j: dixed cost for the jth warehouse

C_{ijl}: unit cost for shipping one unit of product l from warehouse j to customer i

T_{jkl}: unit cost for shipping one unit of product l from plant k to warehouse j

$Prob_s$: probability of demand scenario s

Out_{il}: unit outsourcing cost of product l for customer i

dd_{ils}: demand of customer i for product l under scenario s

W_j: capacity of warehouse j

D_k: capacity of plant k

W: maximum number of open warehouses

P: maximum number of open plants

Decision Variables:

P_k: binary variable for plant k; 1 if plant k open and 0 otherwise

Z_j: binary variable for warehouse j; 1 if warehouse j open and 0 otherwise

X_{ijls}: number of units of product l shipping from warehouse j to customer i under demand scenario s

Y_{jkls}: number of units of product l shipping from plant k to warehouse j under demand scenario s

O_{ils}: number of outsourcing units of product l for customer i under demand scenario s

Formulation:

$$\min \quad z = \sum_k f_k P_k + \sum_j g_j Z_j + \sum_i \sum_j \sum_l \sum_s C_{ijl} X_{ijls} Prob_s + \sum_j \sum_k \sum_l \sum_s T_{jkl} Y_{jkls} Prob_s + \sum_i \sum_l Out_{il} \sum_s O_{ils} Prob_s \tag{8.5}$$

$$s.t. \quad dd_{ils} \leq O_{ils} + \sum_j X_{ijls} \quad \forall i, l, s \tag{8.6}$$

$$\sum_i \sum_l X_{ijls} \leq Z_j W_j \quad \forall j, s \tag{8.7}$$

$$\sum_j Z_j \leq W \tag{8.8}$$

$$\sum_i X_{ijls} \leq \sum_k Y_{jkls} \quad \forall j, l, s \tag{8.9}$$

$$\sum_j \sum_l Y_{jkls} \leq D_k P_k \quad \forall k, s \tag{8.10}$$

$$\sum_k P_k \leq P \tag{8.11}$$

$$P_k, Z_j \in \{0, 1\} \quad \forall k, j \tag{8.12}$$

$$Y_{jkls}, X_{ijls}, O_{ils} \geq 0 \quad \forall i, j, k, l, s \tag{8.13}$$

The objective function (8.5) minimizes the sum of total fixed cost, total expected transportation cost from warehouses to customers, total expected transportation cost from plants to warehouses and total expected outsourcing cost. Constraint set (8.6) guarantees that demand can be satisfied by shipments from open warehouses and outsourcing. Constraint set (8.7) ensures that the number of product units shipped out of each open warehouse does not exceed the corresponding warehouse capacity. Constraint (8.8) ensures that the number of open warehouses does not exceed a pre-specified limit. Constraint set (8.9) ensures that the total number of product units shipped out of warehouses to customers does not exceed the total number shipped into the warehouses from the plants. Constraint set (8.10) enforces the capacity restriction on each open plant. Constraint (8.11) restricts the total number of open plants.

This formulation is slightly different from a typical deterministic supply chain design formulation, where demand is assumed to be satisfied with shipments from warehouses. In the above formulation, demand may not be always satisfied due to uncertainty. If demand cannot be satisfied with shipments from warehouses, we assume it can be satisfied with outsourcing at a much higher cost.

To strengthen the formulation (8.5)-(8.13), notice that we may add two extra constraints:

$$\sum_k D_k P_k \geq \sum_{s=Low} \sum_i \sum_l dd_{ils} \tag{8.14}$$

$$\sum_j W_j Z_j \geq \sum_{s=Low} \sum_i \sum_l dd_{ils} \tag{8.15}$$

Constraint (8.14) guarantees that the total capacity of open plants exceeds the total customer demand when the low demand scenarios for all customers and all products are used. Similarly, constraint (8.15) makes sure that the total capacity of open warehouses exceeds the total customer demand when the low demand scenarios for all customers and all products are used. These two constraints are usually satisfied when the outsourcing cost (or penalty cost) is high. This is similar to the feasibility constraint in a typical deterministic formulation, where total shipments to the customers are assumed to exceed total customer demand.

3.3 An Enumeration Based Stochastic Programming Approach

We propose an alternative formulation, where we use one binary variable to represent a complete supply chain configuration in the first stage. Notice that in formulation (8.5)-(8.13), we use two binary variables to model a supply chain configuration, one for each plant and one for each warehouse. In order to use this alternative formulation, we need a complete enumeration of the supply chain configurations before we start solving the stochastic programming problem.

Index Sets:

$\mathcal{I}$: customer zones $\mathcal{I} = \{i \mid 1, \ldots, I\}$

$\mathcal{J}$: warehouses $\mathcal{J} = \{j \mid 1, \ldots, J\}$

$\mathcal{L}$: products $\mathcal{L} = \{l \mid 1, \ldots, L\}$

$\mathcal{K}$: plants $\mathcal{K} = \{k \mid 1, \ldots, K\}$

$\mathcal{S}$: demand scenarios $\mathcal{S} = \{s \mid 1, \ldots, S\}$

$\mathcal{C}$: supply chain configuration $\mathcal{C} = \{c \mid 1, \ldots, C\}$

Parameters:

f_k: fixed cost for the kth plant

g_j: dixed cost for the jth warehouse

TC_{ijl}: unit cost for shipping one unit of product l from warehouse j to customer i

TW_{jkl}: unit cost for shipping one unit of product l from plant k to warehouse j

$Prob_s$: probability of demand scenario s

Out_{il}: unit outsourcing cost of product l for customer i

dd_{ils}: demand of customer i for product l under scenario s

W_j: capacity of warehouse j

D_k: capacity of plant k

P_{ck}: open plant in configuration c; 1 if plant k is in configuration c, 0 otherwise.

Z_{cj}: open warehouse in configuration c; 1 if warehouse j is in configuration c, 0 otherwise

$Cost_c$: Total facility fixed cost of configuration c

$Cost_c$ is derived as follows: $Cost_c = \sum_k P_{ck} f_k + \sum_j Z_{cj} g_j$

Decision Variables:

$Config_c$: binary variable for configuration c; 1 if configuration c is selected and 0 otherwise

X_{ijls}: number of units of product l shipping from warehouse j to customer i under demand scenario s

Y_{jkls}: number of units of product l shipping from plant k to warehouse j under demand scenario s

O_{ils}: number of outsourcing units of product l for customer i under demand scenario s

Formulation:

$$\min \quad z = \sum_c Cost_c Config_c + \sum_i \sum_j \sum_l \sum_s TC_{ijl} X_{ijls} Prob_s + \sum_j \sum_k \sum_l \sum_s TW_{jkl} Y_{jkls} Prob_s + \sum_i \sum_l Out_{il} \sum_s O_{ils} Prob_s \tag{8.16}$$

$$s.t. \quad dd_{ils} \leq O_{ils} + \sum_j X_{ijls} \quad \forall i, l, s \tag{8.17}$$

$$\sum_i \sum_l X_{ijls} \leq Z_{cj} Config_c W_j \quad \forall j, s, c \tag{8.18}$$

$$\sum_i X_{ijls} \leq \sum_k Y_{jkls} \quad \forall j, l, s \tag{8.19}$$

$$\sum_j \sum_l Y_{jkls} \leq P_{ck} Config_c D_k \quad \forall k, s, c \tag{8.20}$$

$$\sum_c Config_c = 1 \tag{8.21}$$

$$Config_c \in \{0, 1\} \quad \forall c \tag{8.22}$$

$$Y_{jkls}, X_{ijls}, O_{ils} \geq 0 \quad \forall i, j, k, l, s \tag{8.23}$$

The objective function (8.16) minimizes the sum of total fixed cost, total expected transportation cost from warehouses to customers, total expected transportation cost from plants to warehouses and total expected outsourcing cost. Constraint set (8.17) guarantees that demand can be satisfied by shipment from open warehouses and outsourcing. Constraint set (8.18) ensures that for each supply chain configuration, the number of product units shipped out of each open warehouse in the configuration do not exceed the corresponding warehouse capacity. Constraint set (8.19) ensures that the total number of product units shipped out of warehouses to customers does not exceed the total number shipped into the warehouses from the plants. Constraint set (8.20) enforces the capacity restriction on each opening plant. Constraint (8.21) ensures that only one supply chain configuration is selected.

The advantage of formulation (8.16)-(8.23) is its potential reduction in problem size if only a small number of candidate supply chain configurations need to be considered. The number of candidate supply chain configurations may depend on the restrictions an industry has on the plants or warehouses only.

4. A SAMPLING BASED APPROACH

As evidenced in both stochastic programming formulation (8.5)-(8.13) and (8.16)-(8.23), if the total number of demand scenarios is large, it is difficult to solve the mixed integer programming problem in a reasonable amount of time. We suggest a method to reduce the number of demand scenarios evaluated in the model by the use of sampling. Instead of finding the optimal configuration using all possible demand scenarios, we determine the supply chain design configuration based on a sample of size n demand scenarios selected from all possible demand scenarios. Different sampling techniques may be used to find the representative samples of the demand distribution. The sampling technique should be chosen carefully, since it affects the quality of the solution. For example, the uniform design method, which has been applied successfully in the area of experimental design, is a good candidate. This method samples uniformly in the space of all demand scenarios.

Although we divide the solution methods into categories including enumeration and stochastic programming, the procedures used in the enumeration method could also be applied to stochastic programming. To reduce the total number of candidate supply chain configurations to be considered in the enumeration method, we suggest the use of an adjusted random generating procedure that makes use of a decision maker's knowledge regarding each facility to increase or decrease the likelihood of inclusion in a sample. This procedure can also be applied to the stochastic programming method. We may treat this procedure as a way to generate potentially good columns that can be used with the mixed integer program to reduce the problem size.

5. CONCLUSIONS

Supply chain design models are often deterministic, which may dramatically reduce the effectiveness of these models. We consider a major source of supply chain uncertainty-demand uncertainty, as it relates to the strategic design of supply chains. We propose a conceptual framework for designing a robust supply chain under demand uncertainty. A robust supply chain design finds a supply chain configuration (or a group of supply chain configurations) that provide(s) robust and attractive performance over a variety of possible demand scenarios. We define various performance measures of "robustness" and propose three different types of solution methods: explicit enumeration, stochastic programming and an enumeration based stochastic programming. All of the solution methods we propose are intended to find a good solution within a reasonable amount of computational time. A major component of these heuristics is to intelligently sample demand scenarios to represent the entire set of possible demand scenarios.

References

Arntzen, B.C., Brown, G.G., Harrison, T.P., and Trafton, L.L. 1995. Global supply chain management at Digital Equipment Corporation, *Interfaces* **25**, 69-93.

Beamon, B.M. 1998. Supply chain design and analysis: models and methods, *International Journal of Production Economics* **55**, 281-294.

Birge, J.R. and Louveaux, F. 1997. *Introduction to Stochastic Programming*, Springer Series in Operations Research. Springer-Verlag, New York.

Breitman, R. and Lucas, J. 1987. PLANETS: A modeling system for business planning, *Interfaces* **17**, 94-106.

Camm, J., Dull, F., Evans, J., Sweeney, D. and Wegryn, G. 1997. Blending OR/MS, Judgment, and GIS: Restructuring P&G's supply chain, *Interfaces* **27**, 127-142.

Cohen, M.A., and Lee, H.L. 1987. Manufacturing strategy: concepts and methods, in *The management of productivity and technology in manufacturing*, Ed. P. Kleindorfer, Plenum Press, New York 1987, 153-188.

Cohen, M.A. and Lee, H.L. 1988. Strategic analysis of integrated production-distribution systems: Models and methods, *Operations Research* **36**, 216-228.

Cohen, M.A., Kamesam, P.V., Kleindorfer, P., Lee, H. and Tekerian, A. 1990. *Optimizer: IBM's multi-echelon inventory system for managing service logistics*, *Interfaces* **20**, 65-82.

Gutierrez, G.J. and Kouvelis, P. 1995. A robustness approach to international sourcing, *Annals of Operations Research* **59**, 165-193.

Jucker, J.V. and Carlson, R.C. 1976. The simple plant-location problem under Uncertainty, *Operations Research* **24**, 1045-1055.

Kall, P., and Wallace, S.W. 1994. Stochastic Programming, *Wiley-Interscience series in systems and optimization*, John Wiley & Sons, Chichester, England.

Kleijnen, J.P.C. 1974. *Statistical Techniques in Simulation*, Part I, Marcel Dekker Inc., New York.

Laporte, G., Louveaux, F.V. and Hamme, L.V. 1994. Exact Solution to a Location Problem with Stochastic Demands, *Transportation Science* **28**, 95-103.

Lee, H.L. and Billington, C. 1993. Material management in decentralized supply chains, *Operations Research* **41**, 835-847.

Lee, H.L., and Billington, C. 1995. The evolution of supply chain management models and practice at Hewlett-Packard, *Interfaces* **25**, 42-63.

Logendran, R., and Terrell, M.P. 1988. Uncapacitated plant location-allocation problems with price sensitive stochastic demands, *Computer and Operations Research* **15**, 189-198.

Louveaux, F.V. 1993. Stochastic Location Analysis, *Location Science* **1**, 127-154.

Louveaux, F.V., and Peeters, D. 1992. A dual-based procedure for stochastic facility location, *Operations Research* **40**, 564-573.

Mulvey, J.M., Vanderbei, R.J. and Zenios, S.A. 1995. Robust Optimization of Large-Scale Systems, *Operations Research*, **43**, 264-281.

Owen, S.H. and Daskin, M.S. 1998. Strategic facility location: a review, *European Journal of Operational Research* **111**, 423-447.

Pyke, D.F. and Cohen, M.A. 1993. Performance characteristics of stochastic integrated production-distribution systems, *European Journal of Operational Research* **68**, 23-48.

Pirkul, H. and Jayaraman, V. 1998. A multi-commodity, multi-plant, capacitated facility location problem: formulation and efficient heuristic solution, *Computers & Operations Research* **25**, 869-878.

Sabri, E.H. and Beamon, B.M. 2000. A multi-objective approach to simultaneous strategic and operational planning in supply chain design, *Omega* **28**, 581-598.

Van der Vlerk, M.H. 1996. Stochastic Programming Bibliography. World Wide Web, http://mally.eco.rug.nl/biblio/stoprog.html, 1996-2001.

Van Landeghem, H. and Vanmaele, H. 2002. Robust planning: a new paradigm for demand chain planning, *Journal of Operations Management* **20**, 769-783.

Williams, J. 1983. A hybrid algorithm for simultaneous scheduling of production and distribution in multi-echelon structures, *Management Science* **29**, 77-92.

Chapter 9

THE DESIGN OF PRODUCTION-DISTRIBUTION NETWORKS: A MATHEMATICAL PROGRAMMING APPROACH

Alain Martel

Network Organization Technology Research Center (CENTOR), Université Laval, Québec, Canada, G1K 7P4

Abstract This text proposes a mathematical programming approach to design international production-distribution networks for make-to-stock products with convergent manufacturing processes. Various formulations of the elements of production-distribution network design models are discussed. The emphasis is put on modeling issues encountered in practice which have a significant impact on the quality of the logistics network designed. The elements discussed include the choice of an objective function, the definition of the planning horizon, the manufacturing process and product structures, the logistics network structure, demand and service requirements, facility layouts and capacity options, product flows and inventory modeling, as well as financial flows modeling. Major contributions from the literature are reviewed and a number of new formulation elements are introduced. A typical model is presented, and the use of successive mixed-integer programming to solve it with commercial solvers is discussed. A more general version of the model presented and the solution method described were implemented in a commercial supply chain design tool which is now available on the market.

Keywords: Logistics network design, Supply chain engineering, Location-allocation problems, Capacity planning, Technology selection, Mathematical programming.

1. Context

How many production and distribution centers should a company have to satisfy the demand of its targeted markets? Where should they be located and what should their mission be? What supply sources should they use? What technologies should they install for production, storage, shipping and receiving? Which sub-contractors and public warehouses should they do business with? What means of transportation should they choose? All of these questions are related to strategic and tactical logistics network design issues, which are critical for the success of modern manufacturing and distribution companies. This text proposes a mathematical programming approach to analyze several of these logistics network design issues.

The exact nature of the logistics network design problems encountered in practice depends very much on the industrial context in which they occur. For example:

- The design problem to solve for a high volume consumer goods manufacturer is very different than the problem found in a highly customized make-to-order products industry or in a slow moving repair parts distribution context. In a make-to-stock industry, the order-to-delivery time depends on the positioning of finished goods inventories but, in a make-to-order context, it depends on manufacturing lead times and on the depth of penetration of customer orders in the supply chain, i.e., on the positioning of semi-finished product or raw material inventories.

- When manufacturing resource acquisition, deployment and/or allocation decisions are considered, the nature of the production process must also be taken into account. In some industries, manufacturing processes are divergent: several products are made from a common raw material (e.g. pulp and paper industry, meat industry, etc.). In other sectors the manufacturing processes are convergent: several raw-materials and components are assembled into finished products. In some industries, the manufacturing processes may even include feedback loops.

- Networks covering several countries lead to much more complex design problems than single-country networks. Factors such as exchange rates, transfer prices, duties and income taxes must then be taken into account.

The detailed discussion of all these variants is beyond the scope of this paper. In what follows our coverage focuses on the design of international production-distribution networks for make-to-stock products with convergent manufacturing processes.

As can be seen, logistics network design problems, as defined here, integrate several subproblems which have been treated separately in the literature: capital investment planning for the acquisition of new capacity, technology selection, facility location and manufacturing/distribution resource allocation problems. Capacity expansion problems are usually posed as multi-year capital investments problems under uncertainty (Freidenfelds, 1981; Luss, 1982). The financial planning aspects of the problem, such as real options (Trigeorgis, 1996), are predominant in the analysis and the logistics aspects are highly aggregated. Technology selection problems can be seen as an extension of capacity planning where there are several alternative capacity types available (Fine, 1993, Paquet et al., 2004). At the other extreme, resource allocation problems deal with detailed plant loading and inventory placement decisions under the assumption that the plant/warehouse network configuration is fixed (Glover et al., 1979; Cohen and Moon, 1991; Mazzola and Schantz, 1997). They often consider a single year planning horizon divided into several seasons. The literature on basic discrete location models (Francis et al., 1992; Daskin, 1995; Sule, 2001) concentrates on single period, single echelon, geographical deployment problems. A lot of the effort in this field has been devoted to finding efficient solution methods for a set of well defined problems. Some extensions to classical facility location problems are reviewed by Revelle et al. (1996) and by Owen et al. (1998). An abundant literature exists on location, capacity acquisition and technology selection problems. An integrated review of the early work done in these fields is found in Verter and Dincer (1992). Supply chain design models incorporate elements of all the sub-problems discussed previously. Geoffrion and Powers (1995) and Shapiro et al. (1993) discuss the evolution of strategic supply chain design models and Vidal and Goetschalckx (1997) present many of these models. Shapiro (2001) provides an excellent coverage of several supply chain modeling issues.

In this paper, various formulations of the elements of production-distribution network design models are discussed. The emphasis is put on modeling issues encountered in practice which have a significant impact on the quality of the logistics network designed. The elements discussed include the choice of an objective function, the definition of the planning horizon, the manufacturing process and product structures, the logistics network structure, demand and service requirements, facility layouts and capacity options, product flows and inventory modeling, as well as financial flows modeling. Major contributions from the literature are reviewed and a number of new formulation elements are introduced. A typical model is presented, and the use of successive mixed-integer programming to solve it with commercial solvers is discussed. A more general version of the model proposed and the solution method described were implemented in the *Supply Chain Studio*, a commercial supply chain design tool sold by *Modellium.* This tool was used to optimize the production-distribution network of several multinational companies, including Domtar, one of the largest Pulp and Paper Company in North-America.

2. Modeling approach

Performance evaluation

Although most of the logistics network design models presented in the literature adopt a total system cost minimization objective, this does not necessary lead to the creation of a competitive advantage. Low cost is an order winning criteria valued by several customers but it is not the only one (Hill, 1999). Delivery time, quality and flexibility are other valued criteria which are affected by the logistics activities and resources of the firm. In a make-to-stock industry, for example, the order-to-delivery time depends on the positioning of finished goods inventories in the logistics network and it is a criteria as important as cost for the evaluation of network designs. As explained by Porter (1985), it is the additional value given by customers to such an order winning criterion that creates a competitive advantage. Figure 9.1 illustrates the cost accumulation process and the impact of inventory positioning on customer delivery times for a simple multi-echelon (stage) supply chain. As can be seen, costs accumulate as the products pass through the procurement, production and distribution stages, and value is added when the finished products are purchased by customers. The cost of support activities can be interpreted here as all the non-logistic costs incurred by the firm. The response time depends on whether the customers are served from a local or regional warehouse or from a production/distribution center or, more generally, on the distance between a customer and its supply facility. When delivery time is shorter, more revenues are generated through a price premium and/or an increased market share. Total system cost, maximum delivery time and total revenue figures are therefore associated with any logistics network design. In order to evaluate the performance of various designs, their cost and delivery time can be plotted on a graph, as shown in Figure 9.2a). The non-dominated designs are located on an *efficient-frontier*, and any of these designs could constitute a good solution for a firm (Rosenfield, 1985). However, if the impact of delivery time on prices and on demand, and thus on total revenue, is taken into account, as shown in Figure 9.2b), the design maximizing the value added (Total revenue – Total logistics network cost – Cost of support activities) by the logistics network can be identified. Ideally, the objective to pursue should therefore be to find the logistics network design maximizing *net revenues*. In an international context, since different countries have different taxation levels, one should rather seek to maximize after tax global net revenues in a reference currency. Unfortunately, it is not always possible in practice to model the impact of delivery time on price and demand. When this is the case, one should at least sketch the efficient frontier by finding the designs minimizing total system costs for a set of predetermined delivery times. Despite the fact that an abundant literature exists on the impact that quality and flexibility may have on competitiveness, little work has been done to explicitly incorporate them as performance criterion in logistics network design models. By associating different technologies to different quality levels, quality can often be treated in a way similar to delivery times. Some dimensions of flexibility, such as operational flexibility in global networks under exchange rate risk (Kogut and Kulatilaka, 1994; Huchzermeier and Cohen, 1996), have been studied, but more research is needed on the incorporation of the various dimensions of flexibility into network design models. The model presented in what follows seeks to maximize after tax net revenues, taking the impact of delivery times on revenues into account.

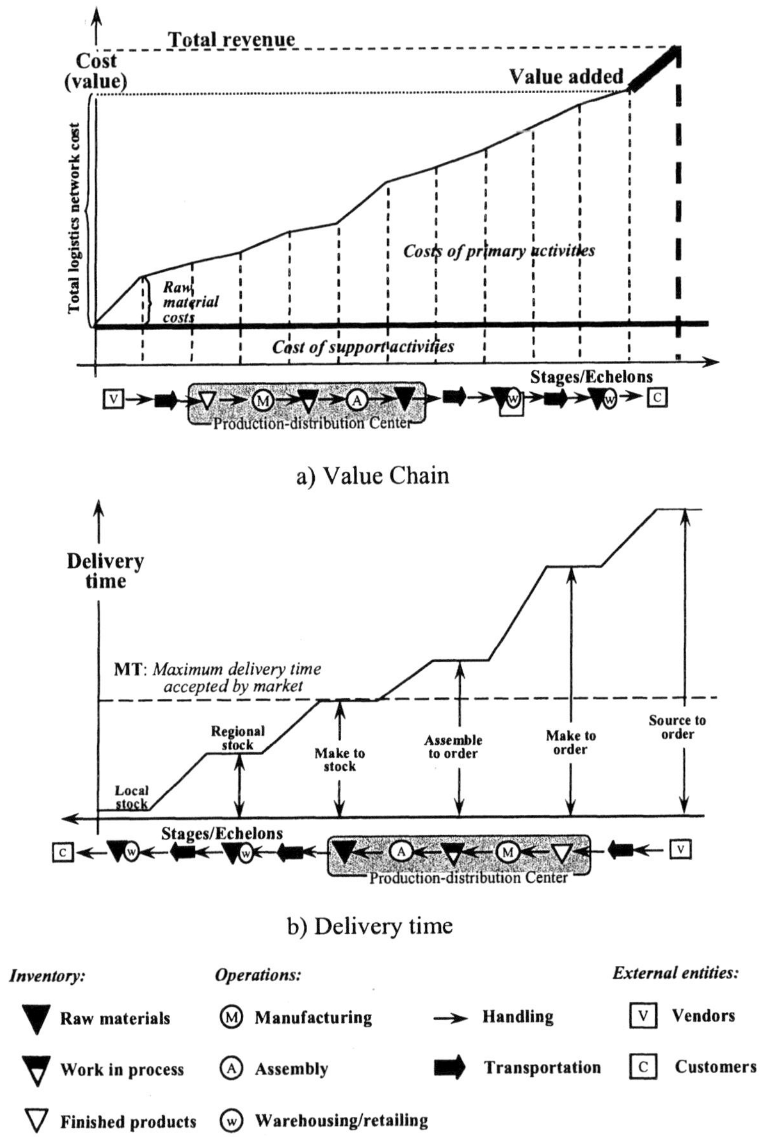

Figure 9.1. Costs, value added and delivery time in the supply chain.

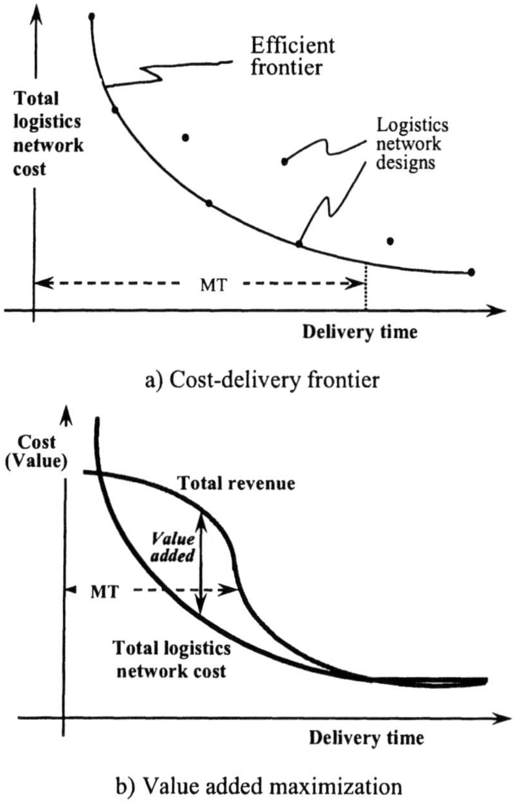

Figure 9.2. Performance evaluation methods.

Planning horizon and uncertainty

In capital intensive industries, capacity expansion decisions may require the explicit consideration of a planning horizon including as much as ten years (Everett et al., 2000, 2001). On the other end, when product supply and/or demand is seasonal, decisions on production and inventory levels for each network location must be made on a quarterly basis or even on a monthly basis. This means that the number of planning periods in logistics network design models could be very large. In addition to the explosion of problem size, using a long planning horizon makes the gathering of meaningful information on the future business environment extremely difficult. Some approach to reduce this complexity must therefore be used in practice.

To clarify this issue, let us first make a distinction between the notions of *season* and *period*. In most design models, 0-1 variables are associated with capacity acquisition and deployment decisions and continuous variables to resource allocation decisions (production and inventory levels, network flows). A *multi-period model* is concerned with the change of state of the network structure (number, location, technology and capacity of facilities) over the long term (typically several one year periods). A *multi-season model* is concerned with the change of mission of the network resources during a planning period (typically months or quarters during a year). Several formulations presented as multi-period models in the literature are in fact single-period multi-season models (Cohen et al., 1989; Arntzen et al., 1995; Dogen and Goetschalckx, 1999). Multi-period models usually concentrate on capacity investment decisions and they limit themselves to single echelon network structures (Shulman, 1991, Everett et al., 2000; Bhutta et al., 2003). Following the pioneering work of Pomper (1976), some authors have also proposed multi-period scenario based stochastic programming models (Eppen et al., 1989; Ahmed et al., 2001; Everett et al., 2001).

Most of the models published in the literature are deterministic single-period mathematical programs (Geoffrion and Graves, 1974; Brown et al., 1987; Cohen and Lee, 1989; Cohen and Moon, 1990; Pirkul and Jayaraman, 1996; Lakhal et al., 2001; Vidal and Goetschalckx, 2001; Cordeau et al., 2002; Paquet et al., 2004). It is understood, however, that since the acquisition and deployment decisions have long-term effects, their analyses must span multiple periods and the model must be either run sequentially over some finite time horizon or, when size permits, expanded to incorporate multiple time periods directly (Cohen and Lee, 1989). Also, the fact that the future is uncertain requires the examination of several scenarios with respect to the firm's strategic options and the evolution of its internal and external environment (Shapiro, 2001) or, when size permits, the transformation of the model into a multi-stage stochastic program with recourse (Birge and Louveaux, 1997). Keeping this in mind, the approach presented in what follows yields deterministic multi-season logistics network design models. The following set is used to denote the planning horizon:

T = Seasons of the planning horizon ($t \in T$).

Modeling process and product structures

In order to arrive at a general production-distribution network design model for a given industrial context, a generic conceptual model of the manufacturing process of the industry must first be elaborated. Such a conceptual model treats products and production stages in an aggregate manner to capture the essence of the manufacturing process, but without concern for operational details (Shapiro, 2001). It can take the form of an activity network or of a bill-of-materials, as illustrated in Figure 9.3 (Lakhal et al., 1999). In these conceptual models, products are grouped into product families and some activities may be an amalgam of several operations. It is common to use process network representations in process manufacturing environments such as petro-chemicals, food, pulp and paper, pharmaceutical, etc. (Brown et al., 1987; Dogan and Goetschalckx, 1999; Philpott and Everett, 2001; Vila et al., 2003). In such contexts, associated with each activity are a number of methods (recipes) that describe how inputs are transformed into outputs using different potential technologies. In discrete parts manufacturing industries, however, a bill-of-materials representation is usually more adequate (Cohen and Moon, 1990; Arntzen et al., 1995; Paquet et al., 2004). This is the approach taken in this paper.

More specifically, the following product structure modeling assumptions are made. Products are classified in families $p \in P$ requiring the same type of production capacity or supplied by the same vendors. Products available only from external suppliers are considered as raw material (RM) and other products can be manufactured in the network plants. The manufactured products (MP) are sub-assemblies (SA) or make-to-stock (MS) finished products. The semifinished products can come partly from external suppliers and partly from the network plants. The aggregated bill-of-materials, illustrated in Figure 9.3b), is an acyclic directed graph. The number associated with the edge $(p'p)$ of the bill-of-materials graph indicates the quantity of the product p' needed to make one product of type p. It is assumed that the vertices of this graph are numbered in topological order, i.e., that for each edge $(p'p)$, we have $p > p'$. A *technology* is defined by the set of products it can manufacture/store, and it is assumed that the bill-of-materials is independent of the technology used. As illustrated in Figure 9.3b), the capacity required to produce one product can be provided by either flexible or dedicated

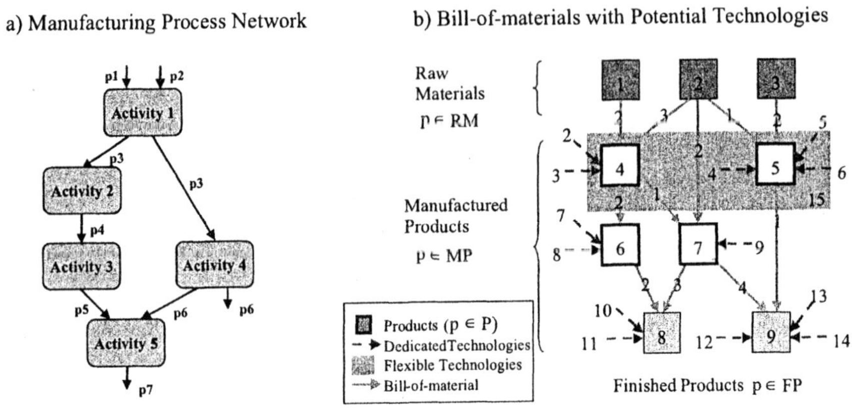

Figure 9.3. Process and product structures.

technologies. Dedicated technologies are associated with only one product family, but flexible technologies can be used to make several product families. Similarly, the capacity needed to stock the finished products can be provided by a set of potential storage technologies. When a facility is used, a technology for the reception and shipping of products must also be implemented. To simplify, it is assumed that this technology can be used for any products and that its capacity can be expressed adequately in terms of the facility outflows.

The notation used to model product structures and technologies is the following:

P = Product families ($p \in P$).

RM = Raw material families ($RM \subset P$).

MP = Manufactured product families, i.e., sub-assemblies and finished products ($MP \subset P$).

SA = Sub-assemblies families ($SA \subset P$).

$g_{pp'}$ = Quantity of product p needed to make one product p'.

KW = Receiving/shipping/handling technologies ($k \in KW$).

KM = Production technologies ($k \in KM$).

KS = Storage technologies ($k \in KS$).

q_{pk} = Technology k capacity consumption rate per unit of product p.

$\overline{w}_p$ = Average weight of family p products in standard weight units.

Network optimization model structure

The structure of logistics networks can be represented by a directed graph. The network nodes correspond to supply sources, to existing facilities, to sites where it would be possible to build or buy a production or distribution center, to the facilities of potential partners (subcontractors, public warehouses, 3PL consolidation centers, etc.) or to demand zones. The network arcs represent the flow of products between the nodes. The specification of the structure of the network and of the mission of its facilities is an important strategic decision. Two approaches to the problem are found in the literature, as illustrated in Figure 9.4. A popular modeling approach has been to assume *a priori* that a multi-echelon structure is required (Geoffrion and Graves, 1974; Cohen and Lee, 1989; Pirkul and Jayaraman, 1996; Vidal and Goetschalckx, 2001). This limits the mission of the facilities to a predetermined role (e.g. intermediate product plant, final product plant, distribution center) and it forbids product flows between facilities on the same echelon. In some contexts, this approach can be far from optimal. In practice, the same facility often has multiple roles: a production-distribution center may produce both intermediate and finished products and serve as a shipping point to some customers; a warehouse close to a supplier may serve as a central warehouse for this supplier's products, but as a local distribution center for other products, etc. For this reason, other authors do not impose any *a priori* echelon structure and expect the optimization model to determine the best structure and mission for the facilities (Arntzen et al., 1995; Paquet et al., 2004).

Two approaches are also used in the literature to model flows in the network. One of them associates decision variables to *paths* in the network (Geoffrion and Graves, 1974; Martel and Vankatadri, 1999). This is particularly appropriate for multi-echelon distribution networks. However, for this approach to work, the product flowing on the path must not change between the first node and the last node, which cannot hold for production facilities. For this reason, models incorporating more than one production echelon either associate decision variables to the *arcs* of the network (Arntzen et al., 1995, Dogan and Goetschalckx, 1999; Vidal and Goetschalckx, 2001; Cordeau, 2002, Paquet et al., 2004), or they

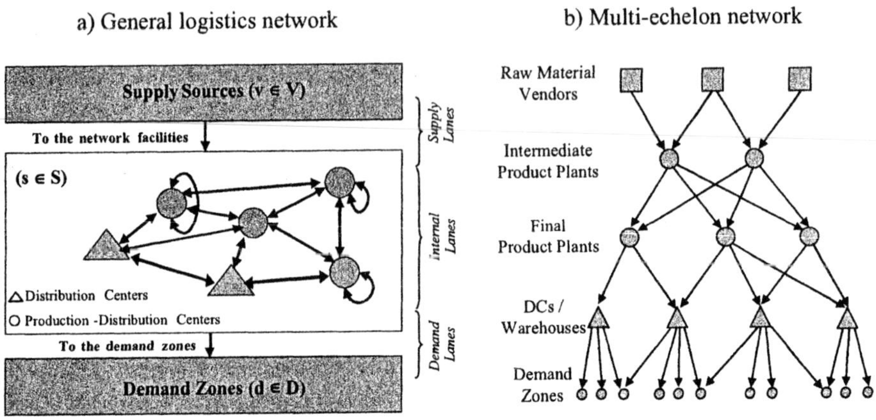

Figure 9.4. Potential logistics network.

use a hybrid approach (Cohen and Moon, 1990). The model presented in this paper is an arc-based formulation for the general logistics network illustrated in Figure 9.4a). Three types of nodes, located in several countries, are present in the network: external vendors ($v \in V$), internal potential facility sites ($s \in S$) and demand zones ($d \in D$). A list of potential internal sites (S) must be identified *a priori* and classified as either production-distribution center sites (Spd) or distribution center sites (Sd). This list usually includes the location of the current facilities, of public warehouses or sub-contractors which could be included in the network, of existing facilities which could be purchased or rented, and of lands where a new facility could be constructed. It is possible also to limit the mission given to potential sites by restricting the set of production (KM_s) and storage (KS_s) technologies which can be implemented in a site, or the set of products (P_s) which can be produced/stored in a site. The network arcs are associated with transportation lanes. Three types of arcs are distinguished: supply arcs, internal arcs and demand arcs. The internal arcs adjacent to a site s are defined by the set of origins of its inbound arcs (S^i_{ps}) and the set of destinations of its outbound arcs (S^o_{ps}). Similar node input and output sets are defined for supply and demand arcs. A continuous decision variable F_{pnst} is associated with the flow of a product p on lane (n, s) in season t. Given that a real logistics network may include several hundred thousand arcs, defining these sets and flow variables in practice is not trivial and it requires the use of an automated arc generation mechanism.

The customer ship-to locations are grouped into demand zones (D). The definition of these demand zones depends on the product-markets (M) of the company and on the geographical dispersion of ship-to points (Ballou, 1994). It is assumed that the company operates national divisions in several countries $o \in O$, and that each of these divisions covers a set of distinct product-markets $m \in M_o$ constituted of several demand zones $d \in D_m$. A market is characterized by a distinct price and service policy. It is assumed that the products shipped to a demand zone can come from more than one distribution center. This is common today because companies tend to operate centralized selling organizations independent of the DC's. Modifying the model however to enforce single DC sourcing is not difficult. Similarly, vendors in close geographic proximity who provide products in the same family can be aggregated into a supply source (V). It is assumed that the seasonal quantity of product which can be supplied by a vendor is bounded. The following sets, indices, parameters and variables are required to define a potential logistics network:

S = Potential network sites ($s \in S$).

O = Countries of the network sites ($o \in O, o(s) =$ country of site s).

S_o = Potential sites in country o.

Sd = Potential distribution center sites ($Sd \subset S$).

Spd = Potential production-distribution center sites ($Spd \subset S$).

Sd_{max} = Upper bound on the number of distribution centers in the network.

Spd_{max} = Upper bound on the number of production-distribution centers in the network.

V_p = Vendors of raw material $p \in RM$ or of manufactured product $p \in MP$.

b_{pvt} = Upper bound on the quantity of raw material p which can be supplied by vendor v in season t.

S^o_{pn} = Set of potential sites (output destinations) which can receive product p from node n.

S^i_{ps} = Set of potential sites (input sources) which can ship product p to site s.

M_o = Potential product-markets in country o ($m \in M = \cup_{o \in O} M_o$).

D_m = Demand zones in product-market m ($d \in D = \cup_{m \in M} D_m$).

D_o = Demand zones in country o ($D_o = \cup_{m \in M_o} D_m$).

$m(d)$ = Product-market of demand zone d.

D^o_{ps} = Set of demand zones (output destinations) which can receive product p from node s.

P_s = Products which can be manufactured/stocked on site s.

P_{ks} = Products which can be manufactured/stocked with technology k on site s.

KM_{ps} = Production technologies which can be used to manufacture product p on site s ($KM_{ps} \subset KM, KM_s = \cup_p KM_{ps}$).

KS_{ps} = Storage technologies which can be used to stock product p on site s ($KS_{ps} \subset KS,\ KS_s = \cup_p KS_{ps}$).

F_{pnst} = Flow of product p between node $n \in V_p \cup S^i_{ps}$ and site s during season t.

The essence of the logistics network design problem boils down to finding an optimal mapping of the product/activity structure onto the potential network structure.

Modeling demand, prices and customer service

Although most of the models available in the literature assume that demand is given and not affected by the logistics network design, this is clearly not realistic. As explained earlier, demand depends on logistics outputs such as delivery times, and the market may be prepared to pay a price premium to obtain these outputs. To take this into account, it is assumed that the company has a choice of marketing policies $i \in I_m$ for each of its product-markets m (Vila et al., 2004). A marketing policy $i \in I_m$ is characterized by the price P_{pdit} the market is prepared to pay for each product $p \in P$ in the demand zones $d \in D_m$ during seasons $t \in T$. It is also characterized by a maximum delivery time and possibly by other value criteria. These value criteria are related to the network design by defining the set of sites in the potential network S^i_{pdi} which could deliver the value characteristics of marketing policy $i \in I_{m(d)}$, for each product p. It is further assumed that the largest demand $\overline{x}_{pdit}$ the company can expect for product p in demand zone d, when marketing policy $i \in I_{m(d)}$ is used, can be estimated, and that the company has minimum market penetration objectives $\underline{x}_{pdt}$ for each of its product-markets.

In this context, the following notation is required to model the demand:

I_m = Marketing policies considered for market m ($i \in I_m$).

S^i_{pdi} = Set of potential sites (input sources) which can ship product p to demand zone d, when marketing policy $i \in I_{m(d)}$ is selected.

P_{pdit} = Amount received for the sales of product p to demand zone d in season t when marketing policy $i \in I_{m(d)}$ is used (in the demand zone country currency).

$\underline{x}_{pdt}$ = Lower bound on the flow of product p to demand zone d in season t imposed by the market penetration objectives of the company.

$\overline{x}_{pdit}$ = Upper bound on the flow of product p to demand zone d in season t imposed by the largest market share the company can expect when marketing policy $i \in I_{m(d)}$ is used.

Y^M_{mi} = Binary variable equal to 1 if marketing policy $i \in I_m$ is used for market m and to 0 otherwise.

F_{psdit} = Flow of product p between site s and demand zone d during season t, when marketing policy $i \in I_{m(d)}$ is selected.

Parallel arcs are defined between the network sites s and the demand zones d to model the flow of products F_{psdit} under the different marketing policies $i \in I_{m(d)}$. Using these flow variables and the marketing policy selection variables Y^M_{mi}, it is seen that the seasonal sale targets

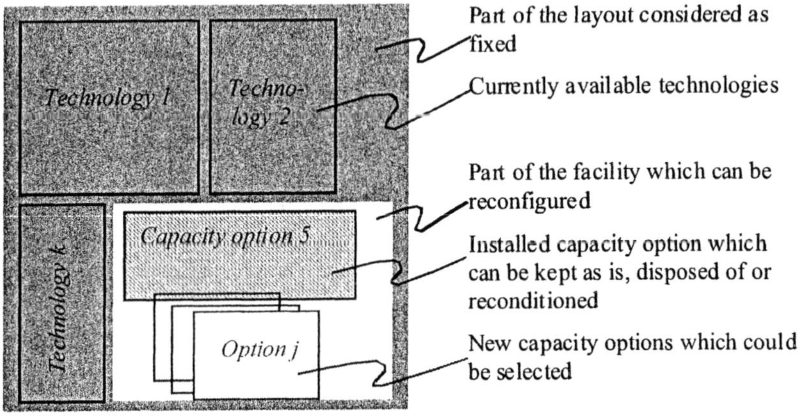

Figure 9.5. Illustration of the Facility Layout Concept.

of the company must respect the following demand and policy selection constraints:

$$\underline{x}_{pdt} Y^M_{m(d)i} \leq \sum_{s \in S^i_{pdi}} F_{psdit} \leq \overline{x}_{pdit} Y^M_{m(d)i} \quad t \in T, p \in P, d \in D, i \in I_{m(d)} \quad (9.1)$$

$$\sum_{i \in I_m} Y^M_{mi} \leq 1 \quad m \in M \quad (9.2)$$

Modeling facility layouts and capacity options

The technical and economic characteristics of the facilities which could be operated on the network sites can be specified with a *facility layout.* The facility layout concept is illustrated schematically in Figure 9.5. A layout $l \in L_s$ for site s is composed of two parts: a fixed part, which cannot be changed and a variable part defining an area which could be reengineered. The technologies implemented in the fixed part are predetermined and they specify the products they can make/stock, the seasonal capacity available b_{lskt}, stated in the units of its technology, and the associated variable costs. The variable part defines an area E_{ls} available for the installation of a set of predetermined *capacity options.* A facility layout may include only a fixed or a variable part. Several layouts can be considered for each site s, including a *status-quo* layout if there is already a facility on the site, and alternative potential layouts corresponding to new construction or reconfiguration opportunities. Numerous capacity options can be available to implement a given technology in the variable part of a layout. An option $j \in J$ can correspond to capacity already in place, to a reconfiguration of an installed equipment to increase its capacity or to the addition of new resources. In this last case, different options can be associated with equipment of different size to reflect economies of scale. Moreover, the simultaneous inclusion of dedicated capacity options and flexible capacity options allow for the modeling of economies of scope. When dealing with a potential equipment replacement/reconfiguration, the options associated with the new potential equipment cannot be selected at the same time as the *status-quo* option, which leads to the definition of mutually exclusive sub-sets of options JR_{ls}^n, $n = 1, ..., N_{ls}$, for some facility layouts. Each option $j \in J$ is characterized by a seasonal capacity, b_{jt}, stated in the units of its technology, by the floor space e_j required to install it, as well as by a fixed cost and a variable cost per product.

The notation required to include layout and option choice decisions in the model is the following:

L_s = Potential facility layouts for site s ($l \in L_s$). By convention, the index $l = 1$ is given to the current layout if there is a facility on site s at the beginning of the horizon.

L_{ks} = Potential facility layouts including fixed technology k capacity for site s ($l \in L_s$).

b_{lskt} = Technology k capacity available for season t in the fixed part of layout l of site s.

E_{ls} = Total area of the variable part of layout l for site s.

Y_{ls} = Binary variable equal to 1 if layout l is used on site s and to 0 otherwise.

Y_{0s} = Binary variable equal to 1 if site s is not used and to 0 otherwise.

J_s = Potential capacity options which can be installed on site s ($j \in J = \cup_{s \in S} J_s$).

J_{ks} = Potential technology k capacity options which can be installed on site s ($J_{ks} \subseteq J_s$).

J_{ls} = Potential capacity options which can be installed on site s when layout l is used ($J_{ls} \subseteq J_s$).

N_{ls} = Number of mutually exclusive option subsets (equipment replacement/reconfiguration) in J_{ls}.

JR^n_{ls} = Mutually exclusive option subsets in J_{ls} ($n = 1, ..., N_{ls}$).

$k(j)$ = Technology of capacity option j.

b_{jt} = Technology $k(j)$ capacity provided by option j for season t.

e_j = Area required to install capacity option j.

Z_j = Binary variable equal to 1 if capacity option j is installed and to 0 otherwise.

Using the layout selection variables Y_{ls}, the following constraints must be included in the model to ensure that at most one layout is selected for each site, and that the total number of facilities used does not exceed the maximum number of distribution and production-distribution centers desired:

$$\textstyle\sum_{l \in L_s} Y_{ls} + Y_{0s} = 1 \qquad s \in S \tag{9.3}$$

$$\sum_{s \in Sd} \sum_{l \in L_s} Y_{ls} \leq Sd_{max} \tag{9.4}$$

$$\sum_{s \in Spd} \sum_{l \in L_s} Y_{ls} \leq Spd_{max} \tag{9.5}$$

Using the capacity option selection variables, Z_j, the following constraints must also be included to ensure that, for a given site, the area required by the selected options does not exceed the area available in the selected layout, and that mutually exclusive options are not selected:

$$\textstyle\sum_{j \in J_{ls}} e_j Z_j - E_{ls} Y_{ls} \leq 0 \quad s \in S, l \in L_s \tag{9.6}$$

$$\textstyle\sum_{j \in JR^n_{ls}} Z_j \leq 1 \qquad s \in S, l \in L_s, n = 1, \ldots, N_{ls} \tag{9.7}$$

Modeling flows and inventories

In addition to deciding the marketing policies, sites, layouts and capacity options to use during the planning horizon, tactical seasonal decisions must be made on the quantity of products to manufacture, the seasonal stocks to accumulate and the flows in the network. This requires the modeling of flows and inventories in the network facilities and the consideration of capacity constraints. Several types of facilities are used in practice and the flow patterns in and between the centers, as well as the nature of the inventory kept, can be quite different from one type of facility to the other. To simplify, it is assumed here that there is a single type of production-distribution center (P-DC) and a single type of distribution center (DC) in the network. The structure of the P-DC's considered is illustrated in Figure 9.6: they include different production technologies and they can manufacture any component or finished products associated with these technologies in the bill-of-materials (Figure 9.3b). When the manufacturing of a product is completed, it is either used to make other products, moved to the facility inventory or shipped to another facility. It is assumed that there is no seasonal inventory of input products and that the plant warehouse contains only products to be shipped directly to the market. All products made for other internal centers are shipped directly to these facilities after production. On the other hand, it is assumed that DC's can receive products from vendors or from any other site, and that they can ship products to the market or to any other site.

The additional notation required to model flows and inventories is the following:

I_{psd} = Sub-set of the demand zone d marketing policies which include site s as a valid supply site for product p $(I_{psd} = \{i | s \in S^i_{pdi}\} \subset I_{m(d)})$.

$\underline{W}_s$ = Lower bound on the seasonal throughput, in standard weight units, required to use distribution center s.

$\overline{X}_{pst}$ = Upper bound on the quantity of family p products which can be manufactured in plant s during season t.

$\underline{X}_{pst}$ = Lower bound on the quantity of family p products which can be manufactured in plant s during season t, when plant s is used.

$\rho^{(0)}_{pst}$ = Number of seasons of product p order cycle and safety stocks kept on average in site s during seasont t (inverse of the inventory turnover ratio).

β_p = Order cycle and safety stocks (max. level)/(avg. level) ratio for product p.

X_{pkst} = Quantity of product p produced in plant s with technology

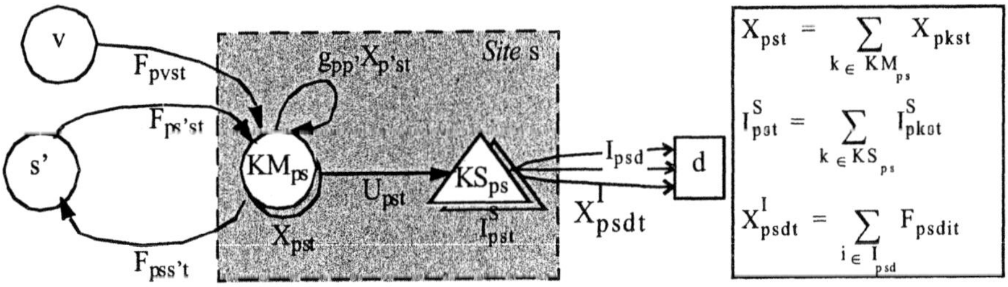

Figure 9.6. Flow of sub-assembly $p \in SA$ in a production-distribution center.

$k \in KM_{ps}$ during season t.

U_{pst} = Quantity of product p transferred to the stock of site s during season t.

I^S_{pkst} = Seasonal inventory of product p stored in site s with technology $k \in KS_{ps}$ at the end of season t.

X^I_{pst} = Throughput of product p in distribution center $s \in S$ for season t.

Any valid network optimization model must ensure that there is equilibrium between the flows of products entering a node, their transformation, stocking and/or consumption in the node and the flows of products exiting the node. The case of a sub-assembly $p \in SA$ manufactured in center $s \in Spd$ is illustrated in Figure 9.6, for a season t. In the part of facility s used by the production technologies (KM_{ps}), the quantity of sub-assembly p manufactured (X_{pst}) must be sufficient to satisfy the needs of the other network sites ($F_{pss't}$), the transfers to the seasonal stock (U_{pst}), and the sub-assembly requirements generated by client products in the bill-of-materials ($g_{pp'}X_{p'st}$, $\forall p' > p$, i.e., by all products p' including subassembly p), taking into account the sub-assemblies coming from other internal sites ($F_{ps'st}$, $\forall s' \neq s$) and from external suppliers (F_{pvst}, $\forall v \in V_p$). In order to have flow equilibrium, the following relations must therefore be satisfied:

$$\sum_{k \in KM_{ps}} X_{pkst} + \sum_{n \in S^i_{ps} \cup V_p} F_{pnst} \geq$$

$$\sum_{s' \in S^o_{ps}} F_{pss't} + \sum_{p' > p} g_{pp'} \sum_{k \in KM_{p's}} X_{p'kst} + U_{pst} \quad t \in T, p \in MP, s \in Spd \quad (9.8)$$

Similarly, in the part of facility s used by the storage technologies (KS_{ps}), additions and withdrawals from the seasonal inventory must be

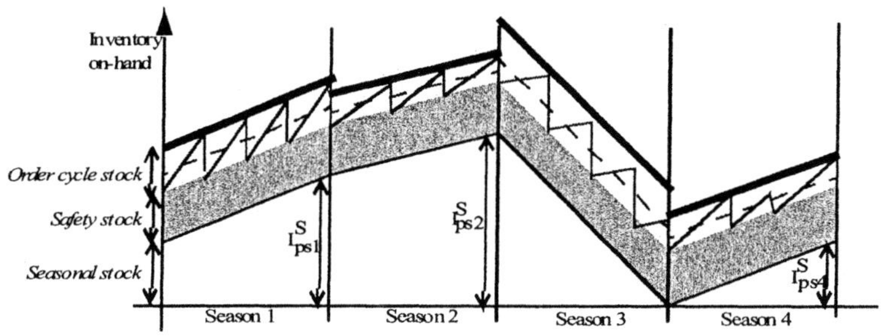

Figure 9.7. Behavior of product p inventory in a distribution center.

accounted for. This yields the following inventory accounting equations:

$$\sum_{k \in KS_{ps}} I^S_{pkst-1} + U_{pst} = X^I_{pst} + \sum_{k \in KS_{ps}} I^S_{pkst} \quad t \in T, p \in MP,$$
$$\left(I^S_{pks0} = I^S_{pks|T|}\right) \qquad s \in Spd \qquad (9.9)$$

where

$$X^I_{pst} = \sum_{d \in D^o_{ps}} \sum_{i \in I_{psd}} F_{psdit} \quad t \in T, p \in P, s \in Spd \qquad (9.10)$$

Seasonal stocks are included in the model to allow the smoothing of production over the planning horizon. As illustrated in Figure 9.7, the seasonal stocks at the beginning and the end of the horizon must therefore be the same, i.e., we must have $I^S_{ps0} = I^S_{ps|T|}$, for all p and s.

The quantity X_{pst} of products which can be manufactured in a given P-DC is limited by the layout and the capacity options selected for that center. This imposes the following capacity constraints:

$$\sum_{p \in P_{ks}} q_{pk} X_{pkst} \leq \sum_{l \in L_{ks}} b_{lskt} Y_{ls} + \sum_{j \in J_{ks}} b_{jt} Z_j \quad t \in T, s \in Spd,$$
$$k \in KM_s \qquad (9.11)$$

In some contexts, it may also be necessary to bound the quantity of products manufactured in a facility, which can be done with the constraints:

$$\underline{X}_{pst} \sum_{l \in L_s} Y_{ls} \leq \sum_{k \in KM_{ps}} X_{pkst} \leq \overline{X}_{pst} \sum_{l \in L_s} Y_{ls} \quad t \in T, p \in P_s,$$
$$s \in Spd \qquad (9.12)$$

To simplify the presentation, it is assumed that offset trade and local content rules do not restrict national production. However, the inclusion of constraints to that effect would not present any difficulty (Cohen et al., 1989; Arntzen et al., 1995). For raw materials, the flow equilibrium constraints required for the P-DC's are:

$$\sum_{n \in V_p \cup S^i_{ps}} F_{pnst} \geq \sum_{s' \in S^o_{ps}} F_{pss't} + \sum_{p'>p} g_{pp'} \sum_{k \in KM_{p's}} X_{p'kst} + X^I_{pst} \qquad t \in T, p \in RM, s \in Spd \quad (9.13)$$

For distribution centers, the flow equilibrium constraints and the inventory accounting equations required are the following:

$$\sum_{n \in V_p \cup S^i_{ps}} F_{pnst} \geq X^I_{pst} \quad t \in T, p \in RM, s \in Sd \qquad (9.14)$$

$$\sum_{k \in KS_{ps}} I^S_{pkst-1} + \sum_{n \in S^i_{ps} \cup V_p} F_{pnst} = X^I_{pst} + \sum_{k \in KS_{ps}} I^S_{pkst} \quad t \in T, p \in MP,$$
$$\left(I^S_{pks0} = I^S_{pks|T|}\right) \qquad s \in Sd \quad (9.15)$$

where

$$X^I_{pst} = \sum_{n \in S^o_{ps}} F_{pnst} + \sum_{d \in D^o_{ps}} \sum_{i \in I_{psd}} F_{psdit} \quad t \in T, p \in P, \; s \in Sd \qquad (9.16)$$

Also, in most contexts, management does not want to operate small DC's. This leads to the imposition of the following lower bounds on DC throughput:

$$\sum_{p \in P_s} \overline{w}_p X^I_{pst} \geq \underline{W}_s \sum_{l \in L_s} Y_{ls} \quad t \in T, s \in Sd \qquad (9.17)$$

The three types of inventories to take into account in the model are represented in Figure 9.7: seasonal stocks, safety stocks and order cycle stocks. The level of order cycle stocks and of safety stocks depends on the inventory management policies and rules used by the company and on the ordering behavior of customers. Using inventory theory it can be shown (Martel, 2002) that, for a given product supply lead time, the relationship between the seasonal flow of a product in a warehouse and the average level of cycle and safety stocks required to support this

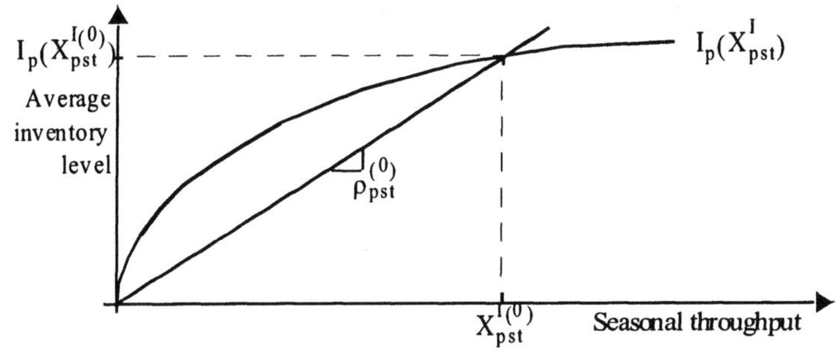

Figure 9.8. Relationship between inventory levels and material flows in a DC.

flow is concave. To simplify things, in what follows, the effect of delivery lead times is assumed to be negligible (see Martel and Vankatadri (1999) for a model incorporating lead times). More specifically, it is assumed that the average inventory level of product p required during season t in warehouse s to support the throughput X_{pst}^I is given by the power function:

$$I_p\left(X_{pst}^I\right) = a_p\left(X_{pst}^I\right)^{b_p} \quad t \in T, p \in P, s \in S \tag{9.18}$$

where a_p and b_p are parameters obtained by regression analysis, from historical or simulation data (Ballou, 1992). The inventory-throughput relationship (9.18) is illustrated in Figure 9.8. Note that, although it is assumed that $I_p()$ is independent of s, in practice it may be more appropriate to use a different function for each type of site (P-DC's, crossdocking centers, local DC's, etc.). Most network design models proposed in the literature do not take the *risk pooling* effects captured by function (9.18) into account: they assume either explicitly (Cohen and Moon, 1990; Arntzen et al., 1995; Dogan and Goetschalckx, 1999) or implicitly that the relationship between inventory levels and throughput is linear. If the historical throughput level and average inventory level observed for product p, in distribution center s, for the most recent season t, are $X_{pst}^{I(0)}$ and $I_p\left(X_{pst}^{I(0)}\right)$, respectively, then the ratio $X_{pst}^{I(0)}/I_p\left(X_{pst}^{I(0)}\right)$ is the familiar inventory turnover ratio, and its inverse

$$\rho_{pst}^{(0)} = I_p\left(X_{pst}^{I(0)}\right)/X_{pst}^{I(0)} \tag{9.19}$$

is the number of seasons of inventory kept in stock. Assuming that the relationship between inventory level and throughput is linear boils down to approximating $I_p\left(X_{pst}^I\right)$ by $\rho_{pst}^{(0)} X_{pst}^I$, as illustrated in Figure 9.8. Such an approximation may not be too bad in the vicinity of $X_{pst}^{I(0)}$, but the DC's throughputs are not known before the optimization model is solved and they can be far from historical values (mainly if a new DC is open or an existing DC is closed), which means that calculating inventory levels with historical inventory turnover ratios can be completely inadequate. An effort is therefore made in this paper to take risk pooling effects into account explicitly.

Function (9.18) provides the average inventory of product p required to support throughput X_{pst}^I. This quantity is needed to calculate inventory holding costs, but it cannot be used directly to calculate the space required to store the products in a warehouse because this space is proportional to maximum inventory levels and not to average inventory levels. For product p, the maximum level of cycle and safety stocks to be stored in a season is obtained by multiplying the average inventory level $I_p\left(X_{pst}^I\right)$ by an amplification factor β_p. In practice, the parameters β_p, $p \in P$, are estimated statistically from the company data on the inventory held in its facilities. From this it is seen that the throughputs and seasonal inventory levels in the DC's must respect the following storage space capacity constraints:

$$\sum_{p \in P_{ks} \cap MP} q_{pk} I_{pkst}^S + \sum_{p \in P_{ks}} q_{pk} \beta_p I_p(X_{pst}^I) \leq \sum_{l \in L_{ks}} b_{lskt} Y_{ls} + \sum_{j \in J_{ks}} b_{jt} Z_j \quad t \in T, s \in S, k \in KS_s \tag{9.20}$$

The flows in all the facilities are also restricted by their receiving and shipping capacity. It is assumed here that this restriction can be properly expressed in terms of the total facility outflows, which leads to the following capacity constraints:

$$\sum_{p \in P_{ks}} q_{pk} \left(\sum_{n \in S_{ps}^o} F_{psnt} + \sum_{d \in D_{ps}^o} \sum_{i \in I_{psd}} F_{psdit} \right) \leq \sum_{l \in L_{ks}} b_{lskt} Y_{ls} + \sum_{j \in J_{ks}} b_{jt} Z_j \quad t \in T, s \in S, k \in KW \tag{9.21}$$

Finally, the limited supply of raw materials and sub-assemblies which can be obtained from external vendors leads to the following inbound flow constraints:

$$\sum_{s \in S_{pv}^{o}} F_{pvst} \leq b_{pvt} \quad t \in T, p \in RM \cup SA, v \in V_p \tag{9.22}$$

Modeling costs

The different costs and revenues associated with the arcs and nodes of a typical multinational logistics network are shown at the top of Figure 9.9, and their correspondence with the decision variables of the optimization model is indicated at the bottom of the figure. Note that several of the costs which are incurred in the network facilities are assigned to the models flow variables. For example, supply-order and receiving costs are assigned to inbound flow variables and customer-order, shipping as well as cycle and safety inventory holding costs are assigned to outbound flow variables. Note also that, in an international context, to take transfer prices and taxes into account correctly, it is necessary to derive an income statement for each network facility. This implies that certain costs associated with the network arcs must be split into the part paid by the origin and the part paid by the destination. For example, for arc (s, s') in Figure 9.9, the origin node pays the customer-order, shipping, transportation, inventory-in-transit and cycle/safety costs but the destination pays the supply-order and receiving costs. In addition, transfer prices are charged to node s' but they are a revenue for node s. Transportation costs are paid by the origin s but they are passed on to the destination s' and duties are paid by the destination. Note finally that, to compute after tax net revenues, the fixed selling costs of the selected markets and the fixed cost of support activities must also be taken into account.

The cost assignments described in Figure 9.9 are based on the following cost modeling assumptions:

- The prices and costs associated with the nodes of the network are given in local currency. The costs associated with the arcs of the network are given in source currency. Exchange rates are known and constant during the planning horizon considered.
- The fixed costs associated with facility layouts reflect potential changes of state (closing an existing facility, building or buying a new facility, changing the layout of a facility, etc.) and fixed

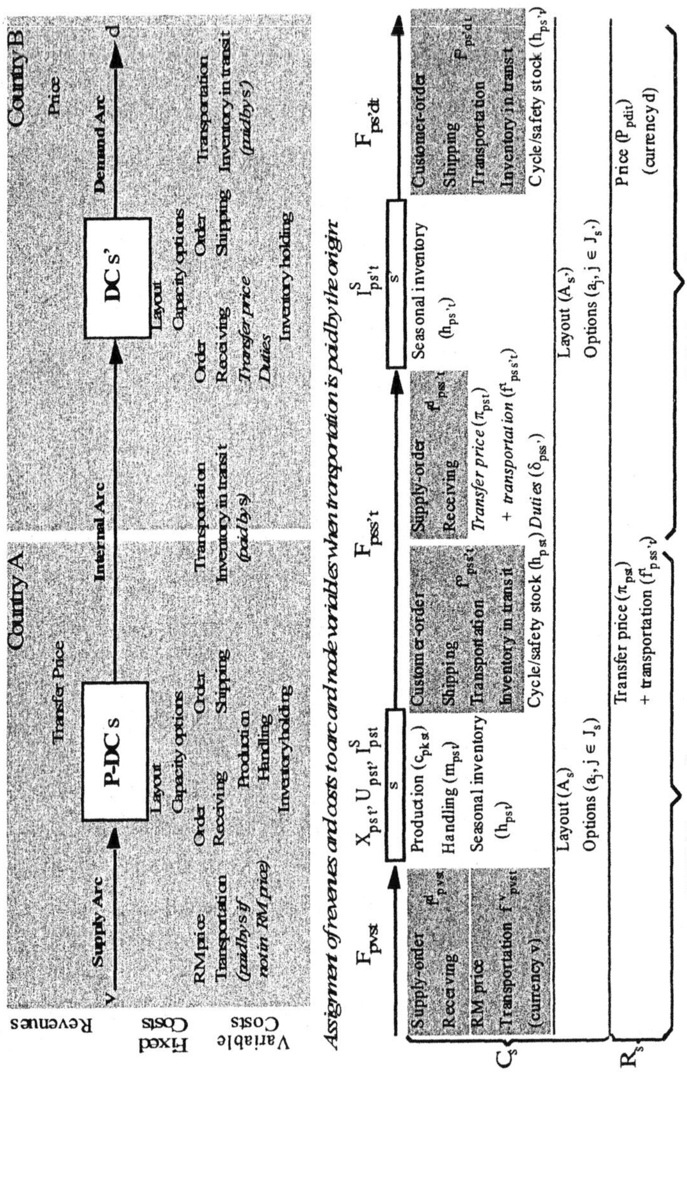

Figure 9.9. Mapping of costs and revenues on arcs and nodes.

operating expenditures, and they depend on the practical context of each potential node. The relevant fixed costs for different contexts are listed in Table 9.1. These costs are based on the engineering economy principles of capital recovery plus return over the planning horizon (Fabrycky and Torgersen, 1966). The fixed costs associated with potential capacity options also cover change of state and operating expenditures. The approach proposed to compute layout fixed costs can also be used to obtain capacity option fixed costs.

- Each time products cross a border; tariffs and duties are charged on the flow of merchandise and these are paid by the importer. In other words, these tariffs are calculated on the inflow to a given site from a foreign country of origin. These tariffs are based on the nature and class of the product. In the majority of countries, border tariffs are calculated on the CIF (Cost, Insurance and Freight) or the FOB (Free on Board) product value. In the model it is assumed that importers in all countries pay border tariffs based on CIF product values. To simplify the presentation, it is also assumed that there are no duty drawback or avoidance possibilities. An approach to model duty drawback and avoidance is presented by Arntzen et al. (1995).

- The transportation costs on the network arcs are paid by the origin. In practice, transportation costs usually display economies of scale with respect to shipment weight and distance, i.e., they can be modeled by a concave function $f(Q, d)$, where Q is the shipment weight and d is the distance between the origin and the destination. Different products can also be included in a given shipment. The flow on a network lane, say $\sum_p F_{pss't}$, corresponds to the sum of all the shipments made on arc (s, s') during season t. If the average weight of the shipments $Q_{ss't}$ on the arc is constant (e.g. truck load), then the shipment frequency during the season is given by $FR_{ss't} = \left(\sum_p \overline{w}_p F_{pss't}\right) / Q_{ss't}$ and the unit shipment cost $f^t_{pss't} = \overline{w}_p f(Q_{ss't}, d_{ss'}) / Q_{ss't}$ is independent of the flow variables $F_{pss't}$. When this is the case, it is reasonable to assume that transportation costs are linear with respect to seasonal flows. On the other end, in practice, it is often the frequency of shipments

Initial state		*Do not use site*		*Use current layout* $(l = 1)$		*Use new layout* $(l > 1)$	
		Decision	Fixed cost (A_s^0)	Decision	Fixed cost (A_{1s})	Decision	Fixed cost (A_{ls})
Current facility	Owned	Close	-Closing cost	*Status-quo*	-Capital recovery -Opportunity cost -Operating cost	Change layout	-Setup cost -Capital recovery -Opportunity cost -Operating cost
	Rented	Close	-Closing cost -Lease penalty	*Status-quo*	-Rent -Operating cost	Change layout	-Setup cost -Rent -Operating cost
	Public	Stop	-Stopping cost	*Status-quo*	-Operating cost	Change layout	-Operating cost
Potential site	New facility or purchased & renovated	Do not use	-Zero			Build/ Buy	-Setup cost -Capital recovery -Opportunity cost -Operating cost
	Rented facility	Do not use	-Zero			Rent	-Setup cost -Rent -Operating cost
	Public	Do not use	-Zero			Use	-Starting cost -Operating cost

Table 9.1. Facility layout fixed costs in different contexts.

$FR_{ss't}$ that is considered constant. When this is the case, the average shipment weight is

$$Q_{ss't} = \left(\sum_p \overline{w}_p F_{pss't}\right) / FR_{ss't} \tag{9.23}$$

and the unit shipment cost $f^t_{pss't}$ is a non-linear function of the seasonal flow variables. A successive linear programming approach to take these non-linearities into account is proposed by Fleischmann (1993). Another possible approach is to discretize the non-linear cost functions by introducing parallel arcs with different transportation costs and bounds on the flow variables. This approach, however, adds a large number of 0-1 variables. To simplify, in what follows, it is assumed that transportation costs are linear.

- Transfer prices for products sent in the internal network are fixed by the accounting department of the company and these do not include transportation costs from the source to the destination. In order to comply with laws and regulations, the transfer price of a given product shipped from a given source must be independent of its destination. In other words, the transfer price from the origin to the destination covers all the accumulated costs up to the shipping of the products from the origin and they include a predetermined margin. An approach to optimize transfer prices is presented by Vidal and Goetschalckx (2001).

- The income taxes paid in a country are calculated on the sum of the net revenues made by all facilities in this country. If a facility reports a loss, this loss is deducted from the total profit of the subsidiary before taxes. It is also assumed that the corporate taxes paid by the parent company are deferred until it pays dividends and that the decision to pay out dividends is independent of the design of the network. The parent company therefore only pays taxes on its local profits and it can be treated in the same way as the subsidiaries.

- The company wishes to maximize its after tax net revenues in a predetermined currency.

To calculate the total revenues and costs of a logistics network design, the following financial parameters and variables are required:

A_{ls} = Fixed cost of using layout l on site s for the planning horizon.

A_s^0 = Fixed cost of not using site s for the planning horizon.

a_j^1 = Fixed cost of using capacity option j for the planning horizon.

a_j^0 = Fixed cost of not using capacity option j for the planning horizon.

A_{mi}^M = Fixed selling cost incurred when marketing policy i is used for product-market m.

A_o^S = Fixed cost of support activities in country o.

c_{pkst} = Unit production cost of product p in production-distribution center s with technology $k \in KM_{ps}$ during season t.

m_{pst} = Unit handling cost for the transfer of product p to the stock of site s during season t.

f_{pvst}^v = Unit cost of the flow of product p between vendor v and site s during season t (this cost includes the product's price and the variable transportation cost).

f_{psnt}^o = Unit cost of the flow of product p between site s and node n paid by origin s during season t (this cost includes the customer-order processing cost, the shipping cost, the variable transportation cost and the inventory-in-transit holding cost).

f_{psnt}^t = Unit transportation cost of product p from site s to node n during season t (this cost is included in f_{psnt}^o).

f_{pnst}^d = Unit cost of the flow of product p between node n and site s paid by destination s during season t(this cost includes the supply-order processing costs and the receiving cost).

h_{pst} = Unit inventory holding cost of product p in facility s during season t.

π_{pst} = Transfer price of product p shipped from site s in season t.

$e_{oo'}$ = Exchange rate, i.e., number of units of country o currency by unit of country o' currency (the index $o = 0$ is given to the base currency, whether it is part of O or not).

δ_{pns} = Import duty rate applied to the CIF price of product p when transferred from the country of node n to the country of site s.

τ_o = Income tax rate of country o.

C_s = Total site s expenses for the planning horizon.

R_s = Total site s revenues for the planning horizon.

OP_o = Operating profit made in country o during the planning horizon.

OL_o = Operating loss made in country o during the planning horizon.

The revenues and expenses of the P-DCs and DCs, in local currency, are outlined in Table 9.2. The expression for the transfer costs of material inflows is obtained by first converting the transfer prices and transportation costs in local currency and then by adding the applicable duties. A similar approach is used to calculate other revenues and expenses.

Using the numbered elements of the expenditures and revenues in Table 9.2, it is seen that:

$$C_s = 1) + 2) + 3 + 5) + 6) + 7) + 9) + 10) \quad s \in Sd. \tag{9.24}$$

$$C_s = 1) + 2) + 3 + 4) + 5) + 6) + 7) + 8) + 9) + 10) \quad s \in Spd. \tag{9.25}$$

$$R_s = 11) + 12) \quad s \in S. \tag{9.26}$$

The operating income for each national division is thus given by:

$$OI_o = \sum_{s \in S_o} (R_s - C_s) - \sum_{m \in M_o} \sum_{i \in I_m} A^M_{mi} Y_{mi} - A^S_o \quad o \in O \tag{9.27}$$

The corporate net revenues before taxes in the reference currency is given by the expression $\sum_{o \in O} e_{0o} OI_o$. To calculate corporate after tax profits, one must first separate the divisions where the margin is positive from the divisions where it is negative because there is no income tax to pay on losses. To do this, OI_o must be separated into its negative and positive parts by defining

$$\text{Operating Income } = OP_o - OL_o, \tag{9.28}$$

where the operating profit $OP_o = OI_o$ if $OI_o > 0$ and the operating loss $OL_o = -OI_o$, otherwise. Clearly, for a given country, the operating profit OP_o and the operating loss OL_o cannot be simultaneously positive. Given this, it is seen that the after tax net revenues

		Distribution Center	Production-Distribution Center
Expenses	1) Inflows transfer cost	$\sum_{t\in T}\sum_{p\in P}\sum_{s'\in S^i_{ps}}(1+\delta_{ps's})e_{o(s),o(s')}(\pi_{ps't}+f^t_{ps'st})F_{ps'st}$	
	2) Raw materials	$\sum_{t\in T}\sum_{p\in RM\cup SA}\sum_{v\in V_p}(1+\delta_{pvs})e_{o(s),o(v)}f^v_{pvst}F_{pvst}$	
	3) Receptions from other sites	$\sum_{t\in T}\sum_{p\in P}\sum_{n\in V_p\cup S^i_{ps}} f^d_{pnst}F_{pnst}$	
	4) Production		$\sum_{t\in T}\sum_{p\in MP}\sum_{k\in KM_{ps}} c_{pkst}X_{pkst}$
	5) Facilities and options cost	$A^0_sY_{0s}+\sum_{l\in L_s}A_{ls}Y_{ls}+\sum_{j\in J_s}[a^1_jZ_j+a^0_j(1-Z_j)]$	
	6) Order cycle and safety stocks	$\sum_{t\in T}\sum_{p\in P}h_{pst}I_p(X^I_{pst})$	
	7) Seasonal stocks	$\sum_{t\in T}\sum_{p\in MP}\sum_{k\in KS_{ps}} h_{pst}I^S_{pkst}$	
	8) Handling		$\sum_{t\in T}\sum_{p\in MP} m_{pst}U_{pst}$
	9) Outflows to other sites	$\sum_{t\in T}\sum_{p\in P}\sum_{s'\in S^o_{ps}} f^o_{pss't}F_{pss't}$	
	10) Outflows to demand zones	$\sum_{t\in T}\sum_{p\in P}\sum_{d\in D^o_{ps}} f^o_{psdt}\sum_{i\in I_{psd}}F_{psdit}$	
Revenues	11) Outflows to other sites	$\sum_{t\in T}\sum_{p\in P}\sum_{s'\in S^o_{ps}}(\pi_{pst}+f^t_{pss't})F_{pss't}$	
	12) Outflows to demand zones	$\sum_{t\in T}\sum_{p\in P}\sum_{d\in D^o_{ps}} e_{o(s),o(d)}\sum_{i\in I_{psd}}P_{pdit}F_{psdit}$	

Table 9.2. Facilities expenses and revenues in local currency.

of the corporation in its reference currency is given by the expression $\sum_{o\in O} e_{0o}\left[(1-\tau_o)OP_o - OL_o\right]$.

3. Optimization model

Based on our previous discussion, the complete mathematical programming model proposed to optimize the structure of a global production-distribution network takes the following form:

$$Z = \max \sum_{o\in O} e_{0o}\left[(1-\tau_o)OP_o - OL_o\right] \quad \text{(MIP)}$$

subject to:
- Demand and marketing policy constraints (9.1) and (9.2)
- Facility layout, space and exclusive options constraints (9.3), (9.6)

and (9.7)
- Upper bound on the number of DCs and P-DCs (9.4) and (9.5)
- Distribution centers throughput definition constraints (9.10) and (9.16)
- Production centers flow equilibrium constraints (9.8)
- Production facilities capacity constraints (9.11) and (9.12)
- Raw materials flow equilibrium constraints (9.13) and (9.14)
- Distribution centers seasonal inventory accounting constraints (9.9) and (9.15)
- Lower bounds on the distribution centers flow (9.17)
- Facilities storage capacity constraints (9.20)
- Facilities shipping (receiving) capacity constraints (9.21)
- External supply constraints (9.22)
- Definitions of the facility total cost (9.24) and (9.25)
- Definitions of facilities total revenue (9.26)
- Definitions of the national divisions operating income

$$\sum_{s \in S_o} (R_s - C_s) - \sum_{m \in M_o} \sum_{i \in I_m} A^M_{mi} Y^M_{mi} - OP_o + OL_o = A^S_o \quad o \in O \quad (9.29)$$

-Non-negativity and binary constraints

$$Y^m_{mi} \in \{0,1\}, m \in M, i \in I_m; Y_{ls} \in \{0,1\}, s \in S, l \in L_s$$
$$Z_j \in \{0,1\}, j \in J$$
$$X_{pkst} \geq 0, \forall(p,k,s,t); U_{pst} \geq 0, \forall(p,s,t); I^S_{pkst} \geq 0, \forall(p,k,s,t);$$

$$F_{pnst} \geq 0, \forall(p,n,s,t); F_{psdit} \geq 0, \forall(p,s,d,i,t);$$
$$R_s \geq 0, C_s \geq 0, s \in S; OP_o \geq 0, OL_o \geq 0, o \in O.$$

This is a large scale non-linear mixed integer programming model. The non-linearities in the model are found in constraints (9.20), (9.24) and (9.25) and they all come from the inventory throughput functions. In order to solve the model efficiently, a method to cope with its size and its non-linearities must be used. Given the power of current MIP commercial solvers, the decision support system developed to generate and solve the model is based on a the solution of successive linear mixed-integer programming problems with a commercial solver, coupled with the use of valid inequalities (cuts) to strengthen the MIP formulation.

Experiments on the solution of particular cases of the model with Benders decomposition were made. It was found however that, to obtain good computation times with Benders decomposition, initial cuts had to be added to the model. It was also found that when these initial cuts were added to the model, the solution times obtained with CPLEX 8.1 were not worst than those obtained with Benders decomposition (Paquet et al., 2004). The approach used does not seek to obtain the global optimum: rather, it is perceived as a practical scenario improvement method based on reasonable approximations of the inventory-throughput functions.

An approach which could be used to linearize the problem is to replace $I_p\left(X^I_{pst}\right)$ by a piecewise linear approximation. This is equivalent to introducing alternative DC's at a given site with different lower and upper bounds on throughput, and adding an additional constraint on layout variables to ensure that only one of the alternative DC's can be used at each site. The problem with this approach is that it increases the number of 0-1 variables in the model significantly. This is why a successive MIP approach was developed. The approximation of the inventory throughput function used at iteration i of the solution method proposed is:

$$I_p\left(X^I_{pst}\right) = \rho^{(i)}_{pst} X^I_{pst} \tag{9.30}$$

where the slope $\rho^{(i)}_{pst}$ is calculated, at each iteration, from the flows of the last solution with the expression:

$$\rho^{(i)}_{pst} = I_p\left(X^{I(i-1)}_{pst}\right) / X^{I(i-1)}_{pst} = a_p\left(X^{I(i-1)}_{pst}\right)^{b_p-1} \tag{9.31}$$

The initial slope $\rho^{(1)}_{pst}$ is obtained by setting:

$$X^{I(0)}_{pst} = \left(\sum_{d\in D} \underline{x}_{pdt}\right) / |S| \quad s \in S, p \in P, t \in T \tag{9.32}$$

or by using historical flows as in (9.19). Although the equal share flows obtained with (9.32) are not necessarily feasible, they yield an initial slope which can be used to start the procedure. An approach based on goal programming to arrive at feasible initial flows is proposed by Martel (2002). The iteration process is continued until the difference between the values of the objective function of two successive solutions is sufficiently small. The successive slope calculation process proposed

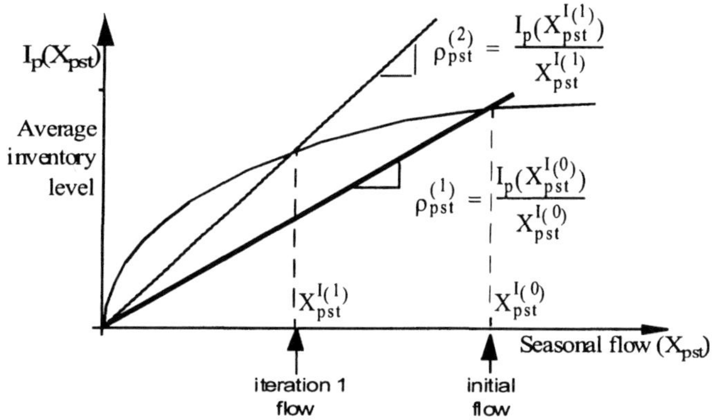

Figure 9.10. Successive linear approximations.

is illustrated in Figure 9.10. When the seasonal throughput obtained for center s during the ith iteration $\left(X^{I(i)}_{pst}\right)$ is positive, the slope can be calculated using relation (9.31). When $X^{I(i)}_{pst} = 0$, however, which necessarily occurs when a site is not used, the slope is not revised and the value obtained at the preceding iteration is retained. A heuristic approach similar to ours is used by Kim and Pardalos (2000) to solve concave piecewise linear network flow problems. To describe the solution algorithm formally, the following notation is needed:

MIP(i) = The mathematical program obtained by replacing $I_p\left(X^I_{pst}\right)$ in the constraints (9.20), (9.23) and (9.24) of MIP by $\rho^{(i)}_{pst} X^I_{pst}$ and by adding appropriate initial cuts.

Sol_i = The solution obtained by solving MIP(i).

$Z_i(Sol)$ = The value of the objective function of MIP(i) for solution Sol.

$Z(i)$ = The exact value of Sol_i, i.e., the value obtained by using the site cost definitions (9.24) and (9.25) to evaluate Sol_i.

Note that, because of the nature of the approximation made, we have:

$$Z(i) = Z_{i+1}(Sol_i)$$

The algorithm used to initialize the solution process and to improve the solutions obtained iteratively is the following:

1) *Initialization:*
   Set $i = 0$
   Obtain *equal-share* initial throughputs for the centers by computing:
   $$X_{pst}^{I(0)} = \left(\sum_{d \in D} \underline{x}_{pdt}\right)/|S| \quad s \in S, p \in P, t \in T$$
2) *Linearization with the last iteration throughputs.*
   $i = i + 1$
   For each product p, each center $s \in S$ and each season $t \in$ T,
   If the throughput $X_{pst}^{I(i-1)}$ is positive, compute the revised inventory duration with:
   $$\rho_{pst}^{(i)} = I_p\left(X_{pst}^{I(i-1)}\right)/X_{pst}^{I(i-1)}$$
   If the throughput $X_{pst}^{I(i-1)}$ is null, keep the inventory duration used at the previous iteration.
   If $i > 1$, set $Z(i-1) = Z_i(Sol_{i-1})$
3) *Check the stopping condition.*
   If $\{i > 2\}$ and $\{[Z(i-1) - Z(i-2)]/Z(i-2) < \epsilon\}$, end.
4) *Solve the mixed-integer programming problem.*
   Find the solution Sol_i of MIP(i)
   Go back to Step 2,

where ϵ is an acceptable tolerance.

Note that if relation (9.23) is added to the model, the solution approach proposed can easily be modified to take concave transportation costs into account. Also, instead of using inventory durations to approximate the inventory-throughput functions, it is possible to use the gradient of $I_p\left(X_{pst}^I\right)$ evaluated at $X_{pst}^{I(i-1)}$ and to limit the throughput change at iteration i to a trust region around $X_{pst}^{I(i-1)}$. This approach, proposed by Martel and Vankatadri (1999), provides a better approximation but it is more difficult to implement and less intuitive. The solution approach proposed here has given very satisfactory results in several real life projects. It was used, for example, to reengineer the North-American production-distribution network of Domtar, one of the largest fine paper producers in the world. The project involved the consideration of 12 paper mills, 13 conversion sub-contractors and 50 distribution centers. More than 100 product families and 1 000 demand zones were taken into account. The problems to solve had about 300 000 variables, including 75 binary variables.

4. Conclusion

This text proposes a mathematical programming approach to design international production-distribution networks for make-to-stock products with convergent manufacturing processes. A more general version of the model proposed and the solution method described were implemented in a commercial supply chain design tool which is now available on the market. The tool was used to solve several real life logistics network design problems. Work is currently in progress to expand the approach to make-to-order contexts and to divergent manufacturing process industries.

References

Ahmed, S., A. King and G. Parija 2001. A multi-stage stochastic integer programming approach for capacity expansion under uncertainty, The Stochastic Programming E-Print Series.

Aikens, C.H. 1985. Facility Location Models for Distribution Planning, EJOR, 22, 263-279.

Arntzen, B., G. Brown, T. Harrison and L. Trafton 1995. Global Supply Chain Management at Digital Equipment Corporation, Interfaces, 21-1, 69-93.

Ballou, R.H. 1992. Business Logistics Management, 3rd ed., Prentice Hall.

Ballou, R.H. 1994. Measuring transport costing error in customer aggregation for facility location, Transportation Journal, Lock Haven.

Bhutta, K., F. Huq, G. Frazier and Z. Mohamed 2003. An Integrated Location, Production, Distribution and Investment Model for a Multinational Corporation, Int. Journal of Production Economics, 86, 201-216.

Birge, J.R. and F. Louveaux 1997. Introduction to Stochastic Programming, Springer.

Brown, G., G. Graves and M. Honczarenko 1987. Design and Operation of a Multicommodity Production/Distribution System Using Primal Goal Decomposition, Management Science, 33-11, 1469-1480.

Cohen, M., M. Fisher and R. Jaikumar 1989. International Manufacturing and Distribution Networks: A Normative Model Framework, in K. Ferdows ed. Managing International Manufacturing, Elsevier, 67-93.

Cohen, M. and H. Lee 1989. Resource Deployment Analysis of Global Manufacturing and Distribution Networks, J. Mfg. Oper. Mgt., 2, 81-104.

Cohen, M. and S. Moon 1990. Impact of Production Scale Economies, Manufacturing Complexity, and Transportation Costs on Supply Chain Facility Networks, J. Mfg. Oper. Mgt., 3, 269-292.

Cohen, M. and S. Moon 1991. An integrated plant loading model with economies of scale and scope, EJOR, 50, 266-279.

Cordeau, J-F., F. Pasin and M. Solomon 2002. An Integrated Model for Logistics Network Design, Les Cahiers du GERAD, G-2002-07.

Daskin, M. 1995. Network and Discrete Location, Wiley Inter-Science.

Dogan K. and M. Goetschalckx 1999. A Primal Decomposition Method for the Integrated Design of Multi-Period Production-Distribution Systems, IIE Trans., 31, 1027- 1036.

Eppen, G., R. Kipp Martin and L. Schrage 1989. A Scenario Approach to Capacity Planning, Operations Research, 37-4, 517-527.

Everett, G., A. Philpott and G. Cook 2000. Capital Planning Under Uncertainty at Fletcher Challenge Canada, Proceedings of 32th Conference of ORSNZ.

Everett, G., S. Aoude and A. Philpott 2001. Capital Planning in the Paper Industry using COMPASS, Proceedings of 33th Conference of ORSNZ.

Fabrycky, W. and P. Torgersen 1966. Operations Economy, Prentice-Hall.

Fine, C.H. 1993. Developments in Manufacturing Technology and Economic Evaluation Models, in: S. Graves, A. Rinnooy Kan and P. Zipkin, eds., Logistics of Production and Inventory, Handbooks in Operations Research and Management Science, vol 4, North- Holland.

Fleischmann, B. 1993. Designing Distribution Systems with Transport Economies of Scale, EJOR, 70, 31-42.

Francis, R.L., L.F. McGinnis and J.A. White 1992. Facility Layout and Location, 2nd ed., Prentice-Hall.

Freidenfelds, J. 1981. Capacity Expansion, North-Holland.

Geoffrion, A. and G. Graves 1974. Multicommodity Distribution System Design by Benders Decomposition, Man. Sci., 20, 822-844.

Geoffrion, A. and R. Powers 1995. 20 Years of Strategic Distribution System Design: An Evolutionary Perspective, Interfaces, 25-5, 105-127.

Glover, F., G. Jones, D. Karney, D. Klingman and J. Mote 1979. An Integrated Production, Distribution, and Inventory-Planning System, Interfaces, 9-5, 21-35.

Hill, T. 1999. Manufacturing Strategy, 3rd ed, McGraw-Hill/Irwin.

Huchzermeier, A. and M. Cohen 1996. Valuing Operational Flexibility under Exchange Rate Risk, Operations Research, 44-1, 100-113.

Kim, D. and P.M. Pardalos 2000. Dynamic Slope Scaling and Trust Interval Techniques for Solving Concave Piecewise Linear Network Flow Problems, Networks, 35-3, 216-222.

Kogut, B. and N. Kalatilaka 1994. Operating Flexibility, Global Manufacturing and the Option Value of a Multinational Network, Management Science, 40-1, 123-139.

Lakhal, S., A. Martel, M. Oral, and B. Montreuil 1999. Network Companies and Competitiveness: A Framework for Analysis, EJOR, 118-2, 278-294.

Lakhal, S., A. Martel, O. Kettani and M. Oral 2001. On the Optimization of Supply Chain Networking Decisions, EJOR, 129-2, 259-270.

Li, S. and D. Tirupati 1994. Dynamic Capacity Expansion Problem with Multiple Products: Technology Selection and Timing of Capacity Additions, Operations Research, 42-5, 958-976.

Luss, H., Operations Research and Capacity Expansion Problems: A Survey, Operations Research, 30, 5, 1982, 907-947.

Martel, A. and U. Vankatadri 1999. Optimizing Supply Network Structures Under Economies of Scale, IEPM Conference Proceedings, Glasgow, Book 1, 56-65.

Martel, A. 2002. Conception et gestion de chaînes logistiques, Manuel de formation, Université Laval.

Mazzola, J. and R. Schantz 1997. Multiple-Facility Loading Under Capacity-Based Economies of Scope, Nav. Res. Log., 44, 1997, 229-256.

Owen, S. and M. Daskin 1998. Strategic Facility Location: A Review, EJOR, 111, 423- 447.

Paquet, M., A. Martel and B. Montreuil 2003. A manufacturing network design model based on processor and worker capabilities, Proceedings of the International Conference on Industrial Engineering and Production Management, Québec.

Paquet, M., A. Martel and G. Desaulniers 2004. Including Technology Selection Decisions in Manufacturing Network Design Models, International Journal of Computer Integrated Manufacturing, 17-2, 117-125.

Philpott, A., and G. Everett 2001. Supply Chain Optimisation in the Paper Industry. Annals of Operations Research, 108 1): 225-237.

Pirkul, H. and V. Jayaraman 1996. Production, Transportation, and Distribution Planning in a Multi-Commodity Tri-Echelon System, Transp. Science, 30-4, 291-302.

Pomper, C. 1976. International Investment Planning: An Integrated Approach, North-Holland.

Porter, M. 1985. Competitive Advantage, Free Press.

Rajagopalan, S. and A. Soteriou 1994. Capacity Acquisition and Disposal with Discrete Facility Sizes, Management Science, 40-7, 903-917.

Revelle, C.S. and G. Laporte 1996. The Plant Location Problem: New Models and Research Prospects, Oper. Res., 44-6, 864-874.

Rosenfield, D., R. Shapiro and R. Bohn 1985. Implications of Cost-Service Trade-offs on Industry Logistics Structures, Interfaces, 15-6, 48-59.

Shapiro, J., V. Singhal and S. Wagner 1993. Optimizing the Value Chain, Interfaces, 23-2, 102-117.

Shapiro, J. 2001. Modeling the Supply Chain, Brooks/Cole Publishing Company.

Shulman, A. 1991. An Algorithm for Solving Dynamic Capacitated Plant Location Problems with Discrete Expansion Sizes, Operations Research, 39-3, 423-436.

Sule, D. 2001. Logistics of Facility Location and Allocation, Marcel Dekker Inc.

Trigeorgis, L. 1996. Real Options, MIT Press.

Verter, V. and C. Dincer 1992. An integrated evaluation of facility location, capacity acquisition, and technology selection for designing global manufacturing strategies, EJOR, 60, 1-18.

Verter, V. and C. Dincer 1995. Facility Location and Capacity Acquisition: An Integrated Approach, Nav. Res. Log., 42.

Vidal, C. and M. Goetschalckx 1997. Strategic Production-Distribution Models: A Critical Review with Emphasis on Global Supply Chain Models, EJOR, 98, 1-18.

Vidal, C. and M. Goetschalckx 2001. A Global Supply Chain Model with Transfer Pricing and Transportation Cost Allocation, EJOR, 129, 134-158.

Vila, D., A. Martel and R. Beauregard 2004. Designing Logistics Networks in Divergent Process Industries: A Methodology and its Application to the Lumber Industry, Working Paper DT-2004-AM-5, Centor, Université Laval, Québec.

Vila, D., A. Martel and R. Beauregard 2005. Taking Market Forces into Account in the Design of Production-distribution Networks: A Positioning by Anticipation Approach, International Conference on Industrial Engineering and Systems Management Proceedings, Marrakech, Morocco.

Chapter 10

MODELING & SOLVING STOCHASTIC PROGRAMMING PROBLEMS IN SUPPLY CHAIN MANAGEMENT USING XPRESS-SP

Alan Dormer, Alkis Vazacopoulos, Nitin Verma, and Horia Tipi
Dash Optimization, Inc, 560 Sylvan Avenue, Englewood Cliffs, NJ 07632, USA

Abstract Supply chains continually face the challenge of efficient decision-making in a complex environment coupled with uncertainty. While plenty of forecasting and analytical tools are available in the market to evaluate and enhance Supply Chain performance, the current functionalities are not sufficient to address issues related to efficient decision making under uncertainty. In this paper we discuss expanding the modeling paradigm to incorporate uncertain events naturally and concisely in a stochastic programming framework, and demonstrate how *Xpress-SP*–a stochastic programming suite–can be used for modeling, solving and analyzing problems occurring in supply chain management.

1. Introduction

A supply chain manager constantly faces the task of making numerous decisions such as the amount of raw material to purchase, routing and shipping finished products to distribution centers, inventory control issues, etc. On top of this, the variability in the underlying uncertain components in the supply chain–be it a sudden increase in demand, a delay in arrival of shipments, or an unusual cut in the supply–can make this job difficult. Stochastic programming techniques (Birge and Louveaux (1997)) are most suitable for supply chain systems because they address the issues of optimal decision-making under uncertainty and prepare the manager by hedging against future risks. However, traditional mathematical programming tools available in the market are not suitable to model and analyze such systems because they either lack appropriate functionalities or their modeling complexity renders them impractical to be used for problems occurring in the enterprise (see Fourer and Gay (1997) and Fragniere and Gondzio (2002)). In the following sections we demonstrate how *Xpress-SP* can be used for modeling, solving and analyzing supply chain optimization problems in an easy-to-use fashion.

We begin by characterizing the uncertainties in supply chain processes in Section 2. Here we identify various risks an organization is exposed to, their impact on revenue, customer satisfaction and various other attributes that indicate the organization's performance. This is followed by a discussion on the benefits of using stochastic programming methodologies for efficiently managing supply chains.

In Section 3, we discuss the basic concepts of stochastic linear programs. We present general formulations of the two-stage and multi-stage stochastic programs. Next we discuss the generation of scenario trees by discretization of random variables, which is followed by a description on forming an extensive deterministic equivalent formulation of stochastic problems.

Section 4 illustrates the framework of the *Xpress-SP* suite. We describe the architecture of the Xpress Stochastic Programming (SP) component and its integration with other components of *Xpress-MP*.

We then highlight several tools and functions available in *Xpress-SP* which facilitate rapid modeling and analysis of stochastic programs. We then consider two examples from the supply chain sector and demonstrate how one can build concise and easy-to-understand stochastic models in *Xpress-SP*. These models are inspired by the operation of assemble-to-order systems and option contracts in supply chains. Section 5 describes a simple base model for a multi-component multi-product assemble-to-order system where demands are random. We discuss the effect of the number of scenarios on the problem, then compare a myopic policy with the optimal policy, and finally study how scenario manipulation affects the accuracy of the solution. The second example is discussed in Section 6. In this example, we demonstrate how models with correlated random variables can be built in a natural fashion in *Xpress-SP*. We also show how to create more complicated asymmetric scenario trees and write stochastic models with global constraints.

2. Uncertainty and its impact on the supply chain

In today's economy, organizations are working harder to reduce development timelines, production costs and lead times, and improve quality. With increasing globalization and ever-growing competition, companies are moving towards achieving better control of their supply chains by implementing better decision support systems, and developing superior business processes. There has also been a lot of emphasis on the free flow of information, transparency, and improving visibility within and across organizations; however companies continue to face stock-outs and mark-downs in their supply chains. A clear cause of these upsets can be attributed to the underlying uncertainties and the risks associated with them (Chopra and Meindl (2001)). Furthermore, the complexity and dependency among various organizational units makes the problem more difficult to handle. With increasing globalization, an organization's exposure to risk is multiplied. At the top-most tier, any enterprise faces essentially three kinds of risks:

1. Financial risk–e.g., excess inventory costs, lost sales.

2. Chaos risk–e.g., fluctuations in demand, supply and availability.

3. Market risk–e.g., missing market opportunities.

The impact of financial risks may be both short-term and long-term. It dramatically changes the allocation of resources for production of goods and services, and jeopardizes the organization's credibility. The volatility of supply and demand interferes with the smooth operation of the organization. Situations such as breakdowns, canceled orders and late deliveries can significantly affect the dynamics of the supply chain. Some extreme scenarios such as natural calamity, political instability, and regulations may severely disrupt the supply chain network. Last but not least, such uncertainties and their consequences often lead to nervousness, overreaction, and mistrust.

A conservative approach to meeting these risks might drastically increase the overhead, such as the amount of resources, man-power, and inventory, and lead to a failure to capitalize on opportunities. Other methodologies may not prepare organizations to hedge against the vagaries of the future sufficiently, because they may be subjective or time consuming. Stochastic programming techniques are efficient and objective methodologies for decision making under uncertainty, and provide an overall optimal solution by balancing risks versus rewards appropriately (Shapiro (2001)). Therefore, there is a clear need and opportunity for a shift in paradigm to stochastic optimization tools for structured and superior decision-making.

3. Stochastic Programming basics

Stochastic programming problems essentially involve sequential decision making in stages, accompanied by random events occurring between consecutive stages (see Birge and Louveaux (1997) and Dupacova et al. (2002)). Such problems occur in supply chains, the financial sector, energy power systems, the transportation industry, etc. (Morton (2004)). The challenge in SP problems is to find decisions at each stage that are the overall best for all possible realizations of events occurring after that stage. These problems can be divided into two-stage and multi-stage problems. From a mathematical programming perspective, in a two-stage problem, the initial decisions are taken first. These are then followed by a random event. Next, the recourse decisions, which are based on this random event, are taken. The multi-stage problem, as the name suggests, consists of multiple stages, with random events occurring between consecutive stages. As an example, retail managers face periodic uncertainty in availability of raw materials and demand of products; however, before these uncertainties are realized, supply chain managers need to determine the long-run production capacity for the system in advance. The recourse action in this context could be for example, a change in short-run production capacity, addition of work-force, or the amount of demand outsourced. In the following sections we discuss the two-stage and multi-stage problems, random events and the associated scenario tree, and the node-based and scenario-based extensive form of the underlying deterministic formulation of stochastic programs.

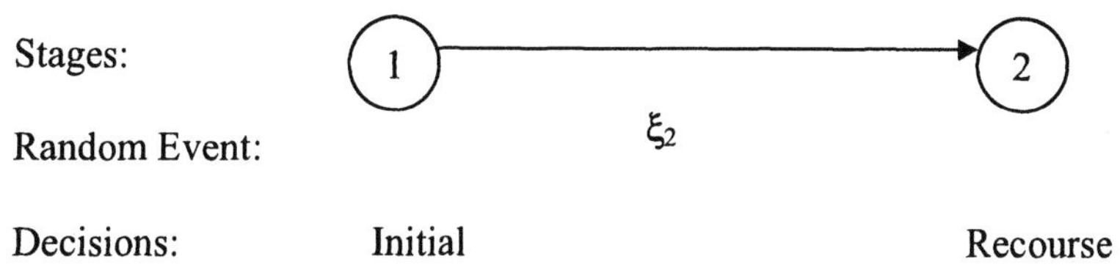

Figure 10.1. Two-stage problem.

3.1 Two-stage stochastic problems

In a two-stage problem, the initial or the first stage decisions (e.g., system design decisions) are made which are followed by random events such as demand, availability, price, or a combination of these. Then the second stage decisions (e.g., operational decisions) are made. The following figure (Figure 10.1) illustrates a two-stage problem.

In standard form, the two-stage stochastic program is:

$$
\begin{array}{ll}
\min & c_1^T x_1 + E_2[\tilde{c}_2^T(\xi_2) x_2(\xi_2)] \\
s.t. & A_{11} x_1 = b_1, \\
& \tilde{A}_{21}(\xi_2) x_1 + \tilde{A}_{22}(\xi_2) x_2(\xi_2) = \tilde{b}_2(\xi_2), \ \forall \xi_2 \in \Xi_2, \\
& l_1 \leq x_1 \leq u_1, \\
& \tilde{l}_2(\xi_2) \leq x_2(\xi_2) \leq \tilde{u}_2(\xi_2), \ \forall \xi_2 \in \Xi_2.
\end{array}
$$

where x_1 and x_2 are the first and the second stage decision variables respectively. $E_2[]$ is the expectation operator with respect to the random event ξ_2. Ξ_2 represents the state space (the set of possible outcomes or the values that ξ_2 can assume) in the recourse stage. There may be separate values for the objective coefficients $\tilde{c}_2'(\xi_2)$, right hand side $\tilde{b}_2(\xi_2)$, and matrix coefficients $\tilde{A}_{21}(\xi_2)$ and $\tilde{A}_{22}(\xi_2)$ for each outcome ξ_2 in Ξ_2, and the recourse decisions $x_2(\xi_2)$ depend on ξ_2, i.e., there are separate sets of recourse decisions for each outcome ξ_2 in the second stage.

3.2 Multi-stage stochastic problems

Multi-stage problems can be viewed as shown in Figure 10.2. Formally, the multi-stage stochastic program is defined as follows:

$$
\begin{aligned}
\min \quad & c_1^T x_1 + E_2[\tilde{c}_2^T x_2 + E_3[\tilde{c}_3^T x_3 + E_4[\cdots + E_T[\tilde{c}_T^T x_T]\ldots]]] \\
s.t. \quad & \sum_{t=1}^{t'} \tilde{A}_{t't} x_t = \tilde{b}_{t'} \ \forall t' \in \{1,\ldots,T\}, \\
& \tilde{l}_t \leq x_t \leq \tilde{u}_t, \ \forall t \in \{1,\ldots,T\}.
\end{aligned}
$$

where

$$
\begin{aligned}
& x_t \equiv x_t(\xi_2, \xi_3, \ldots, \xi_t), \ \forall t \in \{1,\ldots,T\}; \\
& \tilde{c}_t^T \equiv \tilde{c}_t^T(\xi_2, \xi_3, \ldots, \xi_t), \ \forall t \in \{1,\ldots,T\}; \\
& \tilde{b}_{t'} \equiv \tilde{b}_{t'}(\xi_2, \xi_3, \ldots, \xi_t), \ \forall t' \in \{1,\ldots,T\}; \\
& \tilde{A}_{t'p} \equiv \tilde{A}_{t'p}(\xi_2, \xi_3, \ldots, \xi_t), \ \forall t \in \{1,\ldots,t'\}, t' \in \{1,\ldots,T\}; \\
& \tilde{l}_t \equiv \tilde{l}_t(\xi_2, \xi_3, \ldots, \xi_t), \ \forall t \in \{1,\ldots,T\}; \\
& \tilde{u}_t \equiv \tilde{u}_t(\xi_2, \xi_3, \ldots, \xi_t), \ \forall t \in \{1,\ldots,T\}; \\
& E_t[.] \equiv E_{\xi_t|(\xi_2,\ldots,\xi_{t-1})}[.], \ \forall t \in \{2,\ldots,T\}.
\end{aligned}
$$

Each of the above entities depends on the sequence of events $(\xi_2, \xi_3, \ldots)$.[1]

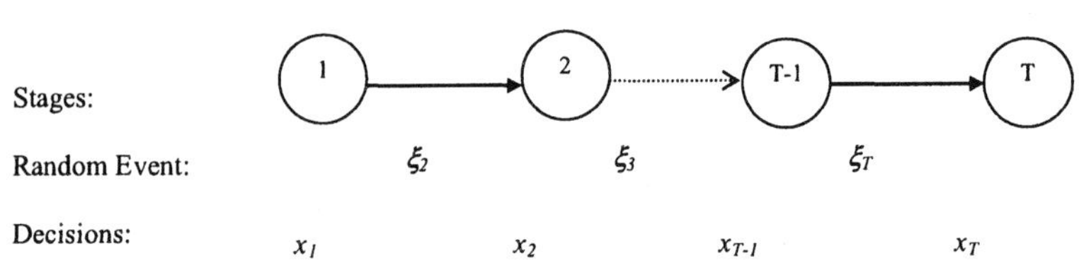

Figure 10.2. Multi-stage problem.

3.3 Scenario generation

If each of the random variables ξ_t assumes discrete values with certain probabilities, then the occurrence of possible outcomes at each stage may be represented as a 'Scenario tree'. An example of a scenario tree with

[1]Note that c_1, A_{11}, b_1, l_1, and u_1 are not random.

T (number of stages) $= 3$ and S (number of scenarios) $= 4$ is shown in Figure 10.3.

In the tth stage there are N_t nodes. For each stage t from 2 to T, each ξ_t has N_t outcomes $(\xi_{t1}\ w.p\ p_{t,1},\ \xi_{t2}\ w.p\ p_{t,2},\ \ldots,\ \xi_{tN_t}\ w.p\ p_{t,N_t})^2$, where p_{tn} is the **conditional probability** of visiting the nth node in the tth stage from its parent node in the $(t-1)^{\text{st}}$ stage. The realizations of $\xi_2, \xi_3, \ldots, \xi_T$ correspond to scenarios (paths in the tree). For each stage t and each node n in that stage, the node has an **unconditional probability** P_{tn} of being visited, which is equal to the product of conditional probabilities along the path to that node. Similarly, each scenario s has a **scenario probability** P_s that is equal to the product of conditional probabilities along the path to that scenario. The following observations can be made about the scenario tree:

1. $\sum\limits_{n':Node(t+1,n')\in Children(Node(t,n))} p_{t+1,n'} = 1,\ \forall n \in \{1,\ldots,N_t\},$ $t \in \{1,\ldots,N-1\}$.

2. $P_{th} = \prod\limits_{(t',n'):Node(t',n')\in path\ to\ Node(t,n)}^{(t,n)} p_{tn},\ \forall n \in \{1,\ldots,N_t\},$ $t \in \{2,\ldots,T\}$.

3. $\sum_{n=1}^{N_t} P_{tn} = 1,\ \forall t \in \{2,\ldots,T\}$.

4. $P_s = \prod\limits_{(t,n):Node(t,n)\in path\ to\ scenario\ s} p_{tn},\ \forall s \in \{1,\ldots,S\}$.

5. $\sum\limits_{s=1}^{S} P_s = 1$.

[2] w.p stands for 'with probability'.

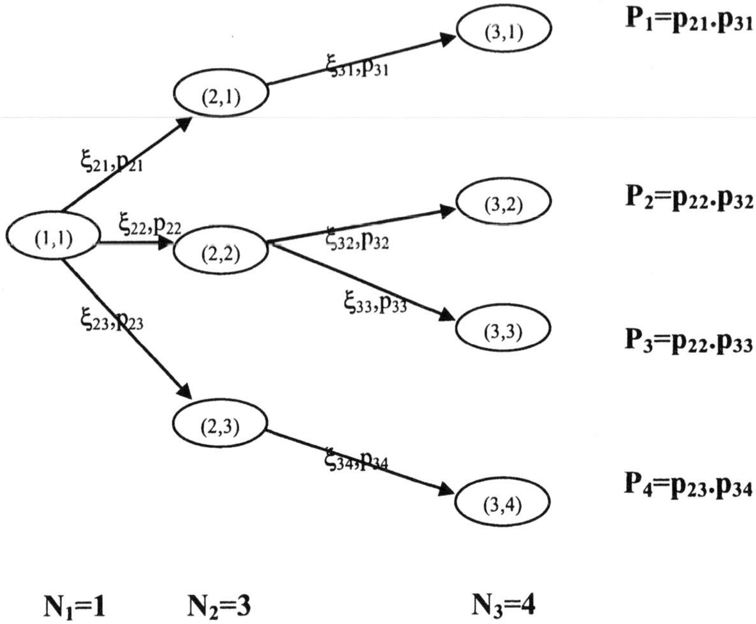

Figure 10.3. Three-stage scenario tree.

3.4 Underlying deterministic model

If we ignored the randomness of the data momentarily, then an underlying deterministic model can be written as follows:

$$
\begin{aligned}
\min \quad & \sum_{t=1}^{T} \tilde{c}_t' x_t \\
s.t. \quad & \sum_{t=1}^{t'} \tilde{A}_{t't} x_t = \tilde{b}_{t'}, \; \forall t' \in \{1, \ldots, T\}, \\
& \tilde{l}_t \leq x_t \leq \tilde{u}_t, \; \forall t \in \{1, \ldots, T\}.
\end{aligned}
$$

3.5 Parsing the underlying deterministic model

Given the dependency of the coefficients $(\tilde{c}_t, \tilde{A}_{tt'}, \tilde{b}_t, \tilde{l}_t, \tilde{u}_t)$ of a stochastic program on the random events (ξ_t), it can automatically be parsed into an extensive form (deterministic equivalent problem) by introducing new variables and constraints. There are basically two ways of creating new variables and constraints: node based and scenario based.

3.5.1 Node based. Given a scenario tree, the underlying deterministic model can be parsed into an extensive mathematical program based on the nodes. The basic idea is to add a subscript of a node number to each of the stochastic decision variables (x_t becomes $x_{t,n}$ for $n = 1, \ldots, N_t$). Then the resulting extensive deterministic model would be:

$$\begin{aligned} \min \quad & \sum_{t=1}^{T} \sum_{n=1}^{N_t} P_{tn} c'_{tn} x_{tn} \\ s.t. \quad & \sum_{(t,n):Node(t,n) \in Path\ to\ Node(t',n')}^{(t',n')} A_{t'tn} x_{tn} = b_{t'n'}, \\ & \qquad \forall n \in \{1, \ldots, N_{t'}\}, t' \in \{1, \ldots, T\}, \\ & l_{tn} \leq x_{tn} \leq u_{tn}, \ \forall n \in \{1, \ldots, N_t\}, t \in \{1, \ldots, T\}. \end{aligned}$$

In this model $c_{tn}, A_{tt'n}, b_{tn}, l_{tn}, u_{tn}$ are the resolved values of $\tilde{c}_t$, $\tilde{A}_{tt'}$, $\tilde{b}_t$, $\tilde{l}_t$, $\tilde{u}_t$ at the node(t, n).

3.5.2 Scenario based. A stochastic model can also be parsed based on scenarios. Here each variable is also subscripted by scenarios (x_t becomes $x_{t,s}$ for $s = 1, \ldots, S$). The parsed mathematical program would look as follows:

$$\begin{aligned} \min \quad & \sum_{s=1}^{S} P_s \sum_{t=1}^{T} c'_{ts} x_{ts} \\ s.t. \quad & \sum_{t=1}^{t'} A_{t'ts} x_{ts} = b_{t's}, \ \forall s \in \{1, \ldots, S\}, t' \in \{1, \ldots, T\}, \\ & l_{ts} \leq x_{ts} \leq u_{ts}, \ \forall s \in \{1, \ldots, S\}, t' \in \{1, \ldots, T\}. \end{aligned}$$

Here $c_{ts}, A_{tt's}, b_{t's}, l_{ts}, u_{ts}$ are realizations of $\tilde{c}_t$, $\tilde{A}_{t't}$, $\tilde{b}_{t'}$, $\tilde{l}_t$, $\tilde{u}_t$ respectively in scenario s.

3.5.2.1. Non-anticipative constraints (NAC) Consider the node $(2, 2)$ in Figure 10.3. When the model is parsed according to scenarios, although we have two separate variables x_{22} and x_{23} corresponding to scenarios 2 and 3 at this node, both of them should assume same value since they cannot depend on the future events ($x_t \equiv x_t(\xi_2, \ldots, \xi_t)$). Therefore, the following set of non-anticipative constraints also needs to be included in the above formulation: $x_{ts} = x_{ts'}$, $\forall n \in \{1, \ldots, N_t\}$, $t \in \{1, \ldots, T-1\}$, $s \neq s' \in \Lambda_{tn}$, where Λ_{tn} is the set of scenarios passing through the node(t, n).

4. Framework of *Xpress-SP*

4.1 Scope

Xpress-SP is used for modeling, solving and analyzing two-stage and multi-stage stochastic linear problems. The design is primarily focused on ease of modeling and analyzing stochastic problems. Users may create scenario trees in one of the following ways:

i. Exhaustively—by specifying the independent or joint discretized distribution of random variables

ii. Symmetrically—by specifying the structure of the tree, and discretized distribution

iii. Explicitly—by specifying the tree structure and assumed values of random variables in the scenario tree

Xpress-SP supports any model of the forms discussed earlier in Section 3.2. Specifically, any of the matrix coefficients, cost coefficients, right hand sides or bounds can be a random variable or a linear/non-linear expression of one or more random variables. *Xpress-SP* version 1.00 supports Xpress-Optimizer's primal simplex, dual simplex and barrier algorithms for solving the extensive form of the underlying stochastic linear problems. The branch and bound algorithm in Xpress-Optimizer is used for solving the extensive forms that are MILPs. One may also solve related problems such as the expected value problem and the perfect information problem (see Sections 5.3.2 and 6.5.1.1).

4.2 Architecture

Xpress-SP is built using Mosel's Native Interface technology (Colombani and Heipcke (2002)). At the modeling level various new 'types', and 'functions' and 'procedures' based on these 'types' are defined in the module 'mmsp'. The library supports several 'control parameters' for controlling the behavior of *Xpress-SP* at the scenario generation, modeling, and solution analysis phases. This library is further integrated

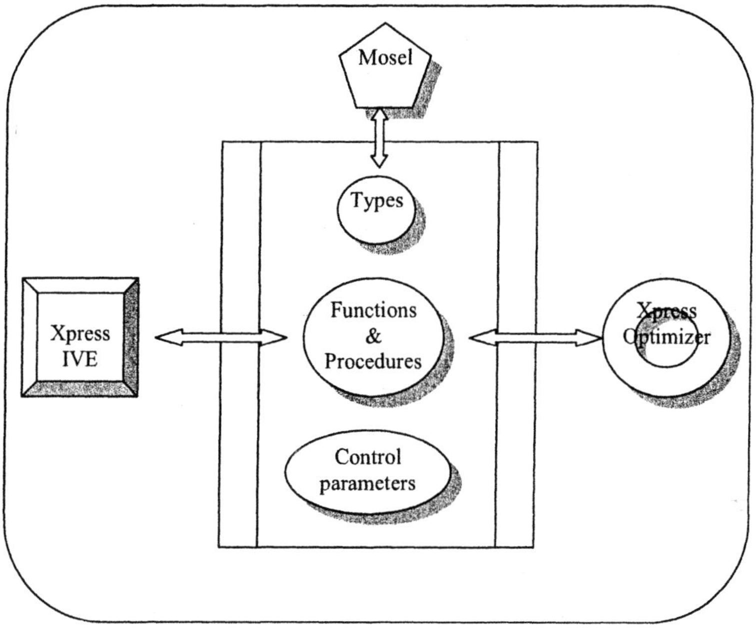

Figure 10.4. Architecture of *Xpress-SP*.

with Xpress-Optimizer for solving the problem and with Xpress-IVE for visualization and analysis. Figure 10.4 gives a pictorial view of the architecture of the Xpress stochastic programming suite.

Mosel supports structures such as 'sets' and 'arrays', programming constructs such as 'forall' and 'while' loops, and other logic building functions. These tools together with the Native Interface technology have facilitated the evolution of *Xpress-SP*'s semantics as a natural extension of modeling conventions. The equivalent extensive form of the stochastic problem is solved using Xpress-Optimizer which is one of the best solvers available in the market for solving large scale LP, MILP, QP and MIQP. The structure of the stochastic problem in the 'mmsp' module is kept simple and easy-to-use by functionally interfacing it with the Xpress Optimizer library. Xpress-IVE is a state-of-the-art Integrated Visual Environment which brings together the Xpress products and facilitates rapid model prototyping and analysis. It is integrated with *Xpress-SP* and provides visualization of stochastic models, scenario trees, and stochastic solutions.

4.2.1 'mmsp' types. The following types–which essentially act as objects for stochastic programs–are defined in 'mmsp':

- sprand: Random variable that assumes different values with certain probabilities, e.g., demand.
- spvar: Stochastic decision variable that assumes values in different scenarios or at different nodes in the scenario tree.
- splinctr: Stochastic constraint built with 'reals', 'sprands', and 'spvars'.
- sprandexp: An expression built with one or more 'sprands' and/or 'reals'.

4.2.2 Functions and Control parameters. *Xpress-SP* provides various functions and control parameters for easy scenario generation and manipulation, rapid modeling of the problem, and for solution analysis. Functions are also provided for evaluating stochastic entities at nodes in the scenario tree or in scenarios. Functions and procedures for accessing Xpress-Optimizer controls are also defined.

4.3 Modeling interface

One can easily write two-stage and multi-stage stochastic linear models in the 'mmsp' module. The syntax of the language is similar to Mosel's modeling language for linear programs. Xpress-IVE provides a modeling interface as shown in Figure 10.5.

4.4 Scenario generation and manipulation

Xpress-IVE provides a 'pie view' and a 'block view' for the visualization of the scenario tree. In the block view the height of the block is proportional to the conditional probability of the node in the scenario tree. The following screen shot (Figure 10.6) shows a scenario tree corresponding to sprands belonging to the second and third stage with distributions $\{2\ w.p\ 0.6,\ 8\ w.p\ 0.4\}$ and $\{3\ w.p\ 0.3,\ 5\ w.p\ 0.6,\ 7\ w.p\ 0.1\}$ respectively.

Similarly, tools are also provided for creating and visualizing asymmetric trees. The following figure (Figure 10.7) shows a scenario tree with three nodes in the second stage. The first node in the second stage

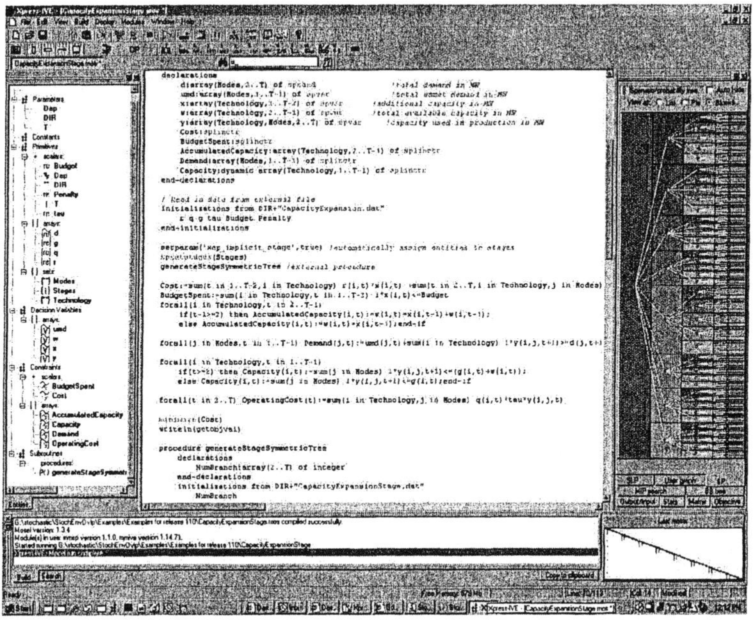

Figure 10.5. *Xpress-SP* model-modeling interface.

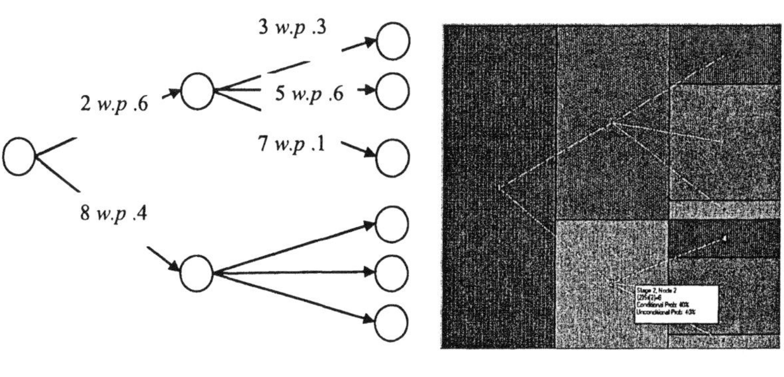

Figure 10.6. Scenario tree visualized in IVE.

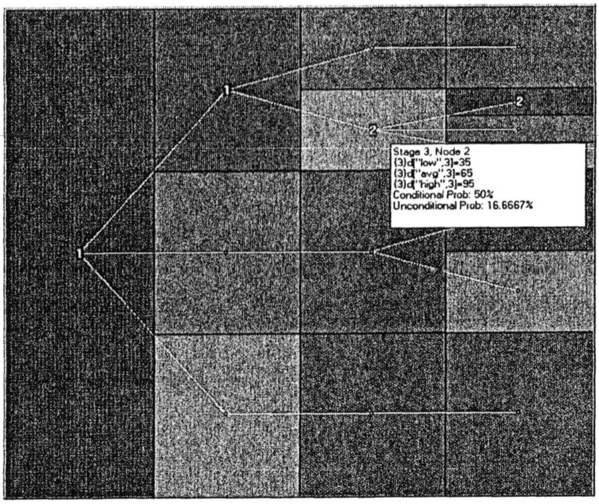

Figure 10.7. Assymetric scenario tree.

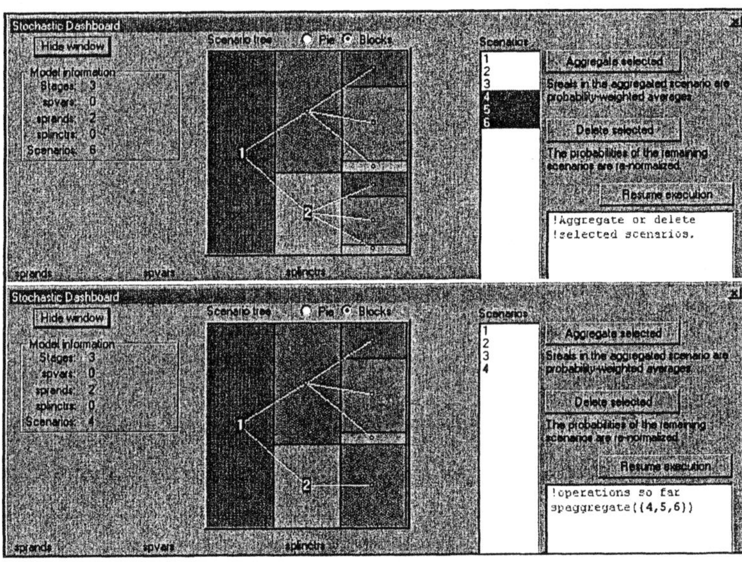

Figure 10.8. Scenario tree: before and after aggregation.

has two branches, whereas other nodes in the second stage have one each.

One may also manipulate the scenario tree after the generation of scenarios in IVE itself. IVE provides scenario aggregation and deletion tools and updates the scenario tree dynamically as shown in Figure 10.8.

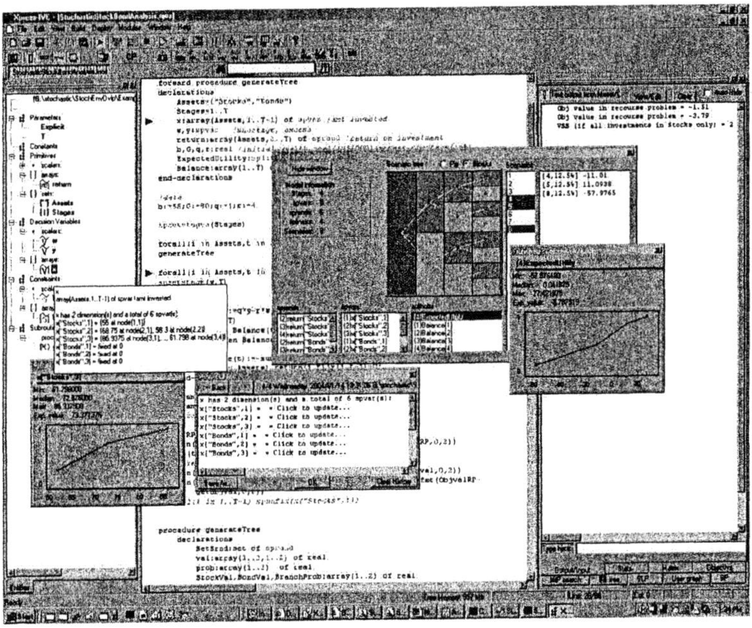

Figure 10.9. Visualization tool in *Xpress-SP*.

4.5 Visualization and analysis

IVE also enables visualization and analysis of stochastic entities, their distributions, the optimal solution, the parsed matrix, and the scenario tree. The following figure (Figure 10.9) gives a pictorial view of these tools.

IVE can easily handle scenario trees with thousands of scenarios. The Hyper version of Xpress optimizer can handle and solve large scale problems. This makes *Xpress-SP* a very robust and scalable tool for stochastic programming. In the next sections we demonstrate the functionalities available in *Xpress-SP* by illustrating the modeling and analysis of two applications in supply chain management.

5. SCM application 1: Two-stage assemble-to-order (ATO) system

In this section we consider an example from assemble-to-order systems, and demonstrate how it can be modeled as a two-stage stochastic linear program and analyzed in *Xpress-SP*.

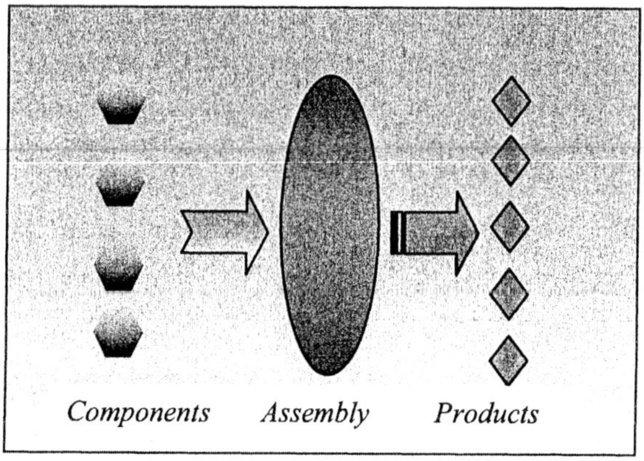

Figure 10.10. Assemble to order systems.

5.1 Problem and applications

In order to manage their inventory efficiently, more companies are moving toward assemble to order systems (Economist (2001)), where typically a set of components are stored in inventory and one or more components are assembled together to produce a range of products. ATO systems involve various challenges such as balancing supply and demand through pricing contracts, dynamically sequencing components, maintaining inventory levels across components with varying lead times, etc. However, their popularity is growing rapidly in various manufacturing sectors because they provide several benefits such as quick response time to order fulfillment, low delivery costs, and a high level of product variety. Some well-known implementations of ATO systems can be seen in companies such as Dell, HP, BMW, and GE (Bylinsky (2000)).

5.1.1 Problem description. We study a simple yet well-known ATO system which is comprised of a set of components and products as described in (Song and Zipkin (2003)). Each product is manufactured by assembling a subset of available components. This can be visualized as shown in Figure 10.10.

We consider a single-period version of the problem, where the demand of products is random, and the inventory level of the components must be such that the expected total cost is minimized. The total cost consists of inventory holding costs and penalties for lost sales of products.

5.1.2 Mathematical formulation.

Model parameters:

- m: total number of components (indexed by i)
- n: total number of products (indexed by j)
- A_{ij}: number of units of component i required to make one unit of product j
- x_0: vector of initial inventory of components
- c: vector of unit cost of procurement of components
- h: vector of unit cost of components
- p: vector of unit costs for shortage of products

Uncertainty:

- $\tilde{d}_j$: random demand for product j

Variables:

- y: vector of inventory position of components
- z: vector of amount of product produced from the components
- x: vector of excess of components left in inventory after the demands are met
- w: vector of lost sales of products

Model:

$$\min_{y \geq x_0} c(y - x_0) + E_{\tilde{d}}\{\min_{x,z,w \geq 0} hx + pw : Az + x = y, z + w = \tilde{d}\}$$

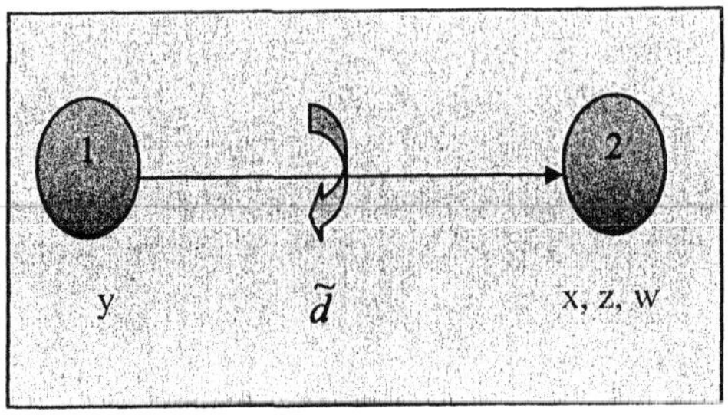

Figure 10.11. sequential event schematics.

5.2 Stochastic framework

In the stochastic programming context, the problem can be visualized as shown in Figure 10.11.

At the first stage, the initial decision y–to position the inventory of components–is taken, which is followed by a random event corresponding to the realization of demand at the second stage. Then the state variables x, z and w are determined.

5.3 SP model

In this section we highlight the key elements of the Mosel stochastic model for the problem. The complete model is shown in Appendix 1. We also compare it with its deterministic equivalent formulation in Mosel.

Given m components and n products, first the stochastic entities are declared.

```
!————————————Declarations—————————————
declarations
  Components=1..m !set of components
  Products=1..n !set of products

  Stages=1..2 !two-stage stochastic problem
  d:array(Products) of sprand !random demand
  x:array(Components) of spvar !excess inventory
  y:array(Components) of spvar !inventory position
  w,z:array(Products) of spvar !lost sales, amount
produced
  TotCost:splinctr !total cost incurred
  SupBal:array(Components) of splinctr !supply balance
end-declarations
```

Demand–being a random variable–is declared as 'sprand' for each product. Stochastic decision variables are declared as 'spvar'. The objective function and other stochastic constraints are declared as 'splinctr'. This is followed by setting the stages and associating each 'sprand' and 'spvar' to a stage as follows:

```
!————————————Stage association———————————————
spsetstages(Stages) !set stages
forall(i in Components) do
spsetstage(y(i),1);spsetstage(x(i),2);
end-do
forall(j in Products) do
spsetstage(d(j),2); spsetstage(w(j),2); spsetstage(z(j),2);
end-do
```

Next, the scenarios are generated by reading in the discretized values and probabilities (not shown here), followed by setting the distributions of demands. Then an exhaustive scenario tree is generated based on these distributions.

```
!————————————Scenario generation———————————————
forall(j in Products) spsetdist(d(j),val,prob) !set
distribution
spgenexhtree !generate exhaustive scenario tree
```

Finally, the model constraints are written, followed by a call to the optimization routine (declarations for c, h and p not shown here).

```
!————————————Model formulation———————————————
TotCost:=sum(i in Components) (c(i)*y(i)+h(i)*x(i))+sum(j in
Products) p(j)*w(j)
forall(i in Components) SupBal(i):=sum(j in Products)
A(i,j)*z(j)+x(i)=
forall(j in Products) DemBal(j):=z(j)+w(j)=d(j)
minimize(TotCost) !Optimization
```

The model formulation for stochastic programming problems in *Xpress-SP* is natural. It is parsed internally into its deterministic equivalent (extensive form). Figure 10.12 highlights the key differences between a

```
model "ATO"
uses 'mmsp'

declarations
d:array(Products) of sprand
y:array(Components) of spvar
x:array(Components) of spvar
w,z:array(Products) of spvar
TotCost:splinctr
SupBal:array(Components) of splinctr
DemBal:array(Products) of splinctr
end-declarations

TotCost:=sum(i in Components)
(c(i)*y(i)+h(i)*x(i))+
sum(j in Products) p(j)*w(j)

forall(i in Components) SupBal(i):=
sum(j in Products) A(i,j)*z(j)+x(i)=y(i)

forall(j in Products) DemBal(j):=
z(j)+w(j)=d(j)

end-model
```

```
model "ATO"
uses 'mmxprs'

declarations
prob:array(Scenarios) of real !probabilities
d:array(Products,Scenarios) of real
y:array(Components) of mpvar
x:array(Components,Scenarios) of mpvar
w,z:array(Products,Scenarios) of mpvar
TotCost:linctr
SupBal:array(Components,Scenarios) of linctr
DemBal:array(Products,Scenarios) of linctr
end-declarations

TotCost:=sum(s in Scenarios) prob(s)*
(sum(i in Components) (c(i)*y(i)+h(i)*x(i,s))+
sum(j in Products) p(j)*w(j,s))

forall(i in Components,s in Scenarios)
SupBal(i,s):=sum(j in Products) A(i,j)*z(j,s)+
    x(i,s)=y(i)

forall(j in Products,s in Scenarios)
DemBal(j,s):=z(j,s)+w(j,s)=d(j,s)

!end-model
```

Figure 10.12. Difference between SP-model and its extensive form.

stochastic model written using *Xpress-SP*, and its extensive form written using Mosel[3].

5.3.1 Analysis of the problem. The special case of ATO systems with one component and one product is indeed the well known newsvendor's problem. In the following sections we compare the results obtained for this problem with the theoretical optimal results for the newsvendor's problem. Since the actual distributions need to be discretized, we first study the loss in accuracy with respect to discretization of distributions. We also analyze the gains from using stochastic programming as opposed to a methodology where one would solve the deterministic problem based on expected demands.

5.3.1.1 The newsvendor's problem The problem discussed in Section 5.1.2 is the generalized case of the newsvendor's problem. Specifically, substituting $m = n = 1$ and $A_{11} = 1$, we obtain the newsvendor's problem with cost c, reward p, and salvage value -h. In this section, we discuss this problem with $p = 100$, $h = -20$, $c = 50$, and $\tilde{d}$ uniformly distributed between a and b, where $a = 50$ and $b = 150$; the optimal inventory is thus given by $y^* = (b-a)\left(\frac{p-c}{p+h}\right) + a = 112.5$

[3]Only relevant part of the code is shown.

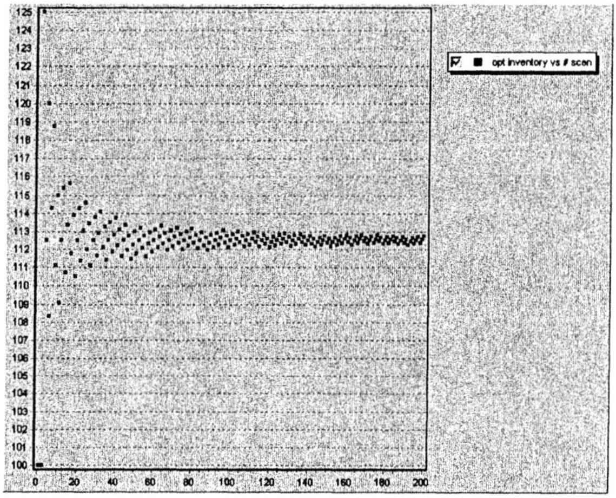

Figure 10.13. IVE plot of optimal inventory with respect to number of scenarios.

Let us first consider the effect of discretization on the optimal results. We divide the interval (a, b) into S equal parts and then solve this problem as a two-stage stochastic linear problem with S scenarios. The following figure (Figure 10.13) shows the IVE-plot of optimal inventory in a two-stage stochastic problem with S scenarios (see Appendix 2 for the Mosel code for generating S scenarios).

From the above figure it is clear that as the number of scenario increases, the optimal inventory tends towards actual optimal solution (112.5). However, the problem size and hence the time required to solve the problem also increases rapidly.

5.3.2 Comparison with myopic (greedy) approaches.

Now consider the original problem with the following parameters:

- $m = 3$; $n = 5$
- $c_1, \ldots, c_3 = 5$; $h_1, \ldots, h_3 = 1$; $p_1, \ldots, p_5 = 20$
- $A = [1, 0, 1, 0, 0; 1, 0, 0, 3, 2; 1, 3, 2, 1, 0]$
- $\tilde{d}_1, \ldots, \tilde{d}_5 = \{5 \; w.p \; 0.2, \;\; 10 \; w.p \; 0.6, \;\; 15 \; w.p \; 0.2\}$

This two-stage problem with $S = 243$ (3^5) scenarios has the optimal solution $y^* = [20, 40, 53.33]$. From the distribution of optimal solution

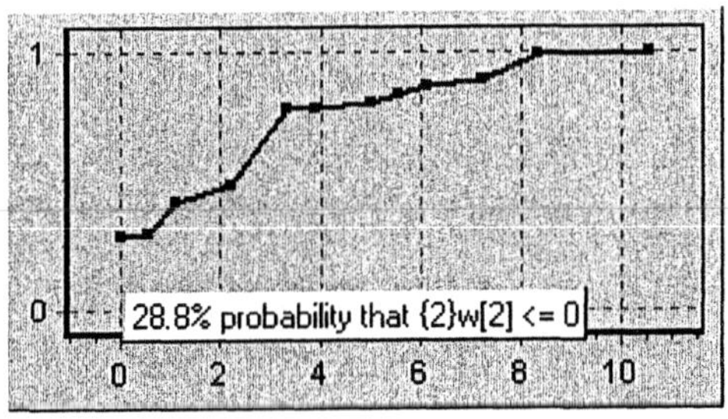

Figure 10.14. IVE plot of probability distribution of lost sales of product 2.

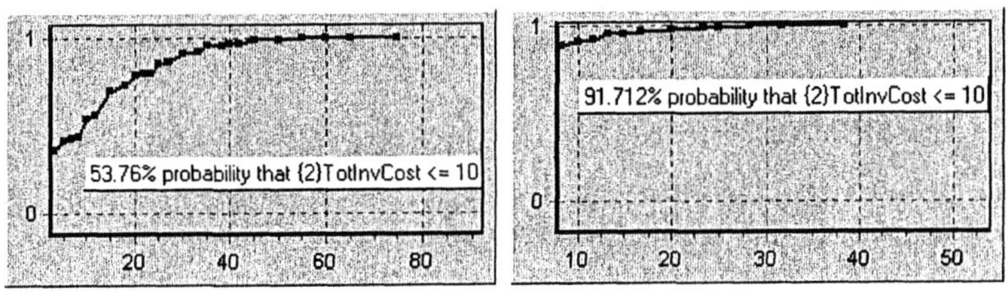

Figure 10.15. Probability distribution of Total inventory cost.

of w_2 (see Figure 10.14) we see that the probability of the sales of product 2 being lost is 0.712 $(1 - 0.288)$.

Now, consider a myopic policy in which the inventory level is decided by considering only the expected demand (10 units) for each product and the objective is to minimize the penalty cost of not meeting the demand. Then the optimal solution is $y = [20, 60, 70]$. Clearly, the increase in inventory would cause a significant change in the total inventory cost in the second stage. This can be observed in Figure 10.15 which shows that the Pr{Total inventory cost>10} is 0.462 as opposed to 0.093 in the optimal solution.

5.3.3 Effects of scenario manipulation. One of the key aspects in an effective SP practice is the generation of scenarios that closely represent the true underlying distribution of random entities, while keeping a reasonably low number of scenarios, so that the problem size is manageable. One may also apply scenario manipulation strategies, e.g., deletion or aggregation of 'extreme' (high/low demand) scenarios. In the following section, we demonstrate how one may aggregate scenarios and further analyze the problem in *Xpress-SP*.

5.3.3.1 Aggregation based on total demand

For the problem with parameters given in Section 5.3.2, one of the criteria for aggregating the scenarios could be total demand, i.e., all scenarios that have same total demand of products would be aggregated together. Using this strategy we obtain 11 scenarios (corresponding to total demands of $25, 30, \ldots, 75$). The demand of each product in the new scenario tree is obtained by aggregating the demands in the original tree weighted by the probabilities of the scenarios aggregated. The optimal solution for this reduced problem is $y = [18, 39, 58]$. In the context of the original problem, its objective value is a lower bound for the optimal expected total cost. In order to evaluate the effect of aggregation, we should test this solution on the original problem, by fixing the y variables and solving the problem. The aggregation can be done in Mosel as follows:

```
!————————————Scen aggregation—————————————
  AggScens !aggregate
  minimize(TotCost) !Optimization of aggregated problem
  writeln("Agg Min Cost=",getojval)
  forall(i in Components) spfix(y(i),getsol(y(i))) !fix
variables
  spgenexhtree !regenerate original tree
  minimize(TotCost) !Optimization of sp problem with fixed
variables
  writeln("RecAggMinCost=",getobjval)
```

In this code, AggScens() is a user-defined procedure for aggregating the scenarios. This procedure is shown in Appendix 3. The 'mmsp' procedure spfix() is used for fixing variables at particular values. This is followed by calling the procedure spgenexhtree() to regenerate the original tree. Now the new stochastic problem is solved again by fixing the y variables at the optimal solution obtained from the reduced problem. It is observed that the objective function value is 0.65% higher than the optimal cost obtained from the original stochastic problem. Considering the reduction of the number of scenarios from 243 to 11, it can be safely concluded that this scenario aggregation strategy is reasonably good. Note that because of the independent and symmetric discretized distributions of demands, the strategy works well for this instance of the problem.

6. SCM application 2: Option contracts in supply chains

Contracts are used by suppliers and buyers to protect themselves against the volatility of demand and market competition. Contracts typically commit involved parties to a pre-defined price or quantity of the underlying commodities. Additionally, they may provide various options such as an option of returning a certain percentage of the unsold items, a provision for securing a pre-determined amount of raw material, etc. From the suppliers' perspective, contracts assure customers and help in consolidating revenues. The format and nature of the contracts in supply chains varies radically across segments of the industry, and the terms and conditions of these contracts also depend on the supplier and the buyer. Both the buyer's and the supplier's problem can be modeled in a multi-stage stochastic framework to achieve new managerial insights and obtain better channels of distribution in the supply chain. The following example is inspired by the work of van Delft and Vial (2003) in modeling an option contract in a supply chain as stochastic program in AMPL. In the following sections we demonstrate how to model and solve this problem using *Xpress-SP*.

6.1 Problem description

We consider the problem of a buyer who is engaged with a supplier in a contract having periodic commitments, for a given number of stages, where the buyer has to make decisions at each stage. At the beginning of the horizon, both parties agree to a fixed amount of product that would be delivered at each stage to the buyer at a specified price. The supplier has a limited amount of options available for each stage, with each unit of an option giving the buyer a right to purchase a unit of product. Whenever this option is exercised, the product is delivered at the next stage. Initially, the buyer may also buy these options for each of the stages at a specified price. At each stage, the buyer creates finished goods from the total amount of product available, and sells them in the market. The demand of finished goods at each stage is uncertain. Hence, at each stage and at a specified exercise price, the buyer may buy additional product (by exercising one or more of the options bought for that stage), in order to meet the demands in the following stages. The unmet demand in each stage is penalized and carried over to the next stage, whereas excess quantity is stored in inventory. If there is excess inventory of products left at the end of the horizon, it is sold at a salvage value; otherwise the unmet demand is lost.

6.2 Mathematical model

Decision variables:

- Q_t: Fixed order of products decided at stage 1, and delivered at stage $t \in \{2, \dots, T\}$
- M_t: Number of options bought at stage 1, which may be exercised at stage $t \in \{2, \dots, T-1\}$
- m_t: Number of options exercised at stage t, and delivered at stage $t+1$, for all $t \in \{2, \dots, T-1\}$

State variables:

- I_t: Finished goods inventory at stage $t \in \{2, \dots, T\}$
- I_t^+: Physical finished goods inventory at stage $t \in \{2, \dots, T\}$
- I_t^-: Backorder of finished goods inventory at stage $t \in \{2, \dots, T\}$

Uncertainty:

- D_t: Demand of finished goods at stage $t \in \{2, \dots, T\}$

Data[4]*:*

- v: Unit salvage value of finished goods (in \$)
- M: Bound on number of options that can be bought at each stage
- o: Unit price for buying an option (in \$)

[4]We assume data is constant across the stages, however this assumption can easily be relaxed.

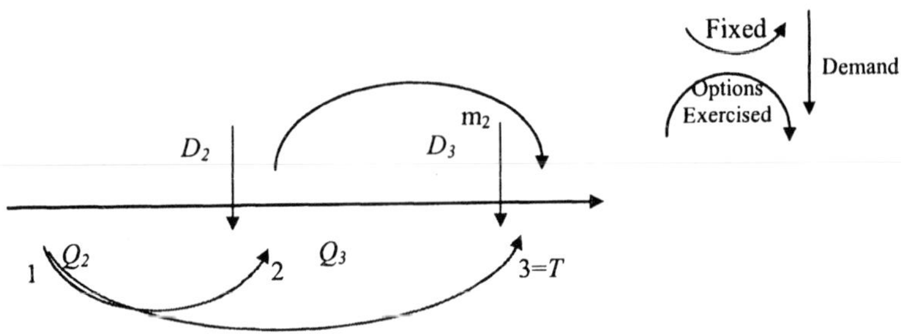

Figure 10.16. Event schematics.

e: Unit price of exercising an already bought option (in \$)

r: Unit selling price of finished goods (in \$)

p: Unit purchasing cost of a product from the supplier (in \$)

s: Unit shortage cost for finished goods (in \$)

h: Unit holding cost for finished goods (in \$)

The process for a problem with $T = 3$ stages can be visualized as shown in Figure 10.16.

In the above figure, the tails of the arcs indicate the stage at which decisions are taken and the head indicates the stage when they come into effect. The inventory level across stages is shown in Figure 10.17.

Model formulation:

The revenue and expenditure functions for a horizon length equal to T are defined as follows:

$$R(I^+, I^-) = r(\tilde{D}_2 - I_2^-) + \sum_{t=3}^{T}(\tilde{D}_t + I_{t-1}^- - I_t^-) + vI_T^+$$

$$E(I^+, I^-, m, Q, M) = \sum_{t=1}^{T-1}(em_t + oM_t) + \sum_{t=1}^{T}(hI_t^+ + sI_t^- + pQ_t)$$

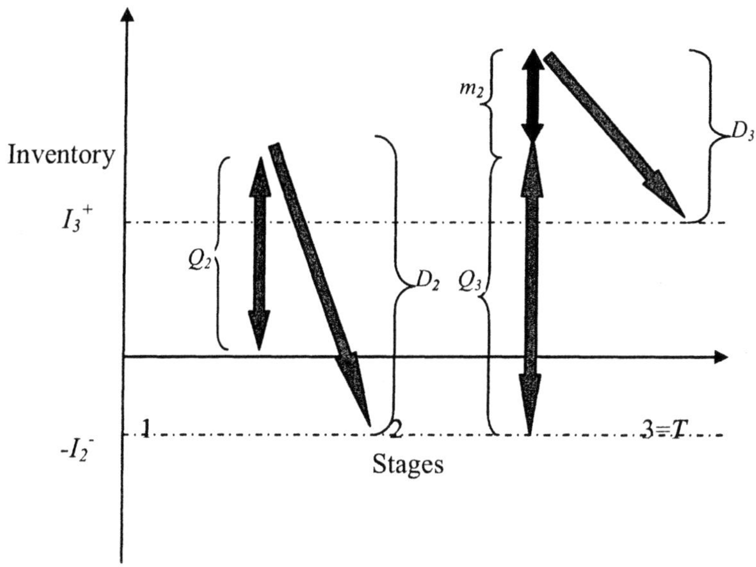

Figure 10.17. Inventory position at each stage.

$$\begin{array}{ll} \max & R(I^+, I^-) - E(I^+, I^-, m, Q, M) \\ s.t. & I_t = I_t^+ - I_t^-, \ \forall t \in \{2, \ldots, T\}, \\ & I_t = \begin{cases} Q_t - \tilde{D}_t & if\ t = 2 \\ I_{t-1} + Q_t + m_{t-1} - \tilde{D}_t & \forall t \in \{3, \ldots, T\}, \end{cases} \\ & 0 \leq m_t \leq M_t \leq \bar{M}, \ \forall t \in \{2, \ldots, T-1\}, \\ & Q_t, I_t^+, I_t^- \geq 0, \ \forall t \in \{2, \ldots, T\}. \end{array}$$

6.3 Demand process

It is assumed that the demand at a stage is correlated with the demand in its previous stage, therefore they:

- form a conditionally heteroskedastic Gaussian process implying:

1. $E(\tilde{D}_{t+1} | \tilde{D}_t = d_t) = \mu + \rho(d_t - \mu)$

2. $Var(\tilde{D}_{t+1} | \tilde{D}_t = d_t) = \sigma^2(1 - \rho^2)$

where μ and σ^2 are the unconditional mean and the variance of $\tilde{D}_t$ respectively, and ρ is the correlation coefficient between $\tilde{D}_t$ and $\tilde{D}_{t+1}$[5]

- are normally distributed, and therefore

$$E(\tilde{D}_{t+1}|\tilde{D}_t = d_t) = \begin{cases} \mu + \rho(d_t - \mu) + \tilde{\epsilon}_{t+1}\sqrt{\sigma^2(1-\rho^2)} \\ \qquad\qquad \forall \in \{2, \ldots, T-1\} \\ \mu + \rho(d_t - \mu) + \tilde{\epsilon}_{t+1}\sigma \qquad t = 1 \end{cases}$$

where $\tilde{\epsilon}_t$ is Normally distributed with mean 0 and variance 1 $\forall t \in \{2, \ldots, T\}$

6.3.1 Discretization. If $\tilde{\epsilon}$ is normally distributed with mean μ_ϵ and variance σ_ϵ^2(in the problem under consideration $\mu_\epsilon = 0$ and $\sigma_\epsilon^2 = 1$), then it is discretized into N_ϵ points by dividing the interval $(\mu_\epsilon + 3\sigma_\epsilon, \mu_\epsilon - 3\sigma_\epsilon)$ into N_ϵ equal sub-intervals, implying

- the length of each sub-interval $\delta_\epsilon = 6/N_\epsilon$
- the nth discretized value $\epsilon_n = (\mu_\epsilon - 3\sigma_\epsilon) + (n - 1/2)\delta_\epsilon$ $\forall n \in \{1, \ldots, N_\epsilon\}$
- the nth discretized probability $p_n = \frac{1}{\sqrt{2\pi\sigma_\epsilon^2}} e^{-\frac{1}{2}\left(\frac{\epsilon_n - \mu_\epsilon}{\sigma_\epsilon}\right)^2} \delta_\epsilon$

6.4 *Xpress-SP* model

In this section we demonstrate the stochastic program in *Xpress-SP*. For clarity, all the data values are explicitly shown in the model itself; however one can easily separate the data and the model. Here we consider the case with[6] $T = 3$, $r = 12$, $v = 2$, $o = 1.5$, $e = 8$, $p = 8$, $s = 6$, $h = 0.5$, $\bar{M} = 10000$, $\mu = 1500$, $\sigma = 330$, $\rho = 0.5$. We also assume that $N_{\epsilon 2} = 41$ and $N_{\epsilon 3} = 31$, which generates a scenario tree with 1, 41, and 1271 (41×31) nodes in the first, second and the third stage of the scenario tree respectively.

[5] We assume that the mean, variance and correlation coefficient are independent of stage, however this assumption may easily be relaxed.
[6] This data is assumed by van Delft and Vial.

SCM.mos

```
model "SCM Option Contract" !model name
uses 'mmsp' !mosel library for stochastic programming

!procedure to generate scenario tree
forward procedure GenScenTree(N:array(range) of integer)
!procedure to get discretized values and probabilities on
Normal distribution
!for a given cardinality of state space
forward procedure GetNormDist(N:integer,x:array(range) of
real,p:array(range) of real)
declarations
  T=3 !number of stages
  Stages=1..T !set of stages

  r=12 !selling price of the finished goods
  h=0.5 !inventory holding cost
  s=6 !shortage cost
  v=2 !salvage value of finished goods
  o=1.5 !unit cost of an option
  e=8 !unit cost of exercising the option
  p=8 !unit purchase cost of product
  Mmax=10000 !upper bounds on the number of option

  mu=1500 !mean demand
  sigma=330 !deviation in demand
  rho=0.5 !correlation coefficient
  eps:array(2..T) of sprand !eps  Normal(0,1)
  Neps:array(2..T) of integer !size of discrete state
  space of eps(t)
  D:array(2..T) of sprandexp !demand in stage t

  M:array(2..T-1) of spvar !# of options bought in 1, to
  exercise in t
  m:array(2..T-1) of spvar !option exercised in t,
  effective in t+1
  Q:array(2..T) of spvar !fixed quantities ordered in 1
  and delivered in t
  I:array(2..T) of spvar !net inventory at stage t
  I_p:array(2..T) of spvar !physical inventory at stage
  t
  I_m:array(2..T) of spvar !backorder at stage t
```

```
    R:splinctr !total revenue
    E:splinctr !total expenses
    P:splinctr !total profit
    NetInv:array(2..T) of splinctr !net inventory balance
    InvBal:array(2..T) of splinctr !inventory balance across
    stages
    MaxNumOpt:array(2..T-1) of splinctr !bound on # of
    options bought
    MaxNumExOpt:array(2..t-1) of splinctr !bound on # of
    options exercised
  end-declarations
  setparam('xsp_implicit_stage',true) !assume last index set
  of array as stage
  spsetstages(Stages) !set the stage set

  !------------Define demand process-------------
  forall(t in 2..T)
    if t>2 then D(t):= mu + rho(D(t-1)-mu) + eps(t)*sigma*(1-rho
  ^2) ^.5
    else D(t):=mu+eps(t)*sigma
    end-if
  Neps:=[41,31] !define discretized grid size
  GenScenTree(Neps) !generate scenario tree
  forall(t in 2..T-1) spsetstage(M(t),1) !explicitly set to
  1-st stage
  forall(t in 2..T) spsetstage(Q(t),1) !explicitly set to
  1-st stage
  forall(t in 2..T) I(t) is_free !set as free variable

  !------------Model formulation-------------
  R:=r*(D(2)-I_(2))+sum(t in 3..T)
  r*(D(t)+I_m(t-1)-I_m(t))+v*I_p(T)
  E:=sum(t in 2..T-1) (e*m(t)+o*M(t))+sum(t in 2..T)
  (h*I_p(t)+s*I_m(t)+p*Q(t))
  P:=R-E
  forall(t in 2..T) NetInv(t):=I(t)=I_p(t)-I_m(t)
  forall(t in 2..T)
    if (t>2) then InvBal(t):=I(t)=I(t-1)+Q(t)+m(t-1)-D(t);
    else InvBal(t):=I(t)=Q(t)-D(t)
    end-if
  forall(t in 2..T-1) MaxNumExOpt(t):=m(t)<=M(t)
  forall(t in 2..T-1) MaxNumOpt(t):=M(t)<=Mmax
  !generate extensive form with entities for each node in
  the event tree
```

```
setparam('xsp_scen_based',false)
maximize(P) !maximize the profit
writeln("Profit=",getobjval) !print obj val

!————————————Procedures————————————
procedure GenScenTree(N:array(range) of integer)
  Nmax:=max(t in 2..T) N(t) !max size of discretized state
space
  declarations
    !store the discretized values and probabilities
    val,prob:array(1..Nmax) of real
  end-declarations
  forall(t in 2..T) do
    forall(n in 1..Nmax) do val(n):=0;end-do
    GetNormDist(N(t),val,prob) !get discretized val &
    probs
    spsetdist(eps(t),val,prob) !set them as distribution
    of eps(t)
  end-do
  spgenexhtree !generate exhaustive tree based on the
  distribution
end-procedure
procedure GetNormDist(N:integer,x:array(range) of
real,p:array(range) of real)
  if(not isodd(N) or N<3) then
    writeln("N must be an odd number and >=3")
    exit(1) !the cardinality of state space must be odd
    and >=3
  end-if
  delta:=6/N !delta element for distribution in (-3,3)
  forall(n in 1..N) do
    x(n):=3+(n-0.5)*delta !n-th discretized value
    n-th discretized probability
    p(n):=(1/(2*M_PI)^0.5)*exp(-(x(n))^2/2)*delta
  end-do
end-procedure
end model
```

The model begins with declaring stages and other entities, which is followed by the setting of stages. When the parameter xsp_implicit_stage is set to true, all the stochastic entities are implicitly associated with one of the stages under the assumption that the last set for indexing the arrays in which they are stored is the stage set; however one can

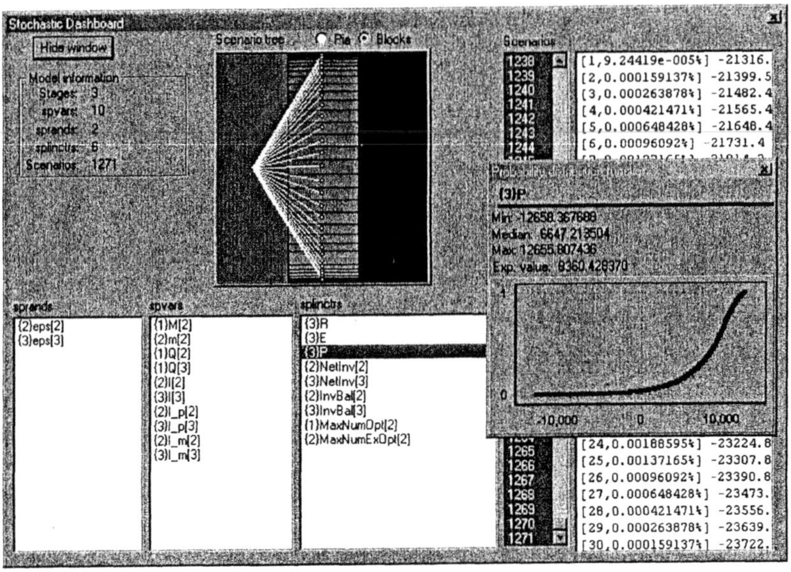

Figure 10.18. Model information.

override this assumption by setting their stages explicitly. Demands are defined as 'sprandexp', and are declaratively assigned their dependence on other 'sprands' and 'sprandexps' in the forall loop, which is followed by scenario generation. In this model, the procedure GenScenTree() is defined for creating the scenario tree. It calls the procedure GetNormDist(), gets the discretized distribution as discussed in Section 6.3.1, and sets the distribution by calling the 'mmsp' procedure spsetdist(). Next, the stages and types of other 'spvars' are defined, the model is formulated, and is then optimized.

6.5 Analysis of stochastic solution

The IVE stochastic dashboard displays the model information (see Figure 10.18). There are 3 stages, 1271 scenarios corresponding to each node in the last stage, 2 random variables, 10 stochastic decision variables, and 6 stochastic constraints. Each of these entities is prefixed by a curly bracket which states its stage number.

The model is parsed internally by creating new variables corresponding to the nodes at the stage of the decision variable in the scenario tree (see Section 3.5.2). The matrix thus generated is ordered according to

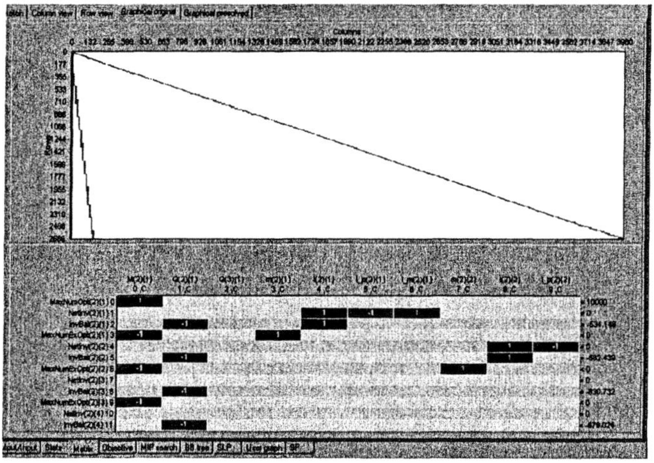

Figure 10.19. Node based matrix.

stages, and within each stage, according to the nodes in that stage. For this problem the Xpress-Optimizer displays the following statistics:

```
Problem Statistics
    2666 (  399 spare) rows
    3980 (    0 spare) structural columns
    9185 ( 1377 spare) non-zero elements
```

The extensive form of the matrix can be visualized in IVE (see Figure 10.19). Note that each stochastic decision variable or constraint is further suffixed by a curly bracket which states the variable's or constraint's node number in the scenario tree.

As mentioned in Section 3.5.2, any stochastic model can also be parsed according to scenarios. This is achieved in *Xpress-SP* by setting the parameter xsp_scen_based to true. The scenario-based problem has the following statistics:

```
Problem Statistics
    16356 (  2453 spare) rows
    12710 (     0 spare) structural columns
    36525 ( 5478 spare) non-zero elements
```

Figure 10.20 shows the matrix structure. In the extensive form obtained from the scenario based parsing, each stochastic decision variable or constraint is indexed by scenarios. In Figure 10.20, the variables

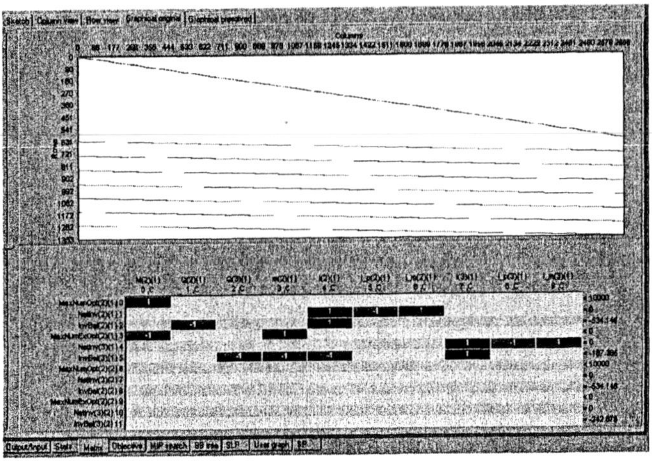

Figure 10.20. Scenario based matrix.

and constraints are suffixed by curly brackets indicating their scenario numbers. Note that the constraints are ordered based on scenarios, so they appear in the block diagonal form. They are followed by the non-anticipative constraints (see Section 3.5.2.1).

6.5.1 Comparison with a related problem. If we order the optimal net profits in various scenarios increasingly and then plot them against cumulative probabilities of corresponding scenarios (this is done automatically in IVE, see Figure 10.18), it is observed that there could occur a loss of as high as \$12,658.36 and a maximum profit of \$12,655.8. The maximum expected profit is \$8360.43 with the probability of loss less than 0.0843. The optimal number of options bought initially for stage 2 is 468 and the fixed orders for stage 2 and 3 are 1982 and 910 units of product respectively.

6.5.1.1. Perfect Information problem

In this section we analyze the problem in the Perfect Information context. Assuming that the future is known with certainty, we solve the problems for each scenario independently. Such problems can be solved in *Xpress-SP* by calling maximize(P,"PI.") where, "PI" refers to the perfect information problem. Each scenario would have its own optimal solution for Q and M. We can aggregate these solutions based on scenario probabilities and obtain a unique implementable solution. The aggregated solution turns out to be $M = 0$ and $Q = [1500, 1500]$ for this problem. Next, we compare this solution with the optimal solution of the original recourse stochastic problem by fixing the first stage variables at the aggregated values obtained, and solving the problem in the stochastic framework. This can be done in *Xpress-SP* by passing the string "PIr." to maximize(). It is observed that the maximum profit for this problem is $923.93 less than the optimal profit[7]; implying a gain of about 11.05% by using stochastic programming for decision-making.

6.6 Reduced scenario tree

van Delft and Vial (2003) studied the effect of discretization of $\tilde{\epsilon}_2$ and $\tilde{\epsilon}_3$ on this problem with the number of discrete values ranging from 5 to 321. Such an analysis for this problem can easily be done in *Xpress-SP* by dynamically reading N_{ϵ_2} and N_{ϵ_3} in the model shown in Section 6.4. Next, we focus on building asymmetric scenario trees in *Xpress-SP*. Although SP provides great insights into the dynamics of a problem with respect to uncertainty and constructing probability distributions of various stochastic entities, it becomes increasingly challenging from a computational point of view to solve stochastic problems as the number of scenarios increases. Hence, in order to prevent the size of the problem from growing too fast, it is important to keep the number of scenarios small, while maintaining the distributions as close to reality as possible. One of the key strategies for restricting the size of the scenario tree is to reduce the number of branches emerging from each node in the later stages. In a sequential decision making process the later decisions depend highly on the initial ones; therefore, it is important to have more realizations for the initial decisions than the later ones. Furthermore, if the probability of visiting a node in a given stage is much smaller as compared to other nodes in that stage, one may curtail the number of branches emerging from that node because the detailed future information collected on that node by having many branches will neither be fruitful nor affect the optimality of the solution. For this purpose van Delft and Vial suggested *fine* and *coarse* branching strategies and

[7]This difference is often referred to as Expected Value of Perfect Information (EVPI).

proposed the following scheme for switching from one to another:

$$Br_{tn} = \begin{cases} N_t^f \ if \ \frac{P_{tn}}{1/N_t} \geq SwitchLevel_t \\ N_t^c \ if \ \frac{P_{tn}}{1/N_t} < SwitchLevel_t \end{cases}$$

$\forall n \in \{1,\ldots,N_t\}, t \in \{1,\ldots,T-1\}$
where

N_t is the number of nodes in the scenario tree at the stage t: $t \in \{1,\ldots,T\}$

P_{tn} is the unconditional probability of occurrence of nth node at t^{th} stage: $n \in \{1,\ldots,N_t\}$, $t \in \{1,\ldots,T\}$

Br_{tn} is the number of branches emerging from n^{th} node at t^{th} stage: $n \in \{1,\ldots,N_t\}$, $t \in \{1,\ldots,T-1\}$

N_t^f is the number of branches emerging from a node in t^{th} stage, if branching is fine at that node: $t \in \{1,\ldots,T-1\}$

N_t^c is the number of branches emerging from a node in t^{th} stage, if branching is coarse at that node: $t \in \{1,\ldots,T-1\}$

$SwitchLevel_t$ is the switching level defined at each stage for switching from fine branching to coarse branching: $t \in \{1,\ldots,T-1\}$

6.6.1 Implementation in *Xpress-SP*. A scenario tree based on such a rule can be easily created in *Xpress-SP*. For the problem under consideration, the procedure 'GenScenTree()' in the Mosel implementation (see Section 6.4) is changed to take the three arguments: N_t^f, N_t^c, and $SwitchLevel_t$, as shown next.

```
!-------------------Procedures-------------------
procedure GenScenTree(Nfine:array(range)) of integer,
  Ncoarse:array(range) of integer,SwitchLevel:array(range) of
  real)
  Brmax:=max (t in 1..T-1) Nfine(t) !max # branches from
  a node
  Nmax:=integer(Brmax ^(T-1)) !max # of nodes in a
  stage
  declarations
    N:array(1..T) of integer !# of nodes in each stage
```

```
      val,prob:array(1..Brmax) of real !assumed vals and
      probs by eps
      fine:dynamic array(1..T,1..Nmax) of boolean !is this a
      fine node
      UnconProb:dynamic array(1..T,1..Nmax) of real
  !unconditional prob
    end-declarations
    !initialize
    N(1):=1;UnconProb(1,1):=1;
    if(SwitchLevel(1)¡1) then fine(1,1:=true)
    else fine(1,1):=false
    end-if
    !begin creating tree
    forall(t in 1..T-1) do
      forall(n in 1..N(t)) do
        if(fine(t,n)) then Br:=Nfine(t);
        else Br:=Ncoarse(t);
        end-if
        spaddchildren(t,n,Br) !create Br # branches from
        node(t,n)
        GetNormDist(Br,val,prob) !get realized vals and
        probs
        forall(b in 1..Br) do
          n_:=spgetchild(t,n,b) !node number in next stage
          spsetrandatnode(eps(t+1),n_,val(b)) !set val at
          node
          spsetprobcondatnode(t+1,n_,prob(b)) !set prob at
          node
          UnconProb(t+1,n_):=UnconProb(t,n)*prob(b)
          !update
       end-do
     end-do
     N(t+1):=spgetnodecount(t+1) !update # nodes in next
     stage
     !update whether next stage nodes are finite or coarse
     if(t+1¡T) then forall(n in 1..N(t+1))
       if(N(t+1)*UnconProb(t+1,n)>=SwitchLevel(t+1))
       then fine(t+1,n):=true
       else fine(t+1,n):=false
       end-if
     end-if
  end-do
  spgentree !generate the tree
end-procedure
```

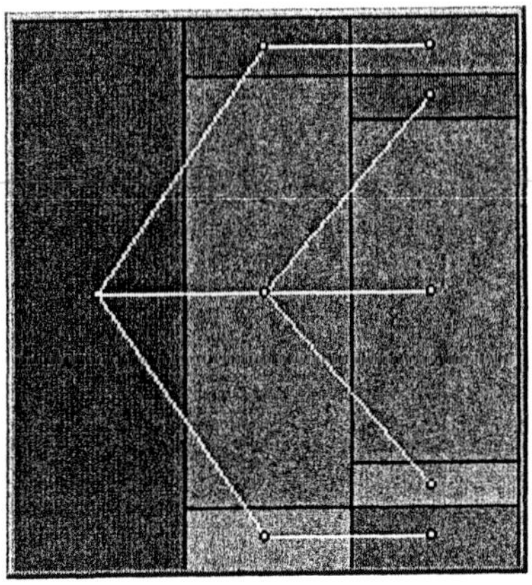

Figure 10.21. Reduced scenario tree.

The above procedure defines a Boolean dynamic array 'fine'. By stating it as 'dynamic', we ensure that Mosel doesn't create any unnecessary entry. If 'fine_{tn}' is *true*, then the number of branches emerging from the nth node in the tth stage is N_t^f; otherwise it is N_t^c. The procedure begins by initializing N_1, and P_{11} to 1, and then N_t, P_{tn} and Br_{tn} are updated within the 'forall' loop. For the purpose of illustration, we set $N^f = [3,3]$, $N^c = [3,1]$[8] and $SwitchLevel = [0.5, 1.5]$, which generates the following scenario tree[9]:

6.7 Global constraints

In this section we demonstrate how global constraints can easily be modeled in *Xpress-SP*. Global constraints are different from the regular constraints in SP, in the sense that instead of having decision variables along a path to a particular scenario or a node in the scenario tree, these constraints chain variables across all the nodes of a particular stage or all the scenarios (depending on whether the problem is 'node based' or 'scenario based'). Such constraints are particularly useful in writing certain financial or managerial constraints. We show how to model 'chance constraints' as global constraints in *Xpress-SP*. Next, we

[8] *Xpress-SP* also supports problems with trap stage scenario trees, e.g., if $N^f = [3,3]$ and $N^c = [3,0]$, then the first and the third node in the second stage will not have any children in the generated scenario tree.

[9] Here we assume that if $N_\epsilon = 1$, then $\epsilon = \{0 \ w.p\ 1\}$.

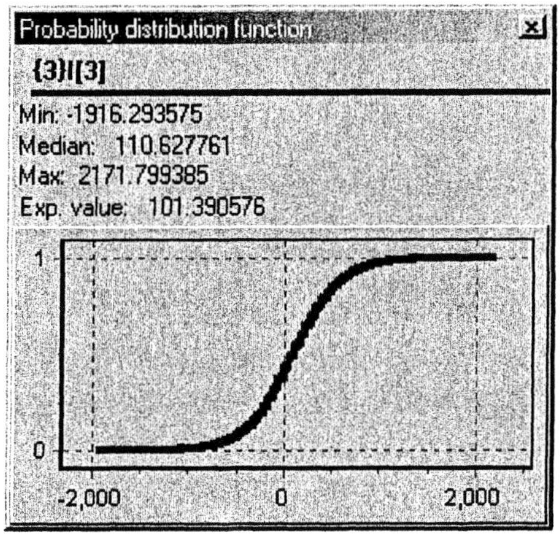

Figure 10.22. Inventory position in the last stage.

also present modeling of global constraints in the context of minimizing the conditional value at risk for this problem, as proposed by van Delft and Vial (2003).

6.7.1 Modeling chance constraints. Chance constraints are quite common in stochastic programming (Birge and Louveaux (1997)). Conceptually, one can think of them as the constraints that need to be satisfied only with certain probability. However, not only are they difficult to model in the traditional algebraic modeling languages, the problem becomes very difficult to solve to optimality. In the context of the current problem, consider the distribution of inventory at the last stage (Figure 10.22).

The probability that net inventory at the last stage (I_T) goes below 0 is 0.42. Now consider a managerial constraint that stipulates that the $Pr\{I_T < 0\} \leq \beta$. Such a constraint can be modeled in SP by adding a:

i. binary variable z belonging to the last stage.

ii. regular constraint $I_T^- \leq \sum_{t=2}^{T} \tilde{D}_t.z$.

iii. penalty θ corresponding to z to the objective function which is small enough so that the optimal solution is not perturbed.

iv. global constraint $z \leq \beta$.

The variable z indicates a negative inventory (lost sales) in the last stage. The regular constraint and the penalty ensures that $I_T^- > 0$ whenever $z = 1$. The global constraint enforces the chance constraint. Note that this constraint is internally parsed as $\sum_{n=1}^{N_T} P_{Tn} z_n \leq \beta$. Hence, the total probability of a negative I_T is enforced to be less than or equal to β. The corresponding Mosel implementation is presented next.

```
!------------Chance constraints------------
declarations
  z:spvar
  theta=5*10 ^(-6) !penalty
  ShortageProb,ShortageBound:splinctr
  beta=0.25 !max shortage prob
end-declarations

spsetstage(z,T)
z is_binary
ShortageBound:=I_m(T<=(sum(t in 2..T) D(t))*z
  !max shortage=D(2)+..+D(T)
ShortageProb:=z<=beta
spsettype(ShortageProb,"global")
!set xprs control parameters for better performance
spsetxprsparam("xprs_cutstrategy",0)
spsetxprsparam("xprs_treegomcuts",10000)
spsetxprsparam("xprs_miprelstop",0.01)

P:=R-E
P-=theta*z
maximize(P)
```

The extensive form of the problem with chance constraint contains 1271 binary variables. The search strategy for finding a good solution is further enhanced by setting Xpress Optimizer control parameters. Using Xpress Optimizer version 14.24, disabling automatic cut generation during the reduction of relaxed LP, and limiting a maximum of 10,000 rounds of Gomory cuts at nodes in the branch and bound tree, a 1% optimal solution with objective value of 8200.5 is obtained after 34 seconds on 2.2 Ghz, P-4 machine with 1 GB RAM. The probability of inventory going below zero is 0.248. The distribution of inventory at the last stage after implementing the chance constraint on it is shown in Figure 10.23:

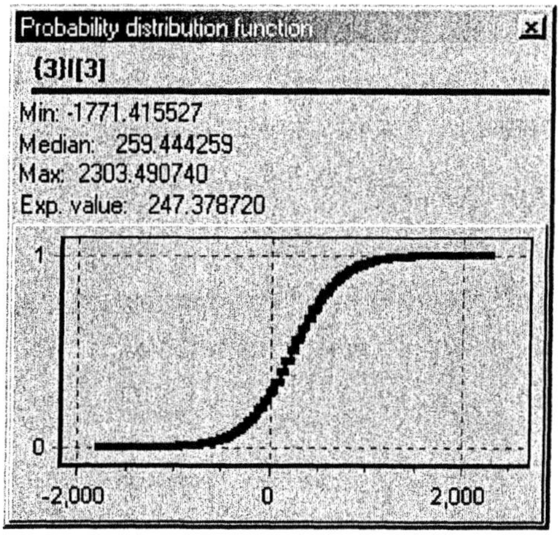

Figure 10.23. Last stage inventory distribution.

6.7.2 Calculating conditional value at risk.

The Conditional Value at Risk measure ($CVaR$) is closely related to the traditionally used Value at Risk measure (VaR) in industry. It is the expected loss under the condition that loss exceeds VaR (Rockafellar and Uryasev (2000)). Formally, $CVaR$ is defined as:

$$CVaR(\alpha) = E[L|L \geq S(\alpha)]$$

where

L is the loss function,

α is the risk level: $0 < \alpha < 1$,

$S(\alpha)$ is the threshhold above which $CVaR$ needs to be calculated.

This can be visualized as shown in Figure 10.24.

Rockafellar and Uryasev (2000) showed that CVaR is a solution to the minimization problem:

$$CVaR(\alpha) = \min_S \{S + E[\max(L - S, 0)]/\alpha\}$$

Hence, in the stochastic programming context, CVaR can be calculated by introducing variables $CVaR$, with S belonging to stage 1, and a non-negative variable z belonging to stage T. The variable z represents $\max(L - S, 0)$; therefore a constraint $z \geq L - S$ must also be introduced.

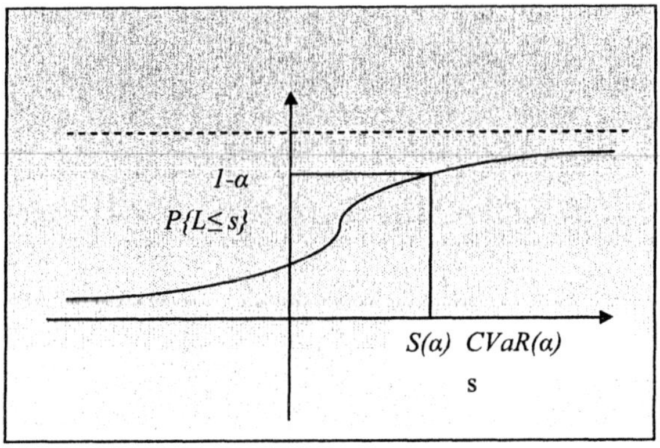

Figure 10.24. Plot of cumulative probability distribution of loss function.

Next, a bound c is introduced on $CVaR$. During minimization, the term $\theta.CVaR$ is added to the objective function, where θ is sufficiently small, so that it does not perturb the optimal solution to the original problem. The constraint $S + z/\alpha \leq CVaR$ is added and set as 'global' implying $S + \sum_{n=1}^{N_T} P_{Tn} z_n/\alpha \leq CVaR$.

The Mosel implementation is shown below:

```
!————————————CVar constraints——————————————
declarations
  z,S,CVar:spvar
  c=6000
  alpha=.01
  theta=5*10 ^(-6)
  CondLoss, CvarLoss,GlbBnd:splinctr
end-declarations
spsetstage(z,T) !by default z>=0
S is_free !S belongs to stage 1 when implicit_stg is
true
CVar is_free
L:=E-R
CvarLoss:=z>=L-S
GlbBnd:=S+z/alpha<=CVar
spsettype(GlbBnd,"global") !specify as global
constraint
CVar<=c !bound
L+=theta*CVar !add this term to objective function
setparam('xsp_scen_based',false)
minimize(L)
```

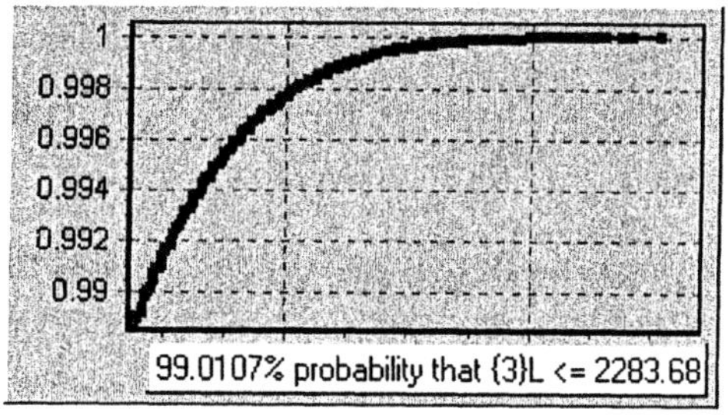

Figure 10.25. Cumulative probability distribution for Loss $\geq S_\alpha$.

After solving the problem for 41 nodes in second stage and 1271 nodes in the third, the minimum value of CVaR obtained is 4053.4 with $S =$ 2283.66. Figure 10.25 shows the distribution of the Loss function for Loss greater than S.

7. Summary and Conclusion

In this paper we have demonstrated how problems that occur in supply chains can be modeled, solved, and analyzed in a stochastic programming framework using *Xpress-SP*. We began by identifying the need and applicability of stochastic programming for efficient supply chain management. Next, we described the basics of stochastic programs, including the structure of a scenario tree and the node-based and scenario-based extensive formulation of the multi-stage stochastic linear problem. Then we summarized the capabilities of the *Xpress-SP* suite by describing its architecture, functionalities and other tools. We then illustrated the utility, flexibility, and scalability of *Xpress-SP* using two examples from supply chain contexts. The first example was taken from assemble to order systems and modeled as a two-stage stochastic program. There we analyzed the variation of the optimal solution value with respect to the number of scenarios, followed by a comparison with a myopic policy, and a brief discussion on the effect of scenario aggregation. The second example was based on supply chain contracts, where we illustrated the functionalities of *Xpress-SP* in greater detail. In that example we showed how scenario trees with correlated random variables can be easily generated. We demonstrated how to analyze the problem and solution visually using Xpress-IVE and how to model chance constraints using global constraints in *Xpress-SP*.

Problems in supply chain management involve a lot of uncertainties and risks. Stochastic programming techniques are promising methodologies for better decision-making under uncertainty, as they provide overall best decisions and balance rewards against risks. *Xpress-SP* provides a range of tools and functions for developing two-stage and multi-stage stochastic linear programs. The syntax of the modeling language available in *Xpress-SP* is natural for problems requiring decision-making under uncertainty. It is concise, flexible, and scalable, and its integration with Xpress IVE makes *Xpress-SP* a state-of -the-art technology for Stochastic Programming.

Appendix

1. Mosel Stochastic model for Assemble to order systems

The following model is built assuming:

- No initial inventory (x0=0)
- Each of the products' demand is independently distributed with a known discretized distribution

ATO.mos

```
model "Assemble to order" !model name
uses 'mmsp' !mosel model library for stochastic programming

parameters !can be changed dynamically at run time
   DatFile="ATO.dat"
   DistFile="ATOdist.dat"
end-parameters

declarations
   m,n:integer !dimensions
end-declarations

initializations from DatFile !read dimensions
   m n
end-initializations

declarations
   Components=1..m !set of components
   Products=1..n !set of products
```

```
    c,h:array(Components) of real !procurement, holding cost
    p:array(Products) of real !penalty cost
    A:array(Components,Products) of real
  end-declarations

  initializations from DatFile
    c h p A
  end-initializations

  declarations
    Stages=1..2 !two-stage stochastic problem
    d:array(Products) of sprand !random demand
    x,y:array(Components) of spvar !excess inventory, inventory position
    w,z:array(Products) of spvar !lost sales, amount produced
    TotCost:splinctr !total cost incurred
    SupBal:array(Components) of splinctr !supply balance
    DemBal:array(Products) of splinctr !demand balance
  end-declarations
  !——————————Stage association——————————
  spsetstages(Stages) !set stages
  forall(i in Components) do
    spsetstage(y(i),1);spsetstage(x(i),2);
  end-do
  forall(j in Products) do
    spsetstage(d(j),2);spsetstage(w(j),2);spsetstage(z(j),2);
  end-do
  !——————————Scenario generation——————————
  declarations
    nVals:integer !number of discretized points
  end-declarations
  initializations from DistFile
    nVals
  end-initializations
  declarations
    val,prob:array(1..nVals) of real !discretized values and probabilities
  end-declarations
  initializations from DistFile
    val prob
  end-initializations
  forall(j in Products) spsetdist(d(j),val,prob)
  !set discretized distribution
  spgenexhtree !generate exhaustive scenario tree
```

```
!——————————Model formulation——————————
TotCost:=sum(i in Components) (c(i)*y(i)+h(i)*x(i)) + sum(j in
Products) p(j)*w(j)
forall(i in Components) SupBal(i):=sum(j in Products) A(i,j)*z(j)+x(i)=y(i)
forall(j in Products) DemBal(j):=z(j)+w(j)=d(j)
minimize(TotCost) !Optimization
end-model
```

2. Two-stage S-scenario tree with uniform distribution

```
declarations
  S=200
  a=50
  b=150
  val,prob:array(1..S) of real
end-declarations

val(1):=a+((b-a)/S)/2
forall(s in 2..S) val(s):=val(s-1)+(b-a)/S
forall(s in 1..S) prob(s):=1/S
forall(j in Products) spsetdist(d(j),val,prob)
spgenexhtree
```

3. Aggregation procedure for ATO model

```
procedure AggScens
  declarations
    TotDem:array(range) of real
    aggSet:set of integer
  end-declarations

  S:=0
  forall(s in 1..spgetscencount) do
    found:=false
    TotScenDem:=sum(j in Products) speval(d(j),s)
    if(S>1) then
      forall(s_ in 1..S) do
        if(TotScenDem=TotDem(s_)) then
          found:=true
          break
        end-if
      end-do
    end-if
```

```
    if(not found) then
      S+=1
      TotDem(S):=TotScenDem
    end-if
  end-do

  forall(s_ in 1..S) do
    aggSet:=
    forall(s in 1..spgetscencount) do
      TotScenDem:=sum(j in Products) speval(d(j),s)
      if(TotScenDem=TotDem(s_)) then
        aggSet+=s
      end-if
    end-do
    if(getsize(aggSet)>1) then
      spaggregate(aggSet)
    end-if
  end-do
end-procedure
```

References

Birge, J.R. and Louveaux, F. 1997. *Introduction to Stochastic Programming*, Springer Series in Operations Research.

Bylinsky, G. 2000. Heroes of U.S. manufacturing, *Fortune* **141**.

Chopra, S. and Meindl, P. 2001. Managing Uncertainty in a Supply Chain: Safety Inventory, *Supply Chain Management; Strategy, Planning, and Operation*, Prentice-Hall Inc.

Colombani, Y. and Heipcke, S. 2002. Mosel: An Overview, May 2002, available at http://www.dashoptimization.com/home/downloads/pdf/mosel.pdf.

Dupacova, J., Hurt, J. and Stephan, J. 2002. *Stochastic Modeling in Economics and Finance* **75**, Kluwer Academic Publishers.

Economist. A long march: Mass customization, July 2001. **360**, Issue 8230.

Fourer, R. and Gay, D.M. 1997. Proposals for Stochastic Programming in the AMPL Modeling Language, Session WE4-G-IN11, *International Symposium on Mathematical Programming*, Lausanne, August 27, 1997, available at http:// iems.nwu.edu/ 4er/ SLIDES/ lsn9708v.pdf.

Fragniere, E. and Gondzio, J. 2002. Stochastic Programming from Modeling Languages, in: *Applications of Stochastic Programming*. Eds. H. Gassmann, S. Wallace and W. Ziemba, SIAM Series on Optimization, available at http://www.maths.ed.ac.uk/ gondzio/ gondzio/ cv-gondzio.html.

Morton, D., 2004. Stochastic Programming Applications, available at http://www.dashoptimization.com/home/downloads/pdf/Stochastic Applications.pdf.

Rockafellar, R.T. and Uryasev, S. 2000. Optimization of conditional value at risk, *Journal of Risk* **2**, pp. 21-41.

Shapiro, J.F. 2001. Decision Trees and Stochastic Programming, *Modeling the Supply Chain*, Duxbury- Thomas Learning.

Song, J-S. and Zipkin, P., 2003. Supply Chain Operations: Assemble-to-Order Systems, in *Handbooks in Operations Research and Management Science* **30**, *Supply Chain Management*, Eds. T. de Kok and S. Graves, North-Holland, Forthcoming

van Delft, Ch. and Vial, J.-Ph. 2003. A practical implementation of stochastic programming: an application to the evaluation of option contracts in supply chains, *Automatica*, Forthcoming.

Chapter 11

DISPATCHING AUTOMATED GUIDED VEHICLES IN A CONTAINER TERMINAL

Yong-Leong Cheng*, Hock-Chan Sen*
Singapore MIT Alliance Program.

Karthik Natarajan*
Department of Mathematics, National University of Singapore.

Chung-Piaw Teo†
SKK Graduate School of Business, Sungkyunkwan University.
Department of Decision Sciences, National University of Singapore.

Kok-Choon Tan
PSA Corporation.
Department of Industrial and Systems Engineering, National University of Singapore.

1. Introduction

The efficiency of a global supply chain network is predicated on the availability of an efficient, reliable global transportation system. No supply chain can operate efficiently if the assets used in the conversion of raw materials, manufacturing and distribution of the product are not being managed effectively. Decreasing costs, lower rates of transport, ris-

*Research supported by a scholarship from the Singapore-MIT Alliance Program in High Performance Computation for Engineered Systems.
†Corresponding author. Research partially supported by a fellowship from the Singapore-MIT Alliance Program in High Performance Computation for Engineered Systems.

ing customer demand, and globalization of trade have caused a steady increase in the use of containers for sea borne cargo. Consequently, container terminals have become an important component of global logistic networks. The repercussions of poor container terminal management and bad planning can prove costly - cargo delayed at port must be accounted for; ships often have to be diverted to other harbors to offload, resulting in added pressure on other ports and additional costs at sea, and delayed delivery.

A shipping container is a box designed to enable goods to be delivered from door to door without the contents being physically handled. The most common sizes are the 20 footers (6.1m long, 2.4m wide and 2.6m tall), and the 40 footers (12.2m long, with the same width and height as the 20 footers). Since its inception, container shipping has become a popular mode to convey products of all types, especially the high-value cargoes. To satisfy customer demand, it is paramount that containers on the ships are unloaded/loaded promptly at the port. According to industry estimates (cf. Chan and Huat (2002)), the typical operating cost for, say a Post Panamax vessel per day can easily come to US$ 30,000 (cf. Table 11.1). Given the high operating cost, it is imperative that vessel operators maximize the yields from the different voyages made by each vessel.

Table 11.1. Operating cost for a typical post Panamax vessel.

	US$/day
Vessel Depreciation Cost (25 years life span)	10,000
Fuel Cost (18 knots cruising speed)	10,000
Wages, Maintenance and Insurance	10,000

The above consideration necessitates the development of highly sophisticated container transportation systems, which allow for efficient movement within the container terminal area. As a result, terminal operators over the world have been increasingly pressurized to provide better and faster service to vessel operators. A major challenge in port management is thus to reduce the turnaround time of the container ships, thus reducing the supply chain cycle time for the shippers. This can be achieved in various ways:

- Deploy more cranes per vessel. This is however constrained by the length of the vessels and the minimum distance required between cranes.
- Improve the handling rate of the individual cranes, by increasing the speeds and semi-automation features of the cranes.
- Improve reliability and maintainability of the cranes to minimize the amount of rework.
- Train and use skilled operators to operate the cranes.
- Provide efficient yard handling and horizontal transportation systems for the loading and discharging/unloading operations.

In this paper, we will focus on the challenges posed by the last method, specifically improving the performance of the horizontal transportation system. Over the last decade, technology and automation have been aggressively introduced into the container terminal business to improve the efficiency of port operations. For example, Automated Guided Vehicle (AGV) systems are used in terminal operations for the retrieval and storage of containers in certain container terminals in Europe. Onboard computers on each AGV communicate using wireless transmissions with the control center to allow the vehicle to navigate to any point within the terminal.

The deployment of AGVs in the horizontal transportation system within the container terminal has given rise to new operational issues. In a manual system, optimizing the deployment and dispatching of trucks to ships has proven to be difficult in the past, due to the lack of control and monitoring mechanisms within the terminal. Most terminal operators simply deployed a fixed number of trucks/drivers to serve a ship, ignoring the real time traffic information and container movement activities within the terminal. In a fully automated system, the entire fleet of AGVs can be mobilized to serve the unloading/loading operations in the most efficient manner. This gives rise to a need to study dispatching decisions in the deployment of the AGV system.

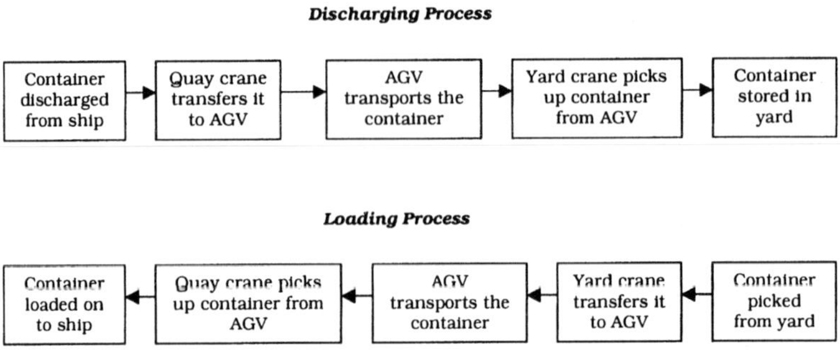

Figure 11.1. Flow of operations.

We start by providing a brief overview of the flow of operations that occur when a ship enters a port (cf. Figure 11.1). When a vessel arrives at the container terminal for transshipment, there are two types of operations that need to be carried out. These are to discharge containers from and/or to load containers onto the vessel. Containers are first discharged from the vessel onto AGVs at the quay side by quay cranes. The AGVs then transport the containers to specific storage locations in the yard area where they are dismounted from the AGVs by yard cranes. Typically, outgoing containers are loaded onto the ship after the majority of incoming containers have been discharged. The outgoing containers from the yard are mounted onto the AGVs using yard cranes. These containers are then carried by AGVs from the yard to the quay area where they are loaded onto the ship by a quay crane.

As mentioned earlier, containers handled by the terminal are typically of two standard sizes: twenty-footer (one TEU) or forty-footer (two TEUs). An AGV may carry a box of one TEU or two TEUs, or carry two boxes of one TEU each. When a container is discharged from a vessel, it is lifted by a proximate quay crane and mounted directly onto an AGV without first landing it on the ground. Landing a container onto the ground necessitates an additional crane operation to lift it from the ground and mount it later onto the AGV, thus reducing the throughput of the whole operation. In order to not delay the progress of the operations, an AGV needs to be readily present near the crane when a container needs to be loaded onto or discharged from a vessel.

The container terminal considered in this paper is based on the layout and operations of a local port operator in Singapore. As one of the world's leading port operators, it plans to automate the container transportation operations by implementing an AGV system in its newest terminal. The scale of the AGV operations in mega container terminals is typically very large, with free ranging AGVs moving in a complicated network of lanes and junctions. A complex layout of the AGV system consists of a network of lanes and junctions shown in Figure 11.2.

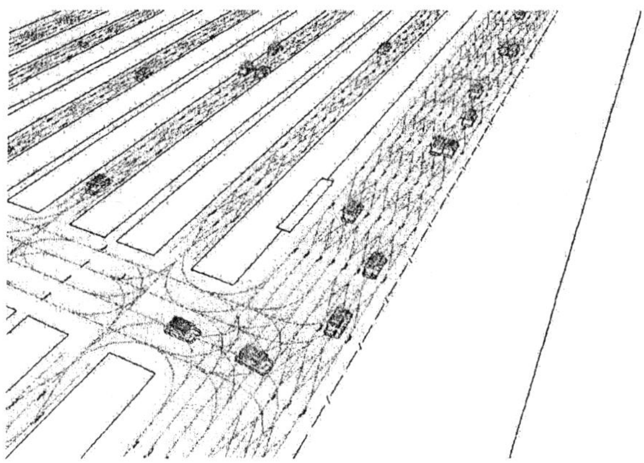

Figure 11.2. Layout of the terminal.

The AGVs transport containers between the quay side and yard side storage areas. These bi-directional AGVs have an advanced navigation system that guides them through the complex network transferring containers from multiple origins to multiple destinations efficiently. Typical operational, planning and control problems in such a system are: dispatching AGVs to transportation jobs, routing of AGVs, and controlling traffic in the network of lanes and junctions. The dispatching module assigns each transportation job to one of the available AGVs. The dispatched AGV will then be instructed to follow the route determined by a routing module, which has details of lanes and junctions to be taken from the origin of the job to its destination.

For the sake of operation safety, the complicated network of lanes and junctions is partitioned into a large number of zones with a restrictive vehicle movement rule. Only one AGV is allowed to occupy a particular zone at any time; thus, any other AGV wishing to use the zone has to wait outside for movement clearance. Typically, the minimum size of a zone is approximately equal to the distance required to stop an AGV from its top speed with the use of a normal controlled braking mechanism. The time required for stopping the AGV is generally less than 10 seconds.

Due to the dynamic nature of terminal operations, breakdowns of AGVs or container handling equipment, unexpected delays in container handling, etc., the planned route of an AGV could interfere with that of another AGV. This in turn leads to a delay in the completion time of transportation jobs involved. For example, when an AGV takes a turn, if there is a vehicle within a certain distance, it may lead to a collision. This is different from routing systems in communication networks where such physical constraints are non-existent. Such issues need to be taken care of by the navigation system along with a host of other conditions that need to be checked by a particular vehicle before it moves.

On top of the complex navigational and control problems faced in the design of such a system, we need to ensure that the AGVs are utilized in a highly efficient manner, to minimize the turnaround time of vessels in the port. Clearly, having too many AGVs roaming in the network is not a cost-effective way to reduce the turnaround time of vessels. Furthermore, due to the added congestion, deploying more AGVs than necessary may in fact slow down the entire system and lead to reduced throughput.

Under this rather complex setting, we focus in this paper on developing efficient dispatching techniques that assign AGVs to container jobs. Our main contributions are as follows:

- By focusing on the work rate optimization issue associated with the quay cranes, we reformulate the AGV dispatching problem as a network flow problem. Our model is similar to the classical tanker-scheduling problem (cf. Ahuja, Magnanti, and Orlin (1993)), and a similar reformulation that has been reported in the literature (cf. Vis et al. (2001)). While earlier models focus on finding the minimum number of AGVs needed to service the vessels (a static problem), the novel feature in our approach is the explicit formulation of waiting time minimization as our primary objective (a dynamic problem). This gives rise to a minimum-cost network flow formulation for the problem of dispatching AGVs to container jobs. For AGVs with unit capacity, solving the minimum-cost-flow network model provides an effective assignment of AGVs to container jobs. Furthermore, this model can be incorporated into a real time dynamic AGV dispatching system, since this problem can be solved efficiently in practice.

- To the best of our knowledge, none of the studies on the AGV dispatching problem in the literature explicitly considers the impact of congestion on the performance of the dispatching algorithm. Overlooking this important aspect may lead to an erroneous conclusion that the performance will improve as more AGVs are deployed. In fact, due to the complicated zone-based navigational routines and space restrictions, the throughput of the terminal is largely dependent on the number of AGVs deployed. Using an AGV deadlock prediction package developed earlier by the group (cf. Moorthy et al. (2003)), we embed the dispatching algorithm within the simulation package to examine the performance of the dispatching algorithm in a dynamic setting. As a benchmark for comparison, we have compared our algorithm with the performance of a widely used greedy dispatching algorithm. Our simulation results show that the proposed method performs significantly better than the existing greedy heuristic used to dispatch AGVs. By carefully taking care of the effect of deadlocks and congestion caused by the AGVs, our simulation system can actually be used to obtain the necessary number of AGVs to be deployed in the system. In fact, the simulation shows that the throughput of the system suffers if too many AGVs are deployed in the system.

Structure of the paper

In Section 2, we review some of the previous work done in the scheduling literature primarily in the seaport context. Section 3 describes our modeling approach to the AGV dispatching problem. In Section 4, we describe a greedy heuristic that has been previously proposed for this class of problems. Section 5 deals with the proposed network flow model for the problem. We discuss the connection between the two algorithms in Section 6. To address issues of network congestion, and to facilitate proper empirical performance comparison, we need to augment the vehicle-dispatching scheme with a deadlock prediction and avoidance mechanism. In Section 7, a deadlock prediction and avoidance scheme that has been implemented is described. In Section 8, we present simulation results to quantify the improvements provided by the new method. Finally, we discuss future research possibilities in Section 9.

2. Literature Review

Over the past few years, there has been a huge amount of research focused on improving the design and operation of container terminals. This is in response to the enormous increase in the number of containers being used in sea transportation and the concomitant increase in the complexity of terminal operations. For excellent reviews on the different strategic and operational issues that arise at container terminals, the reader is referred to the articles by Meersmans and Dekker (2001) and Vis and de Koster (2003). Scheduling AGVs for container transport is one of the key problems identified in these papers.

Bish (2003) considers an integrated problem of determining storage locations for containers along with AGV and crane allocation to minimize the maximum time taken to serve a set of ships. This problem is shown to be NP-hard and a heuristic is proposed for it. In a similar vein, Meersmans and Wagelmans (2001a), and Meersmans and Wagelmans (2001b) consider the AGV and crane allocation problem simultaneously and develop a Beam Search heuristic for this problem. While these approaches focus on joint scheduling problems, we concentrate in this paper on the AGV scheduling problem only.

With specific reference to the scheduling of AGVs, most research has been done in the context of Material Handling Systems. Co and Tanchoco (1991) work with the assignment of transportation equipment to service requests on the shop floor. With assumptions of a fixed shop layout with predetermined material flow paths and fixed fleet sizes, the problem is modeled as a mixed integer program. Egbelu and Tanchoco (1984) develop some heuristic rules for dispatching AGVs in a job shop environment. The heuristics are predominantly either job-based or vehicle-based. Job-based approaches develop heuristics by selecting the nearest vehicle, the farthest vehicle, the longest idle vehicle or the least utilized vehicle to serve the most tightly constrained jobs. Vehicle based approaches on the other hand try to minimize the unloaded travel times in order to maximize the opportunities for jobs to be scheduled. Shortest travel time, longest travel time, maximum outgoing queue size and first-come-first-served are some of the vehicle-based approaches considered.

Kim and Bae (2000) propose mixed integer programming formulations for the AGV dispatching problem under a discrete event time setting. These event times correspond to pickup and delivery times for the containers. For a single quay crane with specified event times to be met, the problem is reduced to an assignment problem. For general cases wherein event times cannot be met, a heuristic is developed. Chen et al. (1998) propose a vehicle-based dispatching strategy for a mega container terminal. The heuristic proposed deploys vehicles to the earliest possible container jobs once the vehicle is free. This vehicle based greedy heuristic does not presuppose any known information on the sequence of jobs available. Bose et al. (2002) obtain an initial solution using either a job-based or vehicle-based approach and subsequently improve on it via an evolutionary algorithm. However, these algorithms only perform well for systems with small numbers of jobs and vehicles. Akturk and Yilmaz (1996) propose an algorithm to schedule vehicles and jobs in a decision-making hierarchy based on mixed integer programming. Their micro-opportunistic scheduling algorithm (MOSA), combines job-based and vehicle-based approaches into a single algorithm. However, the computational time requirements for MOSA become impractical when the job number or the size of vehicle fleet is large. Using neural network models to model the decision processes of expert dispatchers is considered by Potvin, Dufour, and Rousseau (1993) and Potvin, Shen, and Rousseau (1992). Vis et al. (2001) consider the tactical problem of determining the number of AGVs needed at semi-automated container terminals. This paper is most relevant in our context, since they use a network flow formulation to determine the number of AGVs needed at the terminal. In this paper, with suitable modifications to the cost function, we show how the method can be in fact turned into an efficient dispatching scheme.

In practical applications, besides the vehicle-dispatching problem one needs to consider the possible formation of deadlocks in the AGV system. Lee and Lin (1995) and Viswanadham, Narahari, and Johnson (1990) use Petri-net theory to predict deadlock in material handling and AGV systems. The entire network is considered there in a matrix form and the technique requires matrix vector operations in very large dimensions. Hyuenbo, Kumaran, and Wysk (1995) use graph theory to detect impending deadlock situations. To do this a large number of bounded circuits in the AGV system network needs to be found. Yeh and Yeh (1998) develop efficient deadlock prediction strategies for identifying cycles in a dynamic directed graph. Developing on this work, Moorthy et al. (2003) develop a prediction and avoidance scheme for cyclic deadlocks. This scheme is considered in detail in Section 7 since it will be incorporated into the simulation to test the performance of the proposed dispatching scheme. Duinkerken and Ottjes (2000) and Evers and Koppers (1996) perform simulation studies to analyze traffic control issues in AGV systems. To implement effective simulation studies, proper steps need to be taken to ensure the accuracy of results from the model. Systematic approaches to simulation studies have been discussed by Banks et al. (2001) and Law and Kelton (1991).

3. Problem Description

In this paper, we focus exclusively on AGVs with unit capacity. This can be suitably modified in practice, by pairing up consecutive jobs if possible, to address the situation where each AGV can handle up to two 20 TEU containers or one 40 TEU container. This simplification, however, ensures that the problem remains tractable and an efficient dispatching scheme can be devised and implemented in real time. In fact, most of the current literature focuses on AGVs with unit capacity, which is often encountered in container terminals. Henceforth, we will only consider this situation with unit capacity AGVs.

We assume that yard crane resources are always available, i.e., the AGVs will not suffer delays in the storage yard location waiting for the yard cranes. This is not a restrictive assumption in the real implementation, since a good yard storage plan will be able to minimize the amount of congestion in a particular yard location, and hence reduce the amount of delays suffered by the AGVs. Furthermore, yard cranes are relatively much cheaper to acquire than quay cranes. Hence, yard cranes are assumed readily available when necessary.

To maintain a highly efficient automated container terminal, it is crucial to reduce the turnaround time in port of the container ships. Hence, our primary goal is to reduce the time that vessels need to spend in the port (makespan) for their loading and discharging operations. This in turn is equivalent to deploying the AGVs in an effective dynamic manner such that the jobs are executed as soon as they are ready to be processed. Past research in this area has focused on finding dispatching policies so that the containers can be processed as early as possible. This, however, leads to complex scheduling problems that can only be solved for special cases (involving single crane, single job type) or when the number of AGVs to be deployed is small. Instead, we focus on the crane productivity (work rate), which is measured by the number of containers moved per hour.

For each quay crane, there is a predetermined crane job sequence, consisting of loading jobs, or unloading/discharging jobs, or a combination of both. For each loading (discharging) job, there is a predetermined pickup (drop-off) point in the yard, which is the origin (destination) of the job. Given a specified job sequence, the corresponding drop-off (for loading) or pickup (for discharging) times of the jobs at the quay side depends on the work rate of the quay cranes. For example, assuming a work rate of one container every 4 minutes (say), we need the horizontal transportation system to feed a container to the quay crane in every 4 minutes. This allows us to compute the appointment time by the quay side for each container job. To minimize the turnaround time of the vessel, we need to run the cranes at the fastest possible rate such that the AGV deployment system is still able to cope.

Our primary goal in the AGV dispatching problem is in trying to ensure that we can dispatch AGVs such that all the imposed appointment time constraints are met. Namely, we need an AGV to reach the quay crane site in time for a container to be deposited or lifted by the quay crane. If these constraints are satisfied by the deployment scheme, the terminal operates at the desired throughput rate. However, a couple of other factors that need to be taken into account in real AGV deployment systems are:

- **Congestion**: A situation whereby all AGVs queue up at the quay site can lead to traffic congestion. This is undesirable as it reduces the speed at which AGVs travel/operate, especially if there are too many near the quay side. This reduction in speed would cause the AGVs to be late for other jobs that in turn decreases the throughput of the terminal.

To reduce congestion indirectly, we need to try to reduce the idle time of the AGVs at the quay site. This is the time spent waiting for the quay crane to lift/deposit containers from/onto it. Hence, it is desirable to have the AGV arrive at the quay in a just-in-time fashion. This performance measure will indirectly reduce the number of AGVs queuing up at the quay side. Hence, we are interested in finding a feasible AGV deployment that *minimizes the total waiting time* for all the AGVs.

- **Late jobs**: Ideally, solving our model should provide a feasible deployment of AGVs such that all the jobs can be processed exactly at the quay side appointment times. However, in practice this is not possible, due to the limited number of AGVs available and traffic conditions in the network that may force some AGVs to arrive late for the jobs. In this case, we need to allow jobs to be served late. However, capturing the impact of the delays into the appointment times of all future jobs will render the model intractable. This is precisely the bottleneck in earlier approaches to this problem.

  In our model, we will allow jobs to be served late, but we ignore the delays imposed on the appointment time of all future jobs. Instead, we impose a huge penalty for the jobs to be served late, and use dynamic replanning to update the problem status in a rolling horizon format, in order to capture the impact of delays.

Terminology and assumptions

- We consider the unit capacity case wherein each AGV can carry a maximum of one container, regardless of size, at any time.

- Let M denote the set of AGVs available, where $|M|$ is the total number of AGVs. For the AGV dispatching problem, we assume that $|M|$ is fixed and known. In fact, we later show that simulation results can be used to determine the number of AGVs to be deployed in the system.

- Container jobs can be of either the discharging or loading type. Let U and L represent the set of discharging and loading jobs to be served. The total set of container jobs is represented by $N = U \cup L$ where the total number of jobs $|N| = |U| + |L|$.

- For each container job, the time is specified at which it must be either picked up (for discharging) or dropped off (for loading) at the **quay side**. These predetermined times at the quay crane are chosen such that for the set of given jobs, the quay crane operates at its desired productivity level. We denote this pickup/drop-off time for job $i \in N$ by t_i and refer to it henceforth as the *quay side appointment time* for the job. Note that for a loading job i, an AGV must pick up the container from the yard and arrive at the quay before the appointment time t_i, after which it can be deployed to another job. For a discharging job, an AGV must be at the quay before the appointment time t_i, after which it will carry the container to a designated location within the yard. The AGV can only be deployed for another job after completing the delivery.

- Let T_{ij} denote the *travel time* between two distinct jobs i and j measured with respect to the quay side locations of the jobs. The AGVs are assumed to operate at a known average speed throughout the transportation operation. Clearly the computation of T_{ij} depends on the type of job i and j. The travel times can be computed once the distance covered and the type of operations associated with each job (discharging or loading) are known.

- Let $\tilde{T}_{mi}$ denote the travel time from the destination of AGV m to the source of job i at the time of deployment. In a rolling horizon model, this travel time can be similarly calculated based on the destination of the job that AGV m is currently serving and the type and source of job i.

Under these assumptions, that realistically model the container terminal operations, we develop a minimum-cost network flow model to obtain a dispatching strategy that minimizes the total waiting time of the AGVs at the quay cranes in Section 5. It is of paramount importance that such a model should be solvable quickly in practice as is needed by the seaport container terminal in real time operations.

4. Greedy Deployment Scheme

Before we present a comprehensive framework for addressing the AGV deployment problem, we consider a simple and popular heuristic dispatching strategy that has been proposed in the literature (cf. Chen et al. (1998); Egbelu and Tanchoco (1984)) for vehicle dispatching. This strategy, which is easy to implement, has been used in at least one seaport that we are aware of. Its simplicity allows it to be used easily for AGV dispatching in a dynamic fashion. As there is no published benchmark for this class of AGV dispatching problems, we will use the greedy deployment strategy (henceforth called GD) as a benchmark to compare our network flow model with. The GD algorithm is described next.

The goal of the greedy heuristic is to minimize the total time AGVs spend waiting at the quay crane locations to serve their jobs. The jobs are initially arranged in a first-in-first-out manner based on the earliest quay side appointment time t_i at each quay crane. Suppose we have already assigned a set of jobs to the AGV, and the next job in the list is considered. We first choose a list of AGVs that can reach the quay crane location in time after it has completed its previous job. From this list, we pick the AGV that will incur the minimum waiting time at the quay crane location for the job. This process is recursively performed as the jobs are scanned. This job list expands with time as the arrival of new vessels to the terminal necessitates the transportation of more containers. The GD algorithm is best illustrated with the example below.

Example

Consider a terminal with $|M| = 4$ AGVs and $|N| = 4$ container jobs to be processed. The quay side times for the container jobs are displayed in the Table 11.2. From the container job list, the earliest available job

Table 11.2. Quay side time for jobs.

Job i	Appointment time t_i
1	00:30
2	00:31
3	00:32
4	00:36

1, is designated to be served first. To job 1, an AGV is assigned such that the AGV that serves it will incur the minimum waiting time. From

Figure 11.3, we note that based on the current positions of the AGV only three AGVs namely 1, 2 and 4 can reach the quay crane location of container job 1 in time before 00:30. Since the waiting time for AGV 2 is minimum, AGV 2 is assigned to job 1. This procedure is performed recursively to assign the next few available jobs to AGVs. ∎

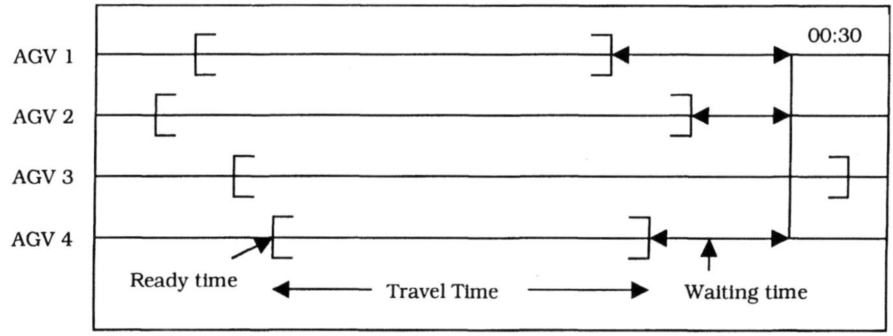

Figure 11.3. Waiting time of AGVs in Example 1.

Dynamic implementation

We now describe how the greedy strategy can be implemented in a dynamic fashion for the AGV dispatching problem. In our implementation, the planning of the time to dispatch each job is done in the following manner: The first k jobs per crane will be assigned an appointed pickup/drop-off time initially. The $(k+1)^{th}$ job for each crane will only be assigned an appointment time when the service of the first job at the quay has actually been completed. The assignment of the $(k+2)^{th}$ job will depend on the completion of the second job and so on. The number k depends on when the re-planning should be done based on historical traffic condition. In our implementation, we used $k = 4$ for dynamically assigning appointment times to jobs (cf. Figure 11.4).

Given the time window W between jobs, the quay side time t_i ($i = 1, \ldots, k$) for the first k jobs is computed as:

$$t_i = \text{Ship-discharge-time} + (i-1) \times W. \tag{11.1}$$

The Ship-discharge-time is the time the ship is ready for discharge at the terminal. The time window W is the interval between successive discharging of containers from the vessels and depends on the quay crane

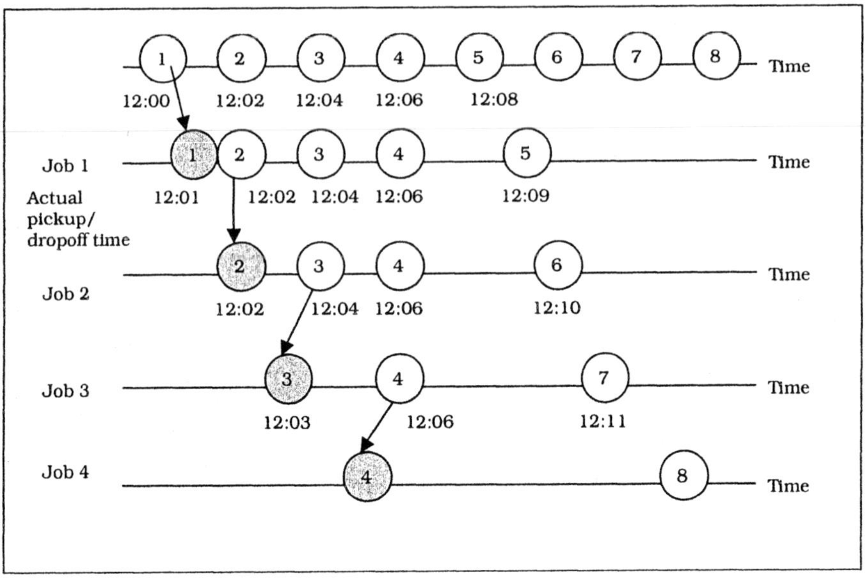

Figure 11.4. Dynamic assignment of appointment time under GD.

work rate (i.e., number of containers moved per hour). By setting a time window of say 4 minutes (i.e., work rate of 15 containers per hour), the operator hopes to work the quay crane at a rate of one container for every 4 minutes.

For any subsequent job i after the first k jobs, the quay side time is computed as:

$$t_i = C_{i-k} + k \times W. \tag{11.2}$$

where C_{i-k} denotes the actual completion time of the $(i-k)^{th}$ job. Note that C_{i-k} is available at the time of deployment of the ith job. Thus given that the ready time for an AGV m is $\tilde{t}_m$, the waiting time for AGV m to serve job i is calculated as:

$$t_i - (\tilde{t}_m + \tilde{T}_{mi}). \tag{11.3}$$

In practice, there will be instances where none of the AGVs can be deployed in time to serve a particular job. In such situations, the AGV that first reaches the quay crane location of a pending container job i

will be selected to serve that job, even though the AGV cannot arrive before the appointment time t_i.

5. Network Flow Method

We now propose a network flow model to solve the AGV dispatching problem. The deployment of the AGVs such that the total waiting time is minimized is found by solving a minimum-cost network flow problem. Such problems can be solved efficiently in practice even for large-scale networks (Ahuja, Magnanti, and Orlin (1993)), making the proposed method extremely attractive. The details of the approach are presented next. Our model can be viewed as an extension of a model proposed by Vis et al. (2001), where our goal is to find a schedule that will minimize the impact of delays and maximize the utilization of the AGVs. In this regard, we find that the objective to minimize the total AGV waiting time at the quay is a reasonably good surrogate for the ultimate goal.

We construct a directed graph $G(V, E)$ to represent the complete network where V denotes the set of nodes and E denotes the set of arcs. The graph is constructed in the following manner. Every container job $i \in N$ (discharging and loading) is represented as a node in G. There is a node corresponding to each AGV $m \in M$, capturing the state of the AGV at the time of deployment. We also insert one sink node s corresponding to the end state of the AGVs, after all jobs have been served. Thus, we have a total of $|N| + |M| + 1$ nodes in the network, denoted as:

$$V = N \cup M \cup \{s\}.$$

To define the (directed) arcs in the network, we introduce the following notation. We call an ordered pair of distinct jobs (i, j) *compatible* (Vis et al. (2001)) if a single AGV can be used to serve job j (arriving at the quay side before t_j) after serving job i. Hence job pair (i, j) is compatible if:

$$t_i + T_{ij} \leq t_j.$$

Similarly an ordered pair (m, i) of AGV m and job i is compatible if:

$$\tilde{t}_m + \tilde{T}_{mi} \leq t_i.$$

The three types of arcs that connect the various nodes in this directed graph are:

- There exists a directed arc from AGV m to a container job i if the AGV and job pair (m, i) is compatible. The cost of such an arc (m, i) corresponds to the waiting time that the AGV m incurs at the quay crane location of job i if it is used to serve job i immediately after finishing the initial job. We assume that the AGV travels at certain predetermined average speed for this computation. Hence:

$$\text{Cost between AGV node } m \text{ and job node } i,$$
$$c_{mi} = t_i - (\tilde{t}_m + \tilde{T}_{mi}) \quad \forall (m, i) \text{ compatible.} \tag{11.4}$$

- There exists a directed arc from container job i to container job j if job pair (i, j) is compatible. The cost of such an arc (i, j) is the waiting time that the AGV incurs if it serves job j immediately after serving job i. Hence:

$$\text{Cost between job node } i \text{ and job node } j,$$
$$c_{ij} = t_j - (t_i + T_{ij}) \quad \forall (i, j) \text{ compatible.} \tag{11.5}$$

- There exists a directed arc from each of the AGV nodes and the container nodes to the sink node s. These arcs signify that an AGV can remain idle after having served any number of container jobs or not having served at all. These arcs are assigned a cost of zero. Hence:

$$c_{ms} = 0 \ \ \forall m \in M, \qquad c_{is} = 0 \ \ \forall i \in N. \tag{11.6}$$

However, in the practical implementation, it may not be possible to process the jobs within the specified time restrictions. To obtain a feasible solution in such cases, we introduce arcs between job pairs (i, j) that are not compatible, i.e., when:

$$t_i + T_{ij} > t_j.$$

Such arcs are highly unattractive as it decreases the quay crane productivity. Hence, we weigh the cost of such arcs with a large penalty value K as:

$$c_{ij} = K(t_i + T_{ij} - t_j) \quad \forall (i, j) \text{ not compatible.} \tag{11.7}$$

Similar arcs with high costs are introduced between AGVs and jobs for pairs that are not compatible where the AGV reaches the quay crane location of the job after the quay side time. The set of arcs in the graph is hence denoted as:

$$\begin{aligned} E &= \{(m,i) : m \in M, i \in N\} \cup \{(i,j) : i,j \in N, i \neq j\} \\ &\quad \cup \{(m,s) : m \in M\} \cup \{(i,s) : i \in N\}. \end{aligned}$$

The AGV dispatching problem then corresponds to finding $|M|$ directed paths in this network (one path starting from each AGV node and ending at node s), visiting all job nodes once, at minimum cost. Let x_{ij} represent the flow on arc (i,j). Mathematically, the problem is formulated as:

$$\begin{array}{llcll} \text{minimize} & \displaystyle\sum_{(i,j)\in E} c_{ij}x_{ij} & & & \\ \text{subject to} & \displaystyle\sum_{i\in V:(m,i)\in E} x_{mi} & = & 1, & \forall m \in M, \\ & \displaystyle\sum_{i\in V:(i,j)\in E} x_{ij} & = & 1, & \forall j \in N, \\ & \displaystyle\sum_{i\in V:(j,i)\in E} x_{ji} & = & 1, & \forall j \in N, \\ & \displaystyle\sum_{i\in V:(i,s)\in E} x_{is} & = & m, & \\ & x_{ij} \in \{0,1\}, & & & \forall (i,j) \in E. \end{array} \tag{11.8}$$

The above problem can be transformed to a network flow problem, using the following well-known node-duplication technique (cf. Ahuja, Magnanti, and Orlin (1993)):

- Split each container job node $i \in N$ into two nodes i' and i'' and add an arc (i', i''). We thus expand the number of container job nodes from $|N|$ to $2|N|$. Let N' and N'' denote these two new sets of container job nodes. Hence the set of nodes in the expanded network is:

$$V' = N' \cup N'' \cup M \cup \{s\}.$$

Correspondingly, the expanded set of the arcs in this network is:

$$\begin{aligned} E' &= \{(m, i') : m \in M, i' \in N'\} \\ &\cup \{(i'', j') : i'' \in N'', j' \in N', i \neq j\} \\ &\cup \{(m, s) : m \in M\} \cup \{(i'', s) : i'' \in N''\} \\ &\cup \{(i', i'') : i \in N\}. \end{aligned}$$

- We set the upper bound and lower bound on the flow traversing through each arc (i', i'') to 1 so that exactly one unit of flow passes through it. The lower and upper bounds on all other arcs are set to 0 and 1 respectively. We let l'_{ij} and u'_{ij} denote the lower and upper bound for each arc (i, j) in the graph.

- The cost of the newly introduced arc (i', i'') is set to zero. Transforming the arc costs from the original model to the new model we obtain:

$$\begin{aligned} c'_{mi'} &= c_{mi}, \quad \forall m \in M, i \in N, \\ c'_{i''j'} &= c_{ij}, \quad \forall i, j \in N, i \neq j, \\ c'_{ms} &= c_{ms}, \quad \forall m \in M, \\ c'_{i''s} &= c_{is}, \quad \forall i \in N, \\ c'_{i'i''} &= 0, \quad \forall i \in N. \end{aligned}$$

The purpose of this transformation, shown in Figure 11.5, will ensure that each job is served by one AGV.

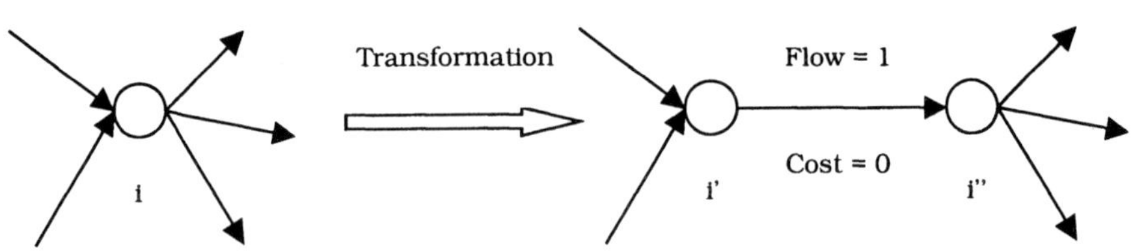

Figure 11.5. Transformation of the network.

Formulation (11.8) is thus reformulated as a minimum cost network flow model as:

$$
\begin{aligned}
\text{minimize} \quad & \sum_{(i,j)\in E'} c'_{ij} x_{ij} \\
\text{subject to} \quad & \sum_{i\in V':(m,i)\in E'} x_{mi} = 1, \quad \forall m \in M, \\
& \sum_{i\in V':(i,j)\in E'} x_{ij} - \sum_{i\in V':(j,i)\in E'} x_{ji} = 0, \quad \forall j \in N' \cup N'', \\
& \sum_{i\in V':(i,s)\in E'} x_{is} = m, \\
& l'_{ij} \le x_{ij} \le u'_{ij}, \quad \forall (i,j) \in E'.
\end{aligned}
\tag{11.9}
$$

The first three constraints in Formulation (11.9) are standard flow conservation constraints while the last constraint provides upper and lower bounds on the flow values. It is well known that the linear programming relaxation of the capacitated minimum cost network flow problem can be solved in polynomial time to yield optimal integral flows. Furthermore, specialized algorithms such as the network simplex method (Lobel (2000)) can be used to solve large-scale problems efficiently. Solving the network flow model generates $|M|$ paths, each of which commences from one AGV node and terminates at the sink node s. In totality the $|M|$ paths cover all the nodes in the network once. Each path from a source AGV node to the sink node prescribes the container job sequence that the AGV should be assigned to. This deployment strategy henceforth is referred to as the Minimum Cost Flow (MCF) algorithm.

Dynamic implementation

In practice, one needs to consider the effects of uncertainty of traffic conditions on the job assignment. In a prescribed job assignment, some of the jobs could be late due to interruptions and this lateness will affect the remaining jobs. Thus, re-planning needs to be done frequently. Here re-planning is done for each crane after every k number of jobs have been deployed. At that instant, a new MCF problem will be formulated based on the number of jobs remaining, the latest status of all jobs and AGVs. Following the GD model, k is selected to be 4. An example of the assignment of the appointed time sequence for 9 jobs for a single crane is illustrated in Figure 11.6.

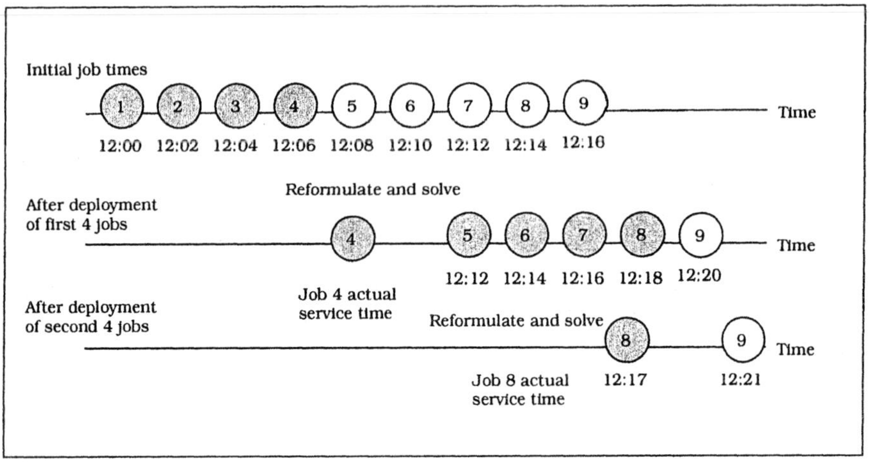

Figure 11.6. Dynamic assignment of appointment time under MCF.

Note that unlike the GD algorithm, the MCF algorithm uses the appointment time information of all future jobs under the assumption of no disruptions to assign the jobs to the AGV. The main advantage is that by doing so, the system is able to anticipate problems that may arise if the AGVs are deployed greedily. However, the solution obtained is dependent on the prescribed appointment time of all future jobs and could be adversely affected if there are delays in serving certain jobs. Hence, it is not a priori clear, whether a dynamic implementation of the MCF algorithm is indeed superior to the GD algorithm.

6. Performance Comparison

6.1 Single Crane Scenario

The GD algorithm can be viewed as a heuristic way to solve the minimum-cost network flow problem, since it tries to design $|M|$ paths from the AGV nodes to the sink node in the network, albeit in a greedy fashion. We first focus on the single crane case when there is only one job type (either discharging or loading jobs, but not mixed). In this case, we show that the GD algorithm actually gives rise to the optimal minimum-cost flow solution.

Theorem 6.1 *For a single quay crane model with sequences of one type of job, either loading or discharging, the solution obtained by GD is as good as the solution obtained by the MCF algorithm.*

Proof. We consider the case when the jobs are all discharging jobs. The case when all jobs are loading jobs can be handled using a similar argument.
Consider the solutions provided by the MCF and GD algorithms for the same set of discharge jobs N. Suppose both algorithms prescribe identical AGV deployment solutions for the first $(k-1)$ container jobs where $(k-1) < |N|$, and suppose they differ in their assignment of the kth job. Let AGV p be assigned to job k by MCF while AGV q is assigned to the same job by GD where p and q are distinct vehicles.

Since MCF has selected AGV p to serve job k, this means that AGV p will reach the source of job k (i.e., the quay side, since all jobs are discharging job) before its appointment time. Furthermore, since GD uses AGV q to serve the same job, we conclude that AGV q will arrive at the quay side of the crane later than AGV p, but before the appointment time of job k.

Let r be the next discharging job, after job k, assigned to AGV q in MCF. Note that now both AGVs arrive before the appointment time of both the jobs k and r. Hence we can interchange the assignments, i.e., using AGV p to serve r and jobs after r served by AGV q previously, and AGV q to serve job k and subsequent jobs served by AGV p, without increasing the total waiting time of the two AGVs. In this way, we obtain a new solution to the minimum-cost network flow problem such that the assignment of the first k jobs are identical to algorithm GD.

By repeating this process for job $(k+1)$ to $|N|$, we can transform a solution for the minimum-cost network flow problem into a solution identical to that given by the GD algorithm without affecting the total waiting time. Hence, GD solves the problem optimally in this special case. ∎

Note that the assumption that the jobs are all of the same type is crucial for the above to hold. For example, if job r is a loading job, then the fact that AGV q can be used to serve r (i.e., bringing the container from yard to the quay side before the appointed time) does not guarantee that AGV p can also be used to serve r, although AGV p will arrive at the quay side of job k earlier than AGV q. This happens if AGV q is nearer to the source of job r (in the yard) than AGV p.

6.2 Multiple Crane Scenario

The performance of GD algorithm however deteriorates considerably in a multiple crane scenario. We demonstrate this with a simple simulation experiment.

Example

Consider a multiple crane AGV dispatching problem with 4 quay cranes. The total number of container jobs is set to 200. Twenty AGVs are used to process the jobs. The quay crane rate is varied between 30 containers per hour and 75 containers per hour. The yard crane rate is set to 24 containers per hour. Each AGV is assumed to travel at a uniform speed. The simulation results are displayed in Table 11.3. For the quay crane

Table 11.3. Effect of quay crane rate on waiting time and late jobs.

Quay Crane Rate (Containers per hour)	Waiting time for GD (Minutes)	Waiting time for MCF (Minutes)
30.00	255	104
33.33	208	108
40.00	212	90
50.00	155	80
54.55	6 late jobs	2 late jobs
60.00	17 late jobs	5 late jobs
66.67	30 late jobs	6 late jobs
75.00	45 late jobs	8 late jobs

rate up to 50 containers per hour, both the MCF and the GD algorithms provide feasible deployment solutions. The waiting time for the MCF algorithm is as much as fifty percent less for these quay crane rates. For the higher quay crane rates, both AGV deployment strategies cause some of the container jobs to be late. Clearly, the number of late jobs for the MCF algorithm is much lesser than the number of late jobs for the GD algorithm. This experiment clearly shows that the performance of the MCF algorithm is significantly better than the GD algorithm in the multiple crane scenario. ∎

7. Deadlock Prediction and Avoidance Algorithms

The deployment schemes obtained from the GD and the MCF algorithms provide dispatching rules for the automated vehicles. In practice, however, constrained space near the quay side locations and fixed paths for AGVs cause deadlocks to occur. Such deadlocks cause part of or the entire system to stall, further delaying job processing. Hence, it is essential to integrate deadlock prediction and avoidance algorithms with the dispatching algorithms.

The AGV system consists of a complicated network of lanes and junctions that is partitioned into zones. For operational safety, at most one vehicle is allowed to occupy each zone. However, it is still possible for a dispatching algorithm to create a cyclic deadlock in the AGV system (Moorthy et al. (2003)). This generic form of deadlock occurs when a chain of vehicles is formed where each vehicle requests for a zone creating a cycle. A cyclic deadlock for four AGVs is shown in Figure 11.7 where an AGV can move to a zone only after the zone is free. To account for such situations, we need to integrate a deadlock prediction strategy and an avoidance algorithm with the proposed AGV dispatching algorithm to test its merits.

7.1 Deadlock Prediction Strategy

In practice, the sample time for the control system of the AGV is very small, in the range of 1.5 to 2 seconds. Hence, we need to ensure that the deadlock prediction strategy is extremely quick and effective. As noted in Section 2, previously proposed methods are either too complicated or computationally expensive to use for this application.

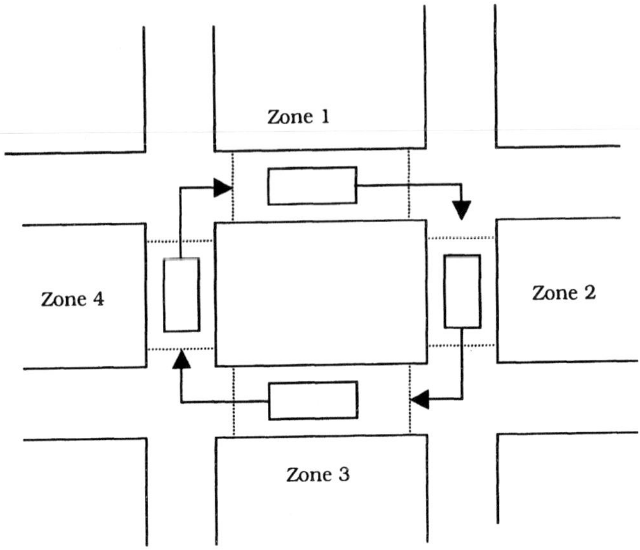

Figure 11.7. Cyclic deadlock in zone control AGV system.

We now describe the basic idea of the deadlock prediction algorithm that is used in the model (Moorthy et al. (2003)). For every sampling instant of about 2 seconds, a check is done to see if a specific vehicle i has moved to a new zone. For each vehicle that changes zones, a deadlock prediction is performed for its next step. If the next zone is occupied by a different vehicle j, then the vehicle must wait for the zone to clear. However if the next zone is not occupied, we still need to make sure that moving into the next zone will not lead to a deadlock that will cause the system to stall. To do this, we look further and examine the next destination of the AGV after the next zone (which is called the next-next zone). If this zone is unoccupied, then the vehicle can proceed as no deadlock can occur. However if the next-next zone is occupied by vehicle k, then the next zone location of vehicle k is found. The same check is performed to see if this next location is occupied. If not, then no deadlock is predicted. Else we proceed and check if we create a cycle and come back to the original next zone of vehicle i. We have a cyclic deadlock in that case. The worst-case complexity of this algorithm is $O(|M|^2)$ where $|M|$ is the total number of AGVs in the network. This worst-case scenario occurs in case there is a huge cyclic deadlock involving all the vehicles. This algorithm is a one-zone-step deadlock prediction method. This technique is seen to be effective in practice from Moorthy et al. (2003).

An example of a cyclic deadlock that is predicted by the algorithm is illustrated in Figure 11.8. Here, AGV 1 request to enter its next zone 2, which is currently free. However by looking at the next-next zone of the AGVs, it can be verified that a cyclic deadlock is formed.

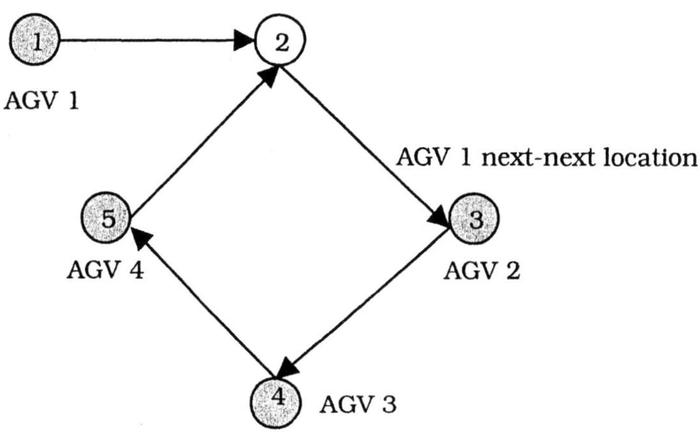

Figure 11.8. Example of cyclic deadlock forming.

7.2 Deadlock Avoidance Strategy

The deadlock avoidance scheme is a wait and proceed strategy. If a deadlock is predicted on a vehicle route, it will stop and wait until at least one vehicle is cleared from the region in which the deadlock is predicted. However, implementing the one-zone step deadlock prediction and the wait and proceed avoidance algorithm might cause other deadlock situations to occur. This form of deadlock is formed because of multiple loops sharing only one unoccupied node (cf. Moorthy et al. (2003)). An example of such a deadlock is provided in Figure 11.9 where a cyclic deadlock is avoided by the wait and proceed strategy but another deadlock is created which has many cycles that share a single empty resource.

Moorthy et al. (2003) proposed a forward-arc strategy to resolve these deadlocks. For detailed simulation studies that test and evaluate the

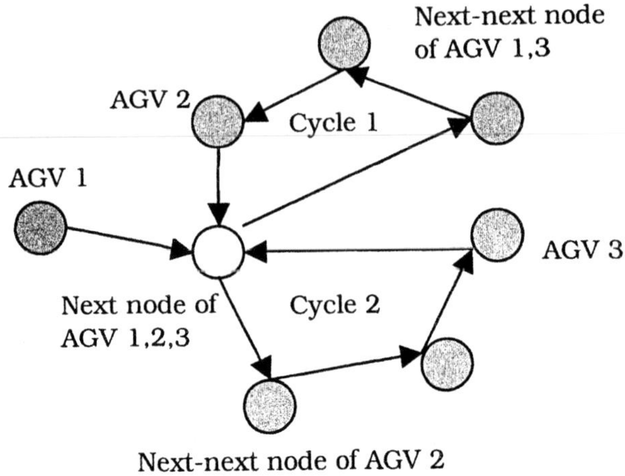

Figure 11.9. Example of multi-cycle deadlock forming.

performance of these deadlock prediction and avoidance strategies, the reader is referred to Moorthy et al. (2003). Our primary interest is to integrate these deadlock prevention strategies with the deployment models to obtain routing strategies for the AGVs.

8. Simulation Study

The simulation study was performed using a discrete event simulation software (AutoMod 9.0). The entire system was modeled in terms of its state at each point in time, entities that pass through the system and events that cause the state to change. For the simulation, the performance of the GD and the MCF algorithm integrated with the one-zone-step deadlock prediction and avoidance algorithms were compared.

Model assumptions

- The layout of the terminal studied consists of 4 berths and 16 quay cranes with 4 quay cranes assigned to each berth.

- We consider AGVs with unit capacity and vary the number of AGVs in the system between 40, 60 and 80. Note that these are the number of AGVs that can be used realistically to support the berthing operation at a small terminal. For larger terminals, the number of AGVs needed could be in the range of 100-200.

- Berths are randomly assigned to incoming ships and the interarrival time of the ships is assumed to follow an exponential distribution with a mean of 60 minutes.

- Each container storage yard is made up of 9 clusters wherein each cluster has 3 control points. At any one time, a quay crane for either discharging or loading process can only use a single cluster. In the real application, it is possible to move the quay cranes but the movement was not simulated here.

- The four quay cranes at each berth process respectively 18%, 25%, 27% and 30% of the number of containers that need to be loaded and unloaded from the ship. This ratio is obtained from empirical data and captures the effect of crane scheduling policy on the final work load allocation for each crane.

- The actual quay crane rate is set to a triangular distribution of (1.375, 1.708, 2.113) minutes. The yard crane rate is set to a triangular distribution of (1.593, 2.172, 2.728) minutes. The crane average operating rate is taken to be the average of the three given values of the triangular distribution. This average operating rate is used in the computation of solutions for both the MCF and the GD algorithms.

- The average velocity of each AGV is important for calculating expected travel times between two points. To realistically determine this average velocity, the simulation model was run a few times before the actual simulation by varying the number of AGVs. The collected statistical data is used to set the average velocity of the AGV in both the models. Based on this data, the average velocities used in meters per second were 3.573, 3.616 and 3.770 for 40, 60 and 80 AGVs respectively.

For a given set of system parameters, the simulation was run for a deterministic period of 4 days. In our first simulation, we evaluate the effect of varying the number of AGVs in the container terminal, maintaining a constant time window between jobs.

Variation in the number of AGVs

The time window between the jobs, W was fixed at 2 minutes. We compare the performance of the MCF and GD algorithms by varying the number of AGVs in the container terminal. The results obtained over a period of 4 days are provided in Table 11.4.

Table 11.4. Effect of varying number of AGVs on performance.

	GD			MCF		
	40 AGV	60 AGV	80 AGV	40 AGV	60 AGV	80 AGV
Number of boxes	17769	20499	21797	18793	21471	22825
Average makespan	7.792	6.213	6.369	7.433	5.910	6.050
Average throughput	53.346	66.297	64.678	58.798	72.660	71.183

The first row in Table 11.4 measures the total number of container jobs that are served over the entire period. Clearly, for an identical number of AGVs, the deployment scheme provided by the MCF algorithm serves more jobs than the GD algorithm. Similarly, with respect to the average makespan of the ship (i.e., the duration that a ship remains in the terminal for loading and unloading operations) and the throughput (i.e., the number of boxes processed per hour), the MCF algorithm significantly outperforms the GD algorithm.

From Table 11.4, it is observed that as the number of AGVs is increased from 40 to 60, there is a significant increase in throughput due to the increase in the amount of available resources. However, increasing the number of AGVs from 60 to 80, in fact decreases the throughput. The deadlock effects caused by AGV congestion for a constant layout of the berths is the primary reason for this. Based on the simulation, in fact we can estimate the number of AGVs to be deployed in the system, by explicitly consider the effects of congestion. For the current system, about 4 to 5 AGVs per crane per berth seems to be optimal.

The observed mean deviation in time from the appointed times is provided in Table 11.5. Ideally, we would like the AGVs to have zero waiting time. If we are late, then we decrease the quay crane productivity and if we are early, we cause congestion. Clearly, from Table 11.5, the MCF algorithm on the average outperforms the GD algorithm with

Table 11.5. Effect of varying number of AGVs on mean time deviation.

	GD			MCF		
	40 AGV	60 AGV	80 AGV	40 AGV	60 AGV	80 AGV
Early Jobs (min)	3.609	2.290	2.235	2.993	2.078	1.850
Late Jobs (min)	6.589	4.848	5.747	4.518	4.026	4.773

respect to the total waiting time. The improvement in deviation of the waiting time for the MCF algorithm over the GD algorithm is in the range of 15 to 30 percent.

Variation in the time window

In the second simulation, we evaluate the effect of varying the time window between jobs, maintaining a constant number of AGVs. The number of AGVs is held constant at 80 over 4 days while the time window W for the jobs is varied. Other specifications for the 4-berth model remain the same as before. The time between jobs is varied between 1.8, 2 and 2.5 minutes. Similar to the previous simulation, we measure the total number of container jobs that are served, average makespan and the throughput of the terminal. The results are displayed in Table 11.6. Clearly, the throughput for the MCF model is about 10% higher

Table 11.6. Effect of varying duration of time window on performance.

	GD			MCF		
	1.8 min	2 min	2.5 min	1.8 min	2 min	2.5 min
Number of boxes	21333	21797	20726	22618	22825	21762
Average makespan	6.889	6.369	6.533	6.577	6.050	6.200
Average throughput	63.264	64.678	63.558	70.186	71.183	70.202

than the GD algorithm. As the time window increases from 1.8 to 2 minutes, the throughput for both the algorithms increases. However, on increasing the time window from 2 to 2.5 minutes, the throughput for both the algorithms decreases. This is because, for the tight time window of 1.8 minutes, more jobs will be late, as the AGVs cannot keep up with the crane productivity level. However, for a loose time window of 2.5 minutes, the AGVs reach the quay crane before the job is ready, causing congestion and hence decreasing the throughput on a whole.

These results clearly indicate that the proposed MCF algorithm outperforms the GD algorithm. An increase of as much as 10% in throughput is observed. Furthermore, the tradeoff between faster processing and increased congestion caused by increasing the amount of AGVs is evident. We can in fact, estimate the number of AGVs to deploy at the container terminal based on this tradeoff.

9. Conclusions and Future Research

A container terminal must operate efficiently to ensure that the time in port for seaport vessels is reduced. This in turn entails formulating efficient dispatching strategies to load and discharge containers from the vessels. In this paper, we focused on finding efficient deployment strategies for AGVs to perform these operations. Current techniques use a simple greedy heuristic to solve this problem. In this paper, a new technique based on a network flow formulation is proposed for dispatching AGVs with unit capacity. This technique has two distinct advantages. One is that it provides an optimal solution to the problem if the goal is to minimize the total waiting time of AGVs. The improvement in the performance is seen to be significant from the simulation study. Furthermore, this solution can be computed extremely efficiently in practice, even for a large AGV network. Thus, the proposed algorithm is extremely suitable for the AGV deployment problem in complex and large seaport container terminals.

The network flow technique proposed is applicable for AGVs with unit capacity, as is the case in many automated terminals. In the particular terminal considered, the AGV has the additional feature of carrying two units of load. An efficient model to obtain dispatching strategies under the multiple load capacity is an interesting and open problem. Extensive simulation studies for the case of multiple units of capacity need to be performed to compare the performance relative to the unit capacity case. Another feature of our model is that it is completely deterministic. However, in practice there may be some randomness involved, especially in travel times of AGVs and the processing times of jobs. Efficient models to solve such stochastic models are another area of future research.

Acknowledgements

The authors would like to thank the referee for helpful comments in improving the overall presentation of the paper.

References

Ahuja, R.K., T.L. Magnanti, J.B. Orlin. 1993. *Network Flows: Theory, Algorithms and Applications.* Prentice Hall.

Akturk, M.S., H. Yilmaz. 1996. Scheduling of automated guided vehicles in a decision making hierarchy. *International Journal of Production Research.* **34**, 2, 577-591.

Banks, J., J.S. Carson, B.L. Nelson, D.M. Nicol. 2001. *Discrete-Event System Simulation.* 3rd Edition, Prentice Hall.

Bish, E.K. 2003. A multiple-crane-constrained scheduling problem in a container terminal. *European Journal of Operational Research.* **1441**, 83-107.

Bose, J., T. Reiners, D. Steenken, S. Vos. 2002. Vehicle dispatching at seaport container terminals using evolutionary algorithms. *Proceedings of the 33rd Annual Hawaii International Conference on System Sciences.* R.H. Sprage (Ed), IEEE, Piscataway, 1-10.

Chan, C.T., L.H. Huat. 2002. *Containers, containerships and quay cranes: a practical guide.* Singapore: Genesis Typesetting & Publication Services.

Co, C.G., J.M.A. Tanchoco. 1991. A review of research on AGVS vehicle management. *Engineering Costs and Production Economics.* **32**, 35-42.

Chen, F.Y., E.K. Bish, Y.T. Leong, Q. Liu, B.L. Nelson, J.W.C. Ng, D. Simchi-Levi. 1998. Dispatching vehicles in a mega container terminal. INFORMS, Montreal, Canada.

Duinkerken, M.B., J.A. Ottjes. 2000. A simulation model for automated container terminals. *Proceedings of the Business and Industry Simulation Symposium.* Washington, ISBN 1-56555-199-0. ISCS.

Egbelu, P.J., J.M.A. Tanchoco. 1984. Characterization of automatic guided vehicle dispatching rules. *International Journal of Production Research.* **22**, 3, 359-374.

Evers, J.J.M., S.A.J. Koppers. 1996. Automated guided vehicle traffic control at a container terminal. *Transportation Research A.* **30**, 1, 21-34.

Hyuenbo, C., T.K. Kumaran, R.A. Wysk. 1995. Graph theoretic deadlock detection and resolution for flexible manufacturing systems. *IEEE Transactions on Robotics and Automation.* **11**, 3, 413-421.

Kim, K.H., J.W. Bae. 2000. A dispatching method for automated guided vehicles to minimize delays of containership operations. *International Journal of Management Science.* **5**, 1, 1-25.

Law, A.M., D.W. Kelton. 1991. *Simulation Modeling and Analysis.* 2nd Edition, McGraw-Hill.

Lee, C.C., J.T. Lin. 1995. Deadlock prediction and avoidance based on Petri nets for zone control automated guided vehicle systems. *International Journal of Production Research.* **33**, 12, 2349-3265.

Lobel, A. 2000. MCF - A network simplex implementation version 1.2. *http://www.zib.de/Optimization/Software/Mcf/index.html.*

Meersmans, P.J.M., R. Dekker. 2001. Operations research support container handling. *Econometric Institute Report EI 2001-22.*

Meersmans, P.J.M., A.P.M. Wagelmans. 2001a. Effective algorithms for integrated scheduling of handling equipment at automated container terminals. *Econometric Institute Report EI 2001-19.*

Meersmans, P.J.M., A.P.M. Wagelmans. 2001b. Dynamic scheduling of handling equipment at automated container terminals. *Econometric Institute Report EI 2001-33.*

Moorthy, R.L., H.G. Wee, W.C. Ng, C.P. Teo. 2003. Cyclic deadlock prediction and avoidance for zone controlled AGV system. *International Journal of Production Economics.* **83**, 3, 309-324.

Potvin, J.Y., G. Dufour, J.M. Rousseau. 1993. Learning vehicle dispatching with linear programming models. *Computers and Operations Research.* **20**, 4, 371-380.

Potvin, J.Y., Y. Shen, J.M. Rousseau. 1992. Neural networks for automated vehicle dispatching. *Computers and Operations Research.* **19**, 3/4, 267-276.

Vis, I.F.A., R. de Koster. 2003. Transshipment of containers at a container terminal: an overview. *European Journal of Operational Research.* **147**, 1-16.

Vis, I.F.A., R. de Koster, K.J. Roodbergen, L.W.P. Peeters. 2001. Determination of the number of automated guided vehicles required at a semi-automated container terminal. *Journal of the Operational Research Society.* **52**, 409-417.

Viswanadham, N., Y. Narahari, T.L. Johnson. 1990. Deadlock prevention and deadlock avoidance in flexible manufacturing systems using Petri net models. *IEEE Transactions on Robotics and Automation.* **6**, 6, 713-723.

Yeh, M.S., W.C. Yeh. 1998. Deadlock prediction and avoidance for zone control AGVs. *International Journal of Production Research.* **36**, 10, 2879-2889.

AutoMod V 9.0 Reference Manual.

Chapter 12

HYBRID MIP-CP TECHNIQUES TO SOLVE A MULTI-MACHINE ASSIGNMENT AND SCHEDULING PROBLEM IN XPRESS-CP

Alkis Vazacopoulos and Nitin Verma

Dash Optimization, Inc, 560 Sylvan Avenue, Englewood Cliffs, NJ 07632, USA

Abstract In this paper we introduce Xpress-CP–a Constraint Programming tool– and demonstrate its modeling and solving capabilities. We consider the multi-machine assignment and scheduling problem (Hooker et al. (1999)), where jobs, with release dates and deadlines, have to be processed on parallel unrelated machines (where processing times depend on machine assignment). Given a job/machine assignment cost matrix, the objective is to minimize the total cost while keeping all machine schedules feasible. We show that by deriving the benefits of MIP and CP techniques simultaneously this problem can be modeled and solved efficiently in a hybrid fashion using Xpress Optimization suite.

1. Introduction

Many real-world problems in supply chains, particularly scheduling, planning and configuration problems involving logic, constraint satisfaction, or discretization are difficult to solve using Mixed Integer Programming (MIP) techniques alone. Due to the large number of integer variables in the MIP formulation, the size of these problems may increase rapidly. Constraint Programming (CP) is an appropriate methodology for such combinatorial decision problems, since it provides an integrated framework for modeling such problems compactly, and supports user-controllable generalized solution procedures. However CP primarily targets constraint satisfaction. It lacks a global perspective, and is efficient only for small-term and medium-term planning problems. Hence, hybrid approaches using both CP and MIP methodologies provide promise for solving very large-scale problems while accounting for certain problem objectives. Such approaches are suitable for various job-shop scheduling, production and distribution planning, and vehicle scheduling and routing problems. The Xpress-MP suite facilitates such approaches. It consists of a framework for both CP and MIP technologies to exist together in Xpress' modeling language–Mosel, and provides tools to coherently exchange information between Xpress-Optimizer and Xpress-CP solver–CHIP. Other optimization packages such as ILOG Optimization Suite also provide similar frameworks. In this paper, we demonstrate the capabilities of Xpress-CP by using it for modeling and solving the Multi-Machine Assignment and Scheduling (MMAS) problem, originally proposed by Hooker et al. (1999). The structure of this problem makes it a suitable candidate for demonstrating Xpress-CP. Although the problem is a generalized and theoretical version of the problems occurring in job-shop, flow-shop and open-shop environments, its variations (see e.g., Pinedo and Chao (1998)) typically occur in production and scheduling problems in large scale supply chains.

1.1 Literature review and applications

The MIP based pure Branch and Bound (B&B) methodologies for solving scheduling problems suffer from the drawback of slow convergence due to weak LP relaxations and a large number of integer variables in the problem. Various specialized algorithms and heuristics have been proposed for single machine and parallel machine problems and their variations (by e.g., Brucker (2001), and Pinedo (1995)). Several constraint programming based techniques and search strategies have also been proposed (see e.g., Baptiste et al. (2001)) for problems based on the job-machine environment. The applicability of other hybrid approaches using CP in conjunction with MIP such as Branch and Infer (Bockmayr et al. (2003)), and Branch and Price (Easton et al. (2003)) have also been researched, and it has been demonstrated that they can be applied for solving supply chain planning and generalized assignment problems, respectively. Pure and hybrid CP applications have been presented in the scheduling sector in sports scheduling by Aggoun and Vazacopoulos (2004), in BASF's plastic production planning by Timpe[1], and integrated lot-sizing and scheduling of Barbot's paint production. Additionally, numerous applications from supply chains have been built using hybrid methods as mentioned by Bockmayr et al. (2003). The MMAS problem itself has been studied by Hooker et al. (1999), Jain and Grossmann (2001), Bockmayr et al. (2003), and recently by Sadykov and Wolsey (2003) in a hybrid optimization framework. Hooker et al. proposed a declarative hybrid modeling framework for the problem and a solution methodology which used CP to generate inequalities for the MIP component of the problem. Jain and Grossmann discussed a generalized class of problems where hybrid approaches can be used, and proposed an Iterative and a Branch and Cut (B&C) scheme for solving the problem. They demonstrated the iterative methodology for the MMAS problem by implementing it on a set of 10 problems. Bockmayr and Pisaruk implemented a B&C scheme on this problem that can be handled by CP using 'separation heuristics' for 'monotone constraints', and tested their implementation on a set of 8 MMAS problems that were minor variations of the 10 problems generated by Jain and Grossmann. Sadykov and Wolsey proposed some more variations of hybrid approaches using combinations of MIP, CP, Column Generation, and strong cuts (these cuts are a tighter version of the Preemptive cuts we propose). They implemented and compared the aforementioned combinations on 9 instances of MMAS problems taken from Bockmayr and Pisaruk, and on additional randomly generated data sets. In this paper, we discuss the architecture of Xpress-CP and demonstrate its capability using the MMAS problem. We propose certain Disjunctive and Preemptive cuts (see Section 3), and compare the results of the implementations of pure (MIP), Iterative (MIP-CP), and B&C (MIP-CP) methodologies. We also generate several random test cases and present the results of this implementation.

[1]Project supported by the European Commission, Growth Program, Research Project LISCOS -Large Scale Integrated Supply Chain Optimization Software, Contract G1RD-CT-1999-00034.

1.2 Outline

In Section 2 we identify the Assignment and the Scheduling components of the problem and present its pure MIP formulation. In Section 3, we discuss cuts for strengthening the LP relaxation of the problem. In Section 4, we review the methodologies proposed by Jain and Grossmann (2001), and Bockmayr et al. (2003) for solving the problem by combining the MIP and CP techniques. Then, in Section 5, we present the Xpress-CP implementation of the hybrid approaches, where we outline the Xpress-CP architecture, illustrate the Xpress-Mosel implementation, and provide our results. Finally in Section 6, we summarize our work and present conclusions.

2. The Multi-Machine Assignment and Scheduling Problem description

We consider a problem with N jobs and M parallel machines, where each machine can process one job at a time. Processing of the ith job can start after the release time r_i and must end before the deadline d_i. The processing time and the cost of processing of ith job on the mth machine are p_{im} and c_{im}, respectively. Following common scheduling terminology, these machines are *unrelated* (Peter (2001)). The goal is to process each job on one of the machines in order to minimize the total processing cost. In the context of supply chains, the jobs can be viewed as various intermediate products that are supplied from an up-stream supplier on various days, and need to be further processed in one of the several factories before their delivery dates to down-stream buyers. The processing time then would be the total time to ship and process the products to factories, while the cost of processing would be the actual cost incurred in transportation and processing. Although all the factories might have the infrastructure for processing the products, the physical location, and the technology used in each factory may affect the total cost and times required to process the products. This problem can be viewed as an assignment and scheduling problem, where each job must be assigned to one of the available machines such that the total assignment cost is minimized, while maintaining the feasibility of the schedule of all jobs assigned to each of the machines (i.e., meeting the release-deadline restrictions of the jobs, while respecting their non-overlapping sequencing on the machines they are assigned to). In the next Section we present the standard MIP formulation of the assignment and the scheduling components of the problem.

2.1 MIP formulation

Sets:

$Jobs = \{1, \ldots, N\}$ (indexed by i, j, k).

$Machines = \{1, \ldots, M\}$(indexed by m, n).

Data:

r_i, d_i: Release and deadline times, respectively.

p_{im}, c_{im}: Processing time and cost of processing, respectively.

Variables:

x_{im}: Binary variable equal to on if job i is assigned to machine m, and 0 otherwise.

s_i: Starting time of processing of job i.

y_{ij}: Binary variable equal to one if job i precedes job $j \neq i$ when both of them are assigned to the same machine, and 0 otherwise.

Basic Formulation:

Assignment

$$\begin{aligned}
\min \quad & \sum_{i \in Jobs, m \in Machines} c_{im} x_{im} \\
s.t. \quad & \sum_{m \in Machines} x_{im} = 1, \ \forall i \in Jobs, \\
& x_{im} \in \{0, 1\}, \ \forall i \in Jobs, \ m \in Machines.
\end{aligned}$$

Scheduling

$$r_i \le s_i \le d_i + \sum_{m \in Machines} p_{im} x_{im}, \ \forall i \in Jobs,$$

$$x_{im} + x_{jm} - 1 \le y_{ij} + y_{ji},$$
$$\forall i, j \in Jobs: \ j > i, m \in Machines.$$

$$s_i + \sum_{m \in Machines} p_{im} x_{im} \le$$
$$s_j + \left(\sum_{k \in Jobs} \max_{m \in Machines} \{p_{km}\} \right) (1 - y_{ij}), \ \forall i, j \in Jobs : i \ne j,$$
$$s_i \ge 0 \ \forall i \in Jobs, \ y_{ij} \in \{0, 1\}, \ \forall i, j \in Jobs: \ i \ne j.$$

The above formulation for the "assignment" portion of the problem is straightforward. The scheduling part is slightly more complex. The first scheduling constraint ensures that the job is scheduled within its time window. The second equation states that when two jobs i and j are assigned to the same machine m, then either i precedes j or j precedes i. The third equation links starting time variables to disjunction variables as follows: Given two jobs i and j, if i precedes j then, between the starting time of i and the starting time of j, there must be enough time to schedule job i, i.e.,

$$s_i + \sum_{m \in Machines} p_{im} x_{im} \le s_j$$

Now, if i does not precede j, the third equation leads to

$$s_i + \sum_{m \in Machines} p_{im} x_{im} \le s_j + \left(\sum_{k \in Jobs} \max_{m \in Machines} \{p_{km}\} \right)$$

which always holds since the right term is a large constant value (i.e., a "big M"). Such values are known to be a major issue for MIP formulations because they lead to very poor relaxations. An alternative for the scheduling part would be to use a time indexed formulation (see for instance Pritsker et al. (1969)) in which a binary variable is associated with each candidate starting time for each activity. As noticed by several researchers, this leads to huge MIP problems that are not likely to be solved, even for medium sized instances.

3. Cuts to tighten the MIP Formulation

The solution time to solve the problem can be improved significantly by addition of the following cuts:

Maximum Duration Cuts

$$\sum_{i \in Jobs} p_{im} x_{im} \leq \max_{i \in Jobs}\{d_i\} - \max_{i \in Jobs}\{r_i\}, \ \forall m \in Machines.$$

The above cuts–originally proposed by Jain and Grossmann (2001)–ensure that the total processing time of the jobs assigned to a machine should not exceed the maximum possible time $(\max_{i \in Jobs}\{d_i\} - \min_{i \in Jobs}\{r_i\})$ available.

Disjunctive Cuts

If the total processing time of a pair of jobs on a machine exceeds the maximum span of time available for processing, then both of them can not be assigned to the machine; hence, we propose the following Disjunctive cuts.

$$x_{im} + x_{jm} \leq 1, \qquad \forall i, j \in Jobs : j > i$$
$$m \in Machines : p_{im} + p_{jm} > \max(d_i - r_j, d_j - r_i).$$

Preemptive Cuts

$$\sum_{\substack{k \in Jobs \\ r_i \leq r_k, d_k \leq d_j}} p_{km} x_{km} \leq d_j - r_i, \quad \forall i, j \in Jobs : d_j > r_i, m \in Machines :$$
$$\sum_{\substack{k \in Jobs \\ r_i \leq r_k, d_k \leq d_j}} p_{km} > d_j - r_i.$$

In the context of MMAS problem we also propose the Preemptive cuts which are similar to the Disjunctive cuts. They impose the restriction that the total processing time of the set of jobs that are released after the release time r_i, are due before d_j, and are assigned to a machine, should not exceed the time-span d_j - r_i. Surprisingly, these cuts are very powerful. They ensure that, once the assignment problem is solved, we have a feasible preemptive schedule on each machine (Carlier and Pinson (1990)). In a recent paper by Sadykov and Wolsey (2003), the authors have also proposed methods which further strengthen the Preemptive cuts, and implemented them in a hybrid fashion using MIP and CP (see MIP$^+$/CP in Sadykov and Wolsey (2003)).

Logical Cuts for the pure MILP formulation

The following set of cuts Jain and Grossmann (2001) can be used in the pure MILP formulation (see Section 2.1).

$$\begin{aligned} y_{ij} + y_{ji} &\leq 1, && \forall i,j \in Jobs : j > i, \\ y_{ij} + y_{ji} + x_{im} + x_{jn} &\leq 2, && \forall i,j \in Jobs : j > i, \\ & && m,n \in Machines : m \neq n. \end{aligned}$$

The first set of cuts ensures that given a pair of jobs assigned to a machine, one of them should precede the other. The second set of cuts prevents the sequencing of a pair of jobs on any machine if they are assigned to different machines. Alternatively, if they are sequenced on a machine, they cannot be assigned to different machines.

4. Hybrid approaches using Constraint Programming

The introduction of binary variables for sequencing the jobs (y_{ij}) on the machines, together with the associated constraints, makes this problem difficult to solve using MIP techniques alone. Since the scheduling component of the problem does not affect the objective function, one can solve this problem faster by applying a hybrid scheme using both MIP and CP. MIP can be used for solving the assignment problem as a master problem, and CP for checking the feasibility of the scheduling problem as a sub-problem. Whenever the sub-problem is infeasible, a cut is added to the master problem. An iterative scheme implemented by Jain and Grossmann (2001) is outlined in Section 4.2, and a B&C scheme implemented by Bockmayr et al. (2003) is discussed in Section 4.3.

4.1 CP scheduling formulation

CP can be used for checking the feasibility of scheduling the processing of jobs on the machine to which they are assigned. Let be the current solution vector to the master problem. Then define $Jobs^m = \{i \in Jobs : x_{im} = 1, x_{jm} \in \{0,1\} \; \forall j \in Jobs\}$ as the set of jobs assigned to machine m based on the current solution.

The CP problem $\forall m \in Machines : Jobs^m \neq \emptyset$ is stated as -

$$
\begin{aligned}
& i.start \in [r_i, d_i - p_{im}] \; \forall i \in Jobs^m \\
& i.duration = p_{im} \; \forall i \in Jobs^m \\
& disjunctive(Jobs^m)
\end{aligned}
$$

The first set of constraints sets the domains of start times of the assigned jobs, while the second set fixes their processing durations on the machine. The disjunction ensures that the sequencing of the jobs is non-overlapping. If the CP problem is infeasible then a "**no good**" cut (Hooker et al. (1999)): $\sum_{i \in Jobs^m} x_{im} \leq x_{im} \leq |Jobs^m| - 1$ may be added to the master problem.

Very efficient constraint propagation techniques, known as "edge-finding", have been developed to solve such scheduling problems (see Carlier and Pinson (1990); Baptiste et al. (2001)). Edge-finding bounding techniques are particular constraint propagation techniques which reason about the order in which several jobs can be processed on a given machine. It consists of determining whether a job can, cannot, or must execute before (or after) a set of jobs which require the same machine. Two types of conclusions can then be drawn: new ordering relations ("edges" in the graph representing the possible orderings of activities) and new time-bounds, i.e., strengthened earliest and latest start and end times of activities.

4.2 Iterative method

In the iterative method, the assignment problem is solved repetitively as an MIP master problem and the "no good" cuts generated from the CP sub-problem are added to it. The loop terminates when the master problem is found to be infeasible or when the MMAS problem is optimal.

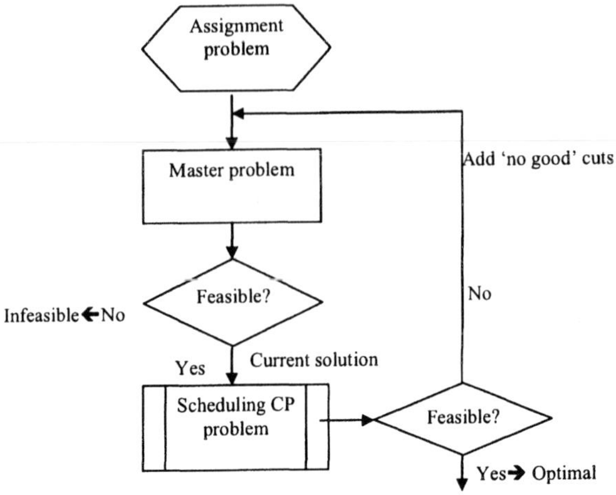

Figure 12.1. Iterative method

4.3 Branch and Cut (B&C) method

The B&C method is similar to the iterative method except that in this method, the master problem is not solved to optimality before adding cuts. Instead, the assignment problem is solved as an MIP problem using the standard B&B method. At each partially feasible node (where one or more machines have been assigned jobs), CP is used to generate the "no good" cuts if possible, thereby ensuring that each integer feasible node is also feasible for the MMAS problem.

5. Implementation in Xpress-CP

The normal approach to solve problems using such a hybrid procedure is to have two models–a planning model and a scheduling model–with the former solved with MIP and the latter solved with CP. The need to use two models and two separate systems increases the complexity, reliability, and lifecycle cost of the system. It also requires some manual intervention to iterate between the two systems, which is expensive and unreliable. Xpress-CP overcomes this limitation by providing a unified framework for modeling problems in a single model, as discussed next.

5.1 Xpress-CP design and capabilities

Xpress-CP combines the mathematical optimization software Xpress-MP and the constraint programming software CHIP in a hybrid optimization framework. An important advantage of Xpress-CP is that the MIP and CP technologies exist within the same software environment and the problem is expressed within a single model. The model is written in Xpress-Mosel and the MIP/CP solver is invoked without the need for complex programming or particular expertise in MIP or CP. Built on Mosel's *Native Interface* technology (Colombani and Heipcke (2002)), Xpress-CP provides a rich collection of *types* and constraint structures to represent problems in supply chain optimization for advanced planning and scheduling. The Mosel module 'xpresscp' provides a high level abstraction by *various types* such as cpdvar, cpoperation, cpnoonoverlap, cplabeling, etc; and supports several functions and *procedures*. Specifically, Xpress-CP provides:

- high level semantic objects such as operations and machines which are part of the language used by the end users to describe problems;
- functionality to get and set the attributes of these objects;
- high level semantic constraints which ease the modelling of the problem's constraints;
- high level primitives to guide during the search procedure;
- various predefined strategies and heuristics.

Various *types*, *procedures* and *functions* defined in Xpress-CP, together with Xpress-CP's superior design enable easy and concise modeling of complex scheduling problems in Mosel

5.2 Mosel model

The Xpress-Optimizer module 'mmxprs' and the Xpress-CP module 'xpresscp' are used for writing the iterative (Section 4.2) and the B&C (Section 4.3) schemes in Xpress-Mosel. In the following section only the relevant parts of the Mosel model are described. The complete model with data files can be obtained from the authors upon request.

First the model entities defined in Section 2.1 are declared as follows:

```
declarations
  Jobs=1,..,N
  Machines=1,..,M
  r,d:array(Jobs) of integer
  p,c:array(Jobs,Machines) of integer
  x:array(Jobs,Machines) of mpvar
end-declarations
```

5.2.1 Assignment.

Next the assignment problem is written as follows in Mosel:

```
TotalCost := sum(i in Jobs,m in Machines) c(i,m)*x(i,m)
forall(i in Jobs) Assignment(i) := sum(m in Machines) x(i,m)=1
forall(i in Jobs,m in Machines) x(i,m) is_binary
```

5.2.2 Cuts.

The Maximum Duration, Disjunctive and Preemptive cuts are written as follows in Mosel:

```
MaxDuration:=max(i in Jobs) d(i)-min(i in Jobs) r(i)
forall(m in Machines)
  MaximumDuration(m):=sum(i in Jobs) p(i,m)*x(i,m)¡=MaxDuration

forall(i,j in Jobs | i>j)
  forall(m in Machines |
p(i,m)+p(j,m)>maxlist(d(i)-r(j),d(j)-r(i)))
    Disjunctive(i,j,m):=x(i,m)+x(j,m)<=1

forall(i,j in Jobs | r(i) <= r(j) and d(i) <= d(j)) do
  S:=union (k in Jobs | r(i)<=r(k) and d(k)<=d(j)) {k}
  forall(m in Machines | sum(k in S) p(k,m)>(d(j)-r(i)))
    Preemptive(i,j,m):=sum(k in S) p(k,m)*x(k,m)<=d(j)-r(i)
end-do
```

5.2.3 CP formulation for feasibility check.

The CP formulation of the problem discussed in Section 4.1 is done by declaring the operations for each job, followed by setting the domains of their starting and processing times.

```
declarations
  o:array(Jobs) of cpoperation
```

```
end-declarations

forall(i in Jobs) do
  max_dur:=max(m in Machines) p(i,m)
  min_dur:=min(m in Machines) p(i,m)
  cpsetstart(o(i),r(i),d(i)-min_dur)
  cpsetduration(o(i),min_dur,max_dur)
end-do
```

The function cpsetstart() sets the domain for the starting times, and the function cpsetduration() sets the bounds on processing times. Now, given a machine m and a set of jobs assigned to it, CP can be used to check the feasibility with respect to scheduling and sequencing of the jobs as follows:

```
function IsFeasible(m:integer,JobsAssigned:set of integer):boolean
  cpstart
  forall(i in JobsAssigned) do
    fes_asgn:=
      if ((r(i)+p(i,m))>d(i),false,
        cpsetminmax(cpgetstart(o(i)),r(i),d(i)-p(i,m)))
    fes_asgn:=
      if (fes_asgn,
        cpsetminmax(cpgetduration(o(i)),p(i,m)),false)
  end-do
  fes_asgn:=if (fes_asgn,CheckFeasibility(JobsAssigned),false)
  cpend
  returned:=fes_asgn
end-function
```

Here, Xpress-CP is invoked by cpstart(), and terminated by cpend(). It is ensured that all the jobs under consideration can be assigned to the machine m simply based on their processing times on the machine and the spans of time available between their release and deadline times. Then, the function cpsetminmax() is used to set the domains of starting and processing times based on the current assignment. The feasibility of current assignment is checked by calling a user-defined function CheckFeasibility() as follows:

```
function CheckFeasibility(JobsAssigned:set of integer):boolean
  declarations
    Oprtns:set of cpoperation
```

```
    Disjunctive:cpnonoverlap
    Label:cplabeling
  end-declarations

  returned:=false
  forall(i in JobsAssigned) Oprtns += {o(i)}
  cpsetops(Disjunctive,Oprtns)
  if cppost(Disjunctive) then
    cpsetops(Label,Oprtns)
    cpsetselectop(Label,"start",1,"smallest")
    cpsetseelectval(Label,"start",1,"indomain_min")
    returned:=cppost(label)
  end-if
end-function
```

In the above functions, all the operations are collected in a set Oprtns, and assigned to a non-overlap type constraint Disjunctive. Next, the feasibility of the current assignment is checked by posting the disjunctive constraint to the CP-solver. The search strategy for finding a solution is defined by creating a primitive Label for the operations, and setting the operation-selection and value-selection criterions. In the MMAS problem we define a high priority search strategy by selecting the smallest values of starting times of operations from their domains. The function CheckFeasibility() returns false if the assigned jobs cannot be sequenced on machines.

5.2.4 Iterative schematics. The Mosel implementation of the iterative scheme discussed in Section 4.2 is as follows:

```
minimize(TotalCost)
while (true) do
  if getprobstat<>XPRS_OPT then break;end-if
  savebasis("previous optimal basis")
  ncut:=0
  forall(m in Machines) do
    JobsAssigned:=union(i in Jobs| getsol(x(i,m))=1) {i}
    NumJobs:=getsize(JobsAssigned)
    if NumJobs>0 then
      if not IsFeasible(m,JobsAssigned,false) then
        TotNumOfCuts+=1
        ncut+1=1
        cuts(TotNumOfCuts):=
```

```
            sum(i in JobsAssigned) x(i,m)<=(NumJobs-1)
          end-if
        end-if
      end-do
      if ncut>0 then
        loadprob(TotalCost)
        loadbasis("previous optimal basis")
        minimize(TotalCost)
      else
        break
      end-if
    end-do
```

The master problem is solved repetitively in the loop which terminates when no more cuts can be added to it or when it is infeasible or unbounded. The feasibility of the sub-problem is checked using CP by calling the function IsFeasible() shown in Section 5.2.3. The global counter TotNumCuts (declaration not shown here) is used to keep track of number of cuts added to the master problem. The 'mmxprs' functions savebasis() and loadbasis() are used to save and load the optimal basis respectively during each iteration which helps in 'warm starting'.

5.2.5 B&C call-back schematics. The logic behind solving the problem using the B&C scheme is similar to that of the iterative *scheme*, except that now the cuts are added during the B&B search. This is achieved in Mosel by turning off the Xpress pre-solver so that the matrix structure is not lost, and directing the Xpress cut-manager to call a user-defined routine at every node of the B&B tree.

The optimizer is intercepted during the B&B search by setting a call-back from the cut-manager as follows:

```
setcallback(XPRS_CB_CUTMGR,"EveryNode")
```

where the function EveryNode() is defined as follows:

```
function EveryNode: boolean
  returned:=false
  setparam("xprs_solutionfile",0)
    forall(i in Jobs,m in Machines) CurrSol(i,m):=getsol(x(i,m))
  setparam("xprs_solutionfile",1)
  loadcuts(NO_GOOD,1)
```

```
  ncut:=0
  forall(m in Machines) do
    if and(i in Jobs) (CurrSol(i,m)=0 or CurrSol(i,m)=1) then
      JobsAssigned:=union(i in Jobs| CurrSol(i,m)=1) {i}
      NumJobs:=getsize(JobsAssigned)
      if NumJobs>0 then
        if not IsFeasible(m,JobsAssigned,false) then
          TotNumOfCuts+=1
          ncut+=1
          cut:=sum(i in JobsAssigned) x(i,m)-(NumJobs-1)
          addcut(NO_GOOD,CT_LEQ,cut)
          returned:=true
        end-if
      end-if
    end-if
  end-do
end-function
```

The routine EveryNode() is called at each node after the LP relaxed problem at that node is solved. The solution at the current node is stored in CurrSol (declaration not shown here) by turning off Xpress' solutionfile temporarily, so that the solution is read from the Optimizer directly instead of the solution file which stores the current best solution. The "no good" cuts are added to the matrix at current node by calling the 'mmxprs' function addcut(). Since these cuts are globally valid, they are loaded by calling the 'mmxprs' function loadcuts(). The constant NO_GOOD (declaration not shown here) is used as an identifier for the cuts.

5.3 Computational Results

In the following sections we present the results of implementation in Xpress-CP. The experiments were done on a P-IV, 2.2GHz machine with 1GB RAM. The default time limit for running the model was set to 1 hour. The Xpress pre-solver is turned off for the B&C method. We used the Maximum-duration, Disjunctive and Preemptive cuts in our implementation. The Xpress components and their version number, used for implementing the hybrid schemes were:

- Mosel (modeling language)- 1.2.4
- IVE (Integrated Visual Environment)- 1.14.70

- mmxprs (Mosel module for MIP)- 1.2.3
- xpresscp (Mosel module for CP)- 0.1.5
- Optimizer (Xpress LP/MIP/QP/MIQP solver)- 14.27
- CHIP (CP solver)- 5.5

5.3.1 Comparison with Jain and Grossmann's results.

Table 5.3.1 shows the comparison with results of Jain and Grossmann's implementation. They used CPLEX 6.5 single processor version on a dual processor SUN Ultra 60 workstation. The problems were originally generated by Jain and Grossmann.

The above table lists the number of major iterations in the Iterative method, the number of cuts added in the B&C approach, the number of nodes in the B&B tree, and corresponding times in seconds to achieve an optimal solution. Since the times required for solving these problems are quite small and the platform used by Jain and Grossmann is different than ours, it is hard to compare the results. As far as the iterative and the B&C methods in Xpress-CP are concerned, the latter seems to solve the problem faster than the former.

5.3.2 Comparison with Bockmayr and Pisaruk's and Sadykov and Wolsey's results.

The following table shows a comparison with results of Bockmayr and Pisaruk's, and Sadykov and Wolsey's implementations. The tested problems are essentially variations of the problems generated by Jain and Grossmann. Bockmayr and Pisaruk used a similar B&C method together with a heuristic for binary variables and a set of cycle cuts, and implemented it using Xpress-MP 2003B, CHIP version 5.3 on a Pentium III machine with 600 MHz with 256 MB memory. Sadykov and Wolsey carried out all the experiments on a PC with P-IV 2 GHz processor and 512 MB RAM. They also used MIP and CP in B&C, together with a tighter version of the Preemptive cuts (MIP^+/CP) and implemented it in Xpress-Mosel 1.3.2, Xpress-Optimizer 14.21 and CHIP 5.4.3.

The above table also lists the best objective value and the best bound for the problems that were not solved to optimality (The corresponding entries for the problem solve to optimality are marked by *). Given their machine specifications, it is observed from Table 5.3.2 that Bockmayr

Problem	obj	Jain & Grossmann: Iterative			Xpress-CP: Iterative			Xpress-CP: B&C		
	val	#iteration	#cuts	time(s)	#iterations	#cuts	time(s)	#cuts	#nodes	time(s)
1a	26	2	1	0.02	1	0	0.11	0	1	0.079
1b	18	1	0	0.01	1	0	0.094	0	1	0.062
2a	60	13	16	0.52	1	0	0.188	1	10	0.219
2b	44	1	0	0.02	1	0	0.109	0	1	0.093
3a	101	31	43	4.18	2	1	0.735	7	219	0.5
3b	83	1	0	0.02	1	2	0.11	0	1	0.078
4a	115	18	26	2.25	3	0	1.172	4	174	0.66
4b	102	1	0	0.04	1	0	0.109	0	1	0.093
5a	158	31	60	14.13	5	6	7.052	11	557	2.344
5b	140	6	6	0.41	1	0	0.156	0	1	0.125

Table 12.1. Comparison with Jain and Grossmann's results.

Problem	obj	Bockmayr & Pisaruk: B&C			Sadykov & Wolsey: MIP^+/CP	Xpress-CP: B&C			
	obj	#cuts	#nodes	time(s)	time(s)	#cuts	#nodes	time(s)	bestobj/bestbound
3_12a	101	93	78	0.47	0.407	7	219	0.5	*
3_12b	104	210	149	0.87	0.328	0	640	1.16	*
5_15a	115	104	144	0.7	0.563	4	174	0.66	*
5_15b	137	164	177	0.9	0.563	3	155	0.86	*
5_20a	159	360	924	4.01	1.313	11	557	2.34	*
5_20b	145	3179	38303	245.77	2.703	136	46338	150.51	*
6_24	231	9901	389372	5762.89	3.672	24	1800	8.31	*
7_30	213	-	-	>10hr	16.656	67	234351	3848.73	214/211.378
8_34[2]	252	NA	NA	NA	3916.42	564	106123	3828.39	255/247.124

Table 12.2. Comparison with Bockmayr and Pisaruk's results.

and Pisaruk's results might be better than ours. This can be attributed to the fact that they use a specialized heuristic and cycle cuts on top of the B&C method which seems to enhance the performance. Similarly, tightening of the Preemptive cuts by Sadykov and Wolsey significantly improves the performance.

5.3.3 Comparison with Randomly generated data.

We generate random data sets in Mosel as follows:

$$c_i \in \{1, \ldots, 20\} \ \forall i \in Jobs$$
$$r_i \in \{1, \ldots, 20\} \ \forall i \in Jobs$$
$$d_i \in \{15, \ldots, 25\} \ \forall i \in Jobs$$
$$p_{im} \in \{1, \ldots, d_i - r_i - 1\} \ \forall i \in Jobs, m \in Machines$$

The problems are generated by varying the number of machines M, the number of jobs N, and the seed for generating the random data. The results of the pure MILP (Xpress pre-solver is on), Iterative (Xpress pre-solver is off), and B&C implementations are tabulated in the following Table. Note that only the non-trivial problems (that require one or more cuts by either of the hybrid methods) are shown below.

From the above results it is observed that the times required for solving the problems using hybrid-schemes are much less than when using pure MILP. Additionally, the B&C method solves most of the problems faster than the Iterative method.

M	N	seed	obj val	MILP bestobj/bestbound	MILP #nodes	MILP time(s)	Iterative #cut	Iterative #iterations	Iterative time(s)	B&C #cut	B&C #nodes	B&C time(s)
5	15	1	95	*	98	5.516	1	2	0.657	4	22	0.281
	20	1	145	*	3352	105.34	3	4	14.485	15	755	4.328
		2	135	*	1170	67.015	0	1	4.797	7	382	2.469
	25	1	192	*	3297	255.47	0	1	6.078	12	1051	7.812
		2	234	*	6425	219.52	0	1	5.954	12	570	5.547
	30	1	247	inf/240	104523	3653.8	5	4	47.296	19	1659	19.016
		2	192	*	29379	1633.6	0	1	12.437	7	494	7.797
10	25	1	86	*	39	27.109	1	2	0.531	1	1	0.234
		2	84	*	4154	359.59	3	4	6.015	5	23	2.203
	30	2	100	*	11596	2310.9	7	4	19.625	12	199	4.563
	40	1	171	214/167	1976	3614.8	5	5	355.64	43	4214	201.25
		2	145	195/142	2635	3618.5	1	2	68.672	7	190	28.687
20	40	1	86	*	1144	2134.6	4	3	23.25	4	11	6.578
	50	1	111	inf/110	173	3600	6	5	261.09	7	17	76.218
	60	1	137	inf/134.3	87	3657.8	5	4	640.59	20	300	248.14
		2	131	165/131	188	3652.6	4	3	108.77	4	13	79.64
	70	1	171	inf/168.558	30	3932.41[c]	8	5	3154.1	10	219	1108.7
		2	178	inf/177.309	21	3693.1	2	2	1197.1	10	203	886.83

Table 12.3. Results for the randomly generated problems.

6. Summary and Conclusion

In this paper we demonstrated the capabilities of Xpress-CP, which provides a natural syntax for expressing scheduling problems, and presented the Xpress-Mosel framework which facilitates the existence of both MIP and CP technologies to co-exist and enable rapid modeling and solving. We considered the Multi-machine assignment and scheduling problem, which, because of its structure, is a perfect candidate for demonstration purposes. We began by presenting the pure MIP formulation of the problem and cuts that could be used for strengthening its linear relaxation. Next, we showed two hybrid approaches to solve the problem, namely Iterative, and B&C, followed by illustrating the implementation of these approaches in Mosel. Finally, we compared our results with those of Jain and Grossmann's, Bockmayr and Pisaruk's, and Sadykov and Wolsey's. We also compared the Iterative and the B&C methods for various problems generated in Mosel randomly. From the results it was observed that the B&C approach solves the problem much faster than the iterative approach in most of the cases, and using stronger cuts further improves the performance significantly.

Acknowledgments

The authors would like to thank Philippe Baptiste for many enlightening discussions on the cuts mentioned in Section 3 for the MMAS problem, and for his help in revising the paper.

References

Aggoun, A. and Vazacopoulos, A. 2004. Solving Sports Scheduling and Time tabling Problems with Constraint Programming, in Economics, Management and Optimization in Sports, Edited by S. Butenko, J. Gil-Lafuente and P.M. Pardalos, Springer.

Baptiste, P., Le Pape, C. and Nuijten, W. 2001. Constraint Based Scheduling. Kluwer.

Bockmayr, A. and Kasper, T. 2003. Branch-and-infer: A framework for combining CP and IP. In Constraint and Integer Programming (Ed. M. Milano), Chapter 3, 59 - 87, Kluwer.

Bockmayr, A. and Hooker, J.N. 2003. Constraint programming. In Handbooks in Operations Research and Management Science: Discrete Optimization (Eds. K. Aardal, G. Nemhauser, and R. Weismantel), Elsevier, To appear.

Bockmayr, A. and Pisaruk, N. 2003. Detecting Infeasibility and Generating Cuts for MIP using CP. 5th International Workshop on Integration of AI and OR Techniques in Constraint Programming for Combinatorial Optimization Problems, CPAIOR'03, Montreal, May 2003.

Brucker, P. 2001. Scheduling Algorithms. Third Edition, Springer.

Carlier, J. and Pinson., E. 1990. A Practical Use of Jackson's Preemptive Schedule for Solving the Job-Shop Problem. Annals of Operations Research 26, 269-287.

Colombani, Y and Heipcke, S. 2002. Mosel: An Overview. May 2002, available at http://www.dashoptimization.com/home/downloads/pdf /mosel.pdf.

Easton, K., Nemhauser, G. and Trick, M. 2003. CP Based Branch-and-Price. In Constraint and Integer Programming (Ed. M. Milano), Chapter 7, 207 - 231, Kluwer.

Hooker, J.N., Ottosson, G., Thorsteinsson, E.S. and Kim, H.J. 1999. On integrating constraint propagation and linear programming for combinatorial optimization. Proceedings of the Sixteenth National Conference on Artificial Intelligence (AAAI-99), AAAI, The AAAI Press/MIT Press, Cambridge, MA. 136-141.

Jain, V. and Grossmann, I.E. 2001. Algorithms for hybrid MIP/CP models for a class of optimization problems. INFORMS J. Computing, 13(4), 258-276, 2001.

Peter, B. 2001. Scheduling Algorithms. Springer Lehrbuch.

Pritsker, A., Watters, L. and Wolfe, P. 1969. Multi-project scheduling with limited resources: a zero-one programming approach. Management Science, 16:93-108.

Pinedo, M. 1995. Scheduling: Theory, Algorithms and Systems. Prentice - Hall, NJ.

Pinedo, M. and Chao, X. 1998. Operations Scheduling with Applications in Manufacturing and services. McGraw-Hill/Irwin.

Sadykov, R. and Wolsey, L. 2003. Integer programming and constraint programming in solving a multi-machine assignment scheduling problem with deadlines and release dates. CORE discussion paper, Nov 2003.